TIME-VARYING IMAGE PROCESSING
AND MOVING OBJECT RECOGNITION, 3

TIME-VARYING IMAGE PROCESSING AND MOVING OBJECT RECOGNITION, 3

Proceedings of the
4th International Workshop
Florence, Italy, June 10–11, 1993

Edited by

V. CAPPELLINI
Department of Electronic Engineering
University of Florence
Florence, Italy

1994

ELSEVIER
Amsterdam – London– New York – Tokyo

ELSEVIER SCIENCE B.V.
Sara Burgerhartstraat 25
P.O. Box 211, 1000 AE Amsterdam, The Netherlands

ISBN: 0 444 81467 1

Transferred to digital printing 2006
Printed and bound by CPI Antony Rowe, Eastbourne

PREFACE

The area of Digital Image Processing is of high actual importance in terms of research and applications. Through the interaction and cooperation with the near areas of Pattern Recognition and Artificial Intelligence, the specific area of "Time-Varying Image Processing and Moving Object Recognition" has become of increasing interest. This new area is indeed contributing to impressive advances in several fields, such as communications, radar-sonar systems, remote sensing, biomedicine, moving vehicle recognition-tracking, traffic monitoring and control, automatic inspection and robotics.

This book represents the Proceedings of the Fourth International Workshop on Time-Varying Image Processing and Moving Object Recognition, held in Florence, June 10-11, 1993. Extended papers reported here provide an authorative and permanent record of the scientific and technical lectures, presented by selected speakers from 15 Nations. Some papers are more theoretical or of review nature, while others contain new implementations and applications. They are conveniently grouped into the following fields:

A.   Digital Processing Methods and Techniques
B.   Pattern Recognition
C.   Image Restoration
D.   Implementation Techniques
E.   Computer Vision
F.   Image Coding and Transmission
G.   Remote Sensing Image Processing
H.   Digital Processing of Biomedical Images
I.   Motion Estimation
J.   Recognition and Tracking of Moving Objects
K.   Traffic Monitoring

New digital image processing and recognition methods, implementation techniques and advanced applications (television, remote sensing, biomedicine, traffic, inspection, robotics, etc.) are presented. New approaches (i.e. digital transforms, neural networks, ...) for solving 2-D and 3-D problems are described. Many papers are concentrated on the motion estimation and recognition - tracking of moving objects. In overall the book presents - for the new outlined area - the state of the art (theory, implementation, applications) with the next-future trends.

This work will be of interest not only to researchers, professors and students in university departments of engineering, communications, computers and automatic control, but also to engineers and managers of industries concerned with computer vision, manufacturing, automation, robotics and quality control.

V. Cappellini

WORKSHOP CHAIRMAN

V. CAPPELLINI, *University of Florence, Florence, Italy*

STEERING COMMITTEE

J.K. AGGARWAL, *University of Texas, Austin, U.S.A.*
M. BELLANGER, *Conservatoire National des Arts et Métiers, Paris, France*
A.G. CONSTANTINIDES, *Imperial College, London, England*
T.S. DURRANI, *University of Strathclyde, Glasgow, England*
G. GALATI, *II University of Rome, Italy*
G.H. GRANLUND, *University of Linkoping, Sweden*
T.S. HUANG, *University of Illinois at Urbana-Champaign, U.S.A.*
M. KUNT, *Ecole Polytechnique Fédérale de Lausanne, Switzerland*
D. LEBEDEV, *Institute for Information Transmission Problems, Moscow, C.S.I.*
S. LEVIALDI, *University of Rome, Italy*
A.R. MEO, *Polytechnic of Turin, Italy*
F. ROCCA, *Polytechnic of Milan, Italy*
J.L.S. SANZ, *IBM Almaden Research Center, San José, U.S.A.*
V. TAGLIASCO, *University of Genoa, Italy*
A.N. VENETSANOPOULOS, *University of Toronto, Canada*
G. VERNAZZA, *University of Genoa, Italy*

SPONSORED BY:

European Association for Signal Processing (EURASIP)
IEEE Middle & South Italy Section
International Center for Signal & Image Processing (ICESP), Florence
Centro d'Eccellenza Optronica (CEO)
Dipartimento di Ingegneria Elettronica, University of Florence
Istituto di Ricerca sulle Onde Elettromagnetiche (IROE) - C.N.R., Florence
Fondazione Ugo Bordoni
Fondazione Scienza per l'Ambiente, Florence
Associazione Italiana di Telerilevamento (A.I.T.)
Gruppo Nazionale Telecomunicazioni e Teoria dell'Informazione (T.T.I.) -
*   C.N.R.*
Ordine degli Ingegneri della Provincia di Firenze
Sezione di Firenze dell'A.E.I.

CO-SPONSORED BY:

Alenia Spazio
AXIS
Esaote Biomedica
Fondazione IBM Italia
Officine Galileo
OTE
Siemens Telecomunicazioni
S.M.A.
Syremont
Telespazio

Cassa di Risparmio di Firenze

CONTENTS

A
DIGITAL PROCESSING METHODS
AND TECHNIQUES

Time-Varying Image Processing and
Moving Object Recognition, 3 – V. Cappellini (Ed.)
© 1994 Elsevier Science B.V. All rights reserved.

Fast Approximate Calculation of the Two-Dimensional Discrete Fourier Transform

Sanjit K. Mitra[a] and Ogjnan V. Shentov[b]

[a] Department of Electrical & Computer Engineering
University of California, Santa Barbara, CA 93106, U.S.A.

[b] Pennie & Edmonds, Inc.
1155 Avenue of the Americas, New York, NY 10036, U.S.A.

The well-known decimation-in-time or decimation-in-frequency FFT algorithm can be viewed as computing the DFT of a longer length sequence as an weighted sum of the DFTs of smaller length sub-sequences generated by a polyphase decomposition of the original sequence followed by down-sampling. A similar approach based on a structural subband decomposition, a generalization of the polyphase decomposition, has resulted in a new FFT algorithm. In the case of two-dimensional (2-D) DFT computation, the algorithm is divided into two distinct stages: a pre-processing 2-D analysis filter bank with down-sampled output and a recombination network. The preprocessing stage generates a set of smaller length 2-D sub-sequences with approximate spectral separations in the 2-D frequency domain from the original 2-D sequence. The overall DFT is then computed by a weighted sum of smaller length DFTs with the weights determined by the recombination network. The shorter length subsequences occupy essentially different portions of the input frequency range. This spectral separation property of the sub-sequences permits elimination of the subsequences having negligible energy contribution resulting in a fast approximate DFT computation method. An example is included illustrating the effectiveness of the proposed method in computing approximately the DFT of digital images.

1. INTRODUCTION

In many practical applications the 2-D signals of interest, such as images, occupy a relatively small portion of their transform domains. This spectral separation property has been exploited very effectively, in a number of image processing applications, like subband image coding [1], [2] etc. However, the property has so far not been exploited in other image processing applications such as discrete transform computation. In this paper, we propose a new 2-D FFT-type algorithm that can be used either for the computation of the full discrete Fourier transform (DFT) or selected DFT samples by making use of the above property. The conventional 2-D FFT algorithm is based on the computation of shorter size 2-D DFTs of subsequences obtained by a polyphase decomposition followed by down-sampling. The proposed 2-D FFT algorithm is based on a similar approach where the polyphase decomposition is replaced by a generalization called the structural subband decomposition. In Section 2 we review the polyphase decomposition and the structural subband decomposition. The new FFT algorithm is derived in Section 3. Section 4 illustrates an image processing application of the new 2-D FFT algorithm. Section 5 contains the concluding remarks.

2. POLYPHASE DECOMPOSITION OF 2-D SEQUENCES AND ITS GENERALIZATION

2.1 The Polyphase Decomposition

Consider a 2-D sequence $x(m,n)$ with a z-transform $X(z_1,z_2)$:

$$X(z_1,z_2) = \sum_{m=-\infty}^{\infty} \sum_{n=-\infty}^{\infty} x(m,n) z_1^{-m} z_2^{-n}. \tag{1}$$

The conventional *polyphase decomposition* of the transform $X(z_1,z_2)$ is given by

$$X(z_1,z_2) = \sum_{r=0}^{M-1} \sum_{s=0}^{N-1} X_{r,s}(z_1^M, z_2^N) z_1^{-r} z_2^{-s}, \tag{2}$$

where

$$X_{r,s}(z_1,z_2) = \sum_{m=-\infty}^{\infty} \sum_{n=-\infty}^{\infty} x(mM+r, nN+s) z_1^{-m} z_2^{-n},$$

$$r = 0, 1, \ldots, M-1; \ s = 0, 1, \ldots, N-1; \tag{3}$$

where functions $X_{r,s}(z_1,z_2)$ are called the *polyphase components* of $X(z_1,z_2)$ [3].

The relation between the sub-sequences $x_{r,s}(m,n)$ given by the inverse transform of $X_{r,s}(z_1,z_2)$, and the original sequence $x(m,n)$ is given by

$$x_{r,s}(m.n) = x(Mm+r, Nn+s); \qquad r = 0, 1, \ldots, M-1; \ s = 0, 1, \ldots, N-1, \tag{4}$$

or in other words, the sub-sequence $x_{r,s}(m,n)$ is simply obtained by down-sampling $x(m,n)$ by a factor of (M,N) with (r,s) indicating the phase of the sub-sampling process as indicated in Figure 1 for $M = 3$ and $N = 2$.

The polyphase decomposition of Eq. (2) can be written in matrix form as:

$$X(z_1,z_2) = \begin{bmatrix} 1 & z_1^{-1} & \cdots & z_1^{-(M-1)} \end{bmatrix} \mathbf{X} \begin{bmatrix} 1 \\ z_2^{-1} \\ \vdots \\ z_2^{-(N-1)} \end{bmatrix}, \tag{5}$$

where

$$\mathbf{X} = \begin{bmatrix} X_{0,0}(z_1^M, z_2^N) & X_{0,1}(z_1^M, z_2^N) & \cdots & X_{0,N-1}(z_1^M, z_2^N) \\ X_{1,0}(z_1^M, z_2^N) & X_{1,1}(z_1^M, z_2^N) & \cdots & X_{1,N-1}(z_1^M, z_2^N) \\ \vdots & \vdots & \ddots & \vdots \\ X_{M-1,0}(z_1^M, z_2^N) & X_{M-1,1}(z_1^M, z_2^N) & \cdots & X_{M-1,N-1}(z_1^M, z_2^N) \end{bmatrix}. \tag{6}$$

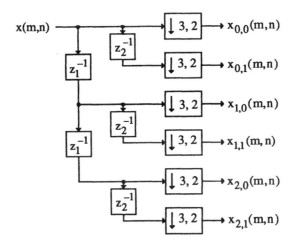

Figure 1 - Conventional polyphase decomposition of a sequence.

2.2 The Structural Subband Decomposition

A generalization of the polyphase decomposition of Eq. (5) is given by

$$X(z_1,z_2) = \begin{bmatrix} 1 & z_1^{-1} & \cdots & z_1^{-(M-1)} \end{bmatrix} A V B \begin{bmatrix} 1 \\ z_2^{-1} \\ \vdots \\ z_2^{-(N-1)} \end{bmatrix}, \tag{7}$$

where $A = [a_{ij}]$ is an $M \times M$ nonsingular matrix, $B = [b_{ij}]$ is an $N \times N$ nonsingular matrix, and

$$V = \begin{bmatrix} V_{0,0}(z_1^M,z_2^N) & V_{0,1}(z_1^M,z_2^N) & \cdots & V_{0,N-1}(z_1^M,z_2^N) \\ V_{1,0}(z_1^M,z_2^N) & V_{1,1}(z_1^M,z_2^N) & \cdots & V_{1,N-1}(z_1^M,z_2^N) \\ \vdots & \vdots & \ddots & \vdots \\ V_{M-1,0}(z_1^M,z_2^N) & V_{M-1,1}(z_1^M,z_2^N) & \cdots & V_{M-1,N-1}(z_1^M,z_2^N) \end{bmatrix}. \tag{8}$$

The expression of Eq. (7) is called the *2-D structural subband decomposition* [4]. The relation between the polyphase components $x_{r,s}(m,n)$ and the generalized polyphase components $v_{r,s}(m,n)$, the inverse transform of $V_{r,s}(z_1,z_2)$ can be established easily from Eqs. (5) and (7), and is given by:

$$V = A^{-1} X B^{-1}. \tag{9}$$

One possible implementation of the above generalized 2-D polyphase decomposition is shown in Figure 2.

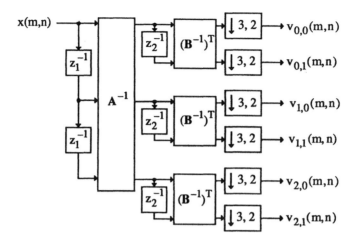

Figure 2 - Generalized polyphase decomposition of a 2-D sequence.

The subband decomposition matrices **A** and **B** can be chosen judiciously to simplify the digital filtering algorithms of interest. Particularly suitable for this purpose is the Hadamard matrix [4]. A K×K Hadamard matrix R_K, with $K = 2^L$, is given by

$$R_K = \underbrace{R_2 \otimes R_2 \otimes \cdots \otimes R_2}_{L-times} \tag{10}$$

where $R_2 = \begin{bmatrix} 1 & 1 \\ 1 & -1 \end{bmatrix}$ and $\otimes$ denotes the Kronecker product. It can be shown that $R_K^{-1} = R_K/K$. Note from Eq. (10) that K must be a power-of-two, i.e., $K = 2^L$. In this case, Eq. (9) reduces to:

$$V = \frac{1}{MN} R_M X R_N. \tag{11}$$

3. EFFICIENT APPROXIMATE 2-D DFT COMPUTATION

The network to the left of the down-samplers in Figure 2 can be considered as a 2-D analysis filter bank. By choosing appropriately the subband decomposition matrices, one can derive at analysis filters with reasonable frequency separation. This property of the analysis filters is exploited here in the fast approximate computation of the 2-D DFT. Recall that the (P,Q)-point DFT $X(k,\ell)$ of a 2-D array $x(m,n)$ of size P×Q is simply given by the equally-spaced frequency samples of its z-transform $X(z_1, z_2)$ on the unit bi-disk, i.e.

$$X(k,\ell) = X(z_1, z_2)\Big|_{z_1 = e^{j2\pi k/P}, z_2 = e^{j2\pi \ell/Q}}, \quad k = 0,1,\ldots,P-1; \ell = 0,1,\ldots,Q-1. \tag{13}$$

Consider first a 4-branch structural subband decomposition of $x(m,n)$ based on the Hadamard transform. Then from Eq. (7), for $M = N = 2$, we get

$$X(z_1,z_2) = \begin{bmatrix} 1 & z_1^{-1} \end{bmatrix} R_2 \begin{bmatrix} V_{0,0}(z_1^2,z_2^2) & V_{0,1}(z_1^2,z_2^2) \\ V_{1,0}(z_1^2,z_2^2) & V_{1,1}(z_1^2,z_2^2) \end{bmatrix} R_2 \begin{bmatrix} 1 \\ z_2^{-1} \end{bmatrix}$$

$$= (1+z_1^{-1})(1+z_2^{-1})V_{0,0}(z_1^2,z_2^2) + (1+z_1^{-1})(1-z_2^{-1})V_{0,1}(z_1^2,z_2^2)$$

$$+ (1-z_1^{-1})(1+z_2^{-1})V_{1,0}(z_1^2,z_2^2) + (1-z_1^{-1})(1-z_2^{-1})V_{1,1}(z_1^2,z_2^2) \tag{14}$$

where $A = B = R_2$, the 2×2 Hadamard matrix. In Eq. (14), the quantities $(1+z_1^{-1})(1+z_2^{-1})$ and $(1-z_1^{-1})(1-z_2^{-1})$ represent, respectively, a separable 2-D lowpass and a highpass filter. The quantity $(1+z_1^{-1})(1-z_2^{-1})$ represents a 2-D filter which is lowpass in the horizontal frequency direction and highpass in the vertical frequency direction, whereas, the quantity $(1-z_1^{-1})(1+z_2^{-1})$ represents a 2-D filter which is highpass in the horizontal frequency direction and lowpass in the vertical frequency direction. Thus, if the 2-D sequence contains primarily low frequency components, we can approximate $X(z_1, z_2)$ as

$$X(z_1,z_2) \cong (1+z_1^{-1})(1+z_2^{-1})V_{0,0}(z_1^2,z_2^2) \tag{15}$$

From Eqs. (13) and (14) it follows that $X(k,\ell)$ can be expressed as[†] :

$$X(k,\ell) = (1+W_P^{-k})(1+W_Q^{-\ell})V_{0,0}(<k>_{P/2},<\ell>_{Q/2})$$
$$+ (1+W_P^{-k})(1-W_Q^{-\ell})V_{0,1}(<k>_{P/2},<\ell>_{Q/2})$$
$$+ (1-W_P^{-k})(1+W_Q^{-\ell})V_{1,0}(<k>_{P/2},<\ell>_{Q/2})$$
$$+ (1-W_P^{-k})(1-W_Q^{-\ell})V_{1,1}(<k>_{P/2},<\ell>_{Q/2}). \tag{16}$$

It should be noted that in Eq. (16), $V_{i,j}(<k>_{P/2},<\ell>_{Q/2})$ denotes the $(P/2,Q/2)$-point DFT of the 2-D sub-sequence $v_{i,j}(m,n)$ generated by the 4-branch structural subband decomposition of $x(m,n)$ based on the Hadamard matrix. This procedure is repeated, by applying the decomposition of Eq. (16) to the computation of the smaller transforms $V_{i,j}(<k>_{P/2},<\ell>_{Q/2})$, and so on until no DFTs are needed. The flow-graph for the first stage in the (P,Q)-point DFT computation is shown in Fig. 3. The (P,Q)-point DFT based on Eq. (16) requires $(P/2)\log_2 P + (Q/2)\log_2 Q$ complex multiplications and $(3P/2)\log_2 P + (3Q/2)\log_2 Q$ additions/subtractions for values of P and Q that are powers of 2. Therefore, the algorithm based on Eq. (16) requires the same number of multiplications as the conventional FFT algorithms, but some more additions.

[†] $<k>_P = k$ modulo P

It follows from our earlier discussions, in the above algorithm, the preprocessing stage has a physical interpretation that is not present in the same advantageous way in other FFT algorithms: Spectral components are somewhat separated in frequency bands by the decompo-

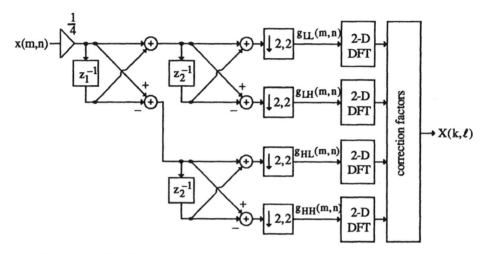

Figure 3 - 2-D DFT calculation using a 4-band structural subband decomposition.

sition block (which is simply a Hadamard subband 2-D filter bank) before the superposition in the weighting factor block yields the exact DFT. In some applications, where signal components are primarily concentrated in a certain frequency range, only parts of the weighting factor block of Fig. 3 may be kept and the remaining portion is removed from computation. The process is equivalent to eliminating some of the smaller DFTs and yields approximate values of the dominant DFT samples.

There are many applications where it is important to have a quick approximate estimate of the signal's dominant spectral samples. Depending on the a-priori information of the analyzed signal, the FFT algorithm outlined above can be preprogrammed to use only a given number of bands (resulting from the subband decomposition of the preprocessing stage) in the computation, increasing iteratively the accuracy by adding more bands if necessary. Alternatively, the energy contents in the different bands can be detected by means of an adaptive scheme to assign priorities in the computation order [5] thus simplifying the DFT computation process.

4. AN IMAGE PROCESSING APPLICATION

In image processing, due to the nature of most natural scenes, most of the energy contents of the corresponding digitized images are concentrated predominantly in the low-low spatial frequency domain. This fact is used in a variety of subband/transform or hybrid coding techniques [6],[7]. The concentration of energy in a localized region of the transform domain makes the approximate DFT computation quite suitable for the calculation of the 2-D image spectra.

Figure 4(a) shows the image "Lena" of size 512×512 with 8 bits per pixel resolution. Figure 4(b) shows its exact 2-D DFT, whereas Figure 4(c) shows the approximate 2-D (512×512)-point DFT computed as follows:

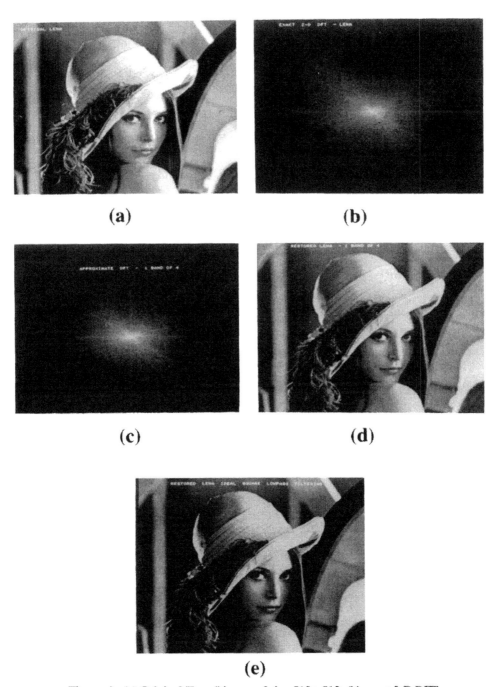

(a)

(b)

(c)

(d)

(e)

Figure 4 - (a) Original "Lena" image of size 512×512, (b) exact 2-D DFT,
(c) approximate 2-D DFT obtained by retaining low-low frequency band,
(d) image reconstructed from the approximate 2-D DFT, and
(e) image reconstructed by ideal 2-D separable lowpass filtering in the transform domain.

$$X(k,\ell) \cong \begin{cases} (1+W_P^{-k})(1+W_Q^{-\ell})V_{0,0}(<k>_{P/2}, <\ell>_{Q/2}), & 0 \le k \le 255, 0 \le \ell \le 255, \\ 0, & \text{otherwise.} \end{cases} \quad (17)$$

i.e., the low-low spatial frequency portion of the DFT is computed using the 256×256 low-frequency sub-samples of the analysis filter bank with the DFT samples not computed set to zero values. Figures 4(b) and (c) show the log-magnitude of the computed spectra to bring out details normally not visible due to the strong dominance of the dc component in the DFT. To evaluate the effectiveness of the approximated 2-D DFT of Figure 4(c) in representing the original image, its 2-D IDFT is shown in Figure 4(d). It can be seen that the subjective degradation in the reconstructed image quality is quite low, given the fact that the computations necessary to compute the approximate spectrum are about 25% of that needed in the exact computation. For comparison, Figure 4(e) shows the image reconstructed from an ideal transform domain filtering for which all components of the exact full size DFT outside the low-low band have been set equal to zero. It is apparent that due to the frequency response of the 2-D lowpass Hadamard filter which tapers off towards the edges of the retained band, the ringing effects in the reconstructed image from the approximated transform are significantly reduced when compared to the ideal filtering case.

5. CONCLUDING REMARKS

A new 2-D DFT computational algorithm based on a subband decomposition of the input 2-D sequence has been outlined. Unlike other well-known algorithms, the proposed method allows physical interpretation of the decomposition process at each step, based on a spectral separation of the subsequences generated by the subband decomposition. This has been shown to be useful in developing fast algorithms for approximate computation of the dominant 2-D DFT samples relying on a-priori knowledge of the input signal spectrum. The proposed approximate DFT algorithm can be used in various other applications, such as approximate convolution and correlation computations.

Acknowledgments

This work was supported in part by a University of California MICRO grant with matching supports from the Digital Instruments, Tektronix, and Rockwell International.

References

[1] N. S. Jayant and P. Noll, *Digital Coding of Waveforms*, Prentice-Hall, Inc., Englewood Cliffs, NJ, 1984.
[2] J. W. Woods and S. D. O'Neil, "Subband coding of images," *IEEE Trans. Acoustics, Speech, Signal Processing*, vol. ASSP-34, pp. 1278-1288, October 1986.
[3] M. Bellanger, G. Bonnerot, and M. Coudreuse, "Digital filtering by polyphase network: Application to sampling rate alteration and filter banks," *IEEE Trans. Acoustics, Speech, Signal Processing*, vol. ASSP-24, pp. 109-114, April 1976.
[4] S. K. Mitra, A. Mahalanobis and T. Saramäki, "A generalized structural subband decomposition of FIR filters and its application in efficient FIR filter design and implementation" *IEEE Trans. on Circuits & Systems*, vol. 40, February 1993 - to be published.
[5] S. K. Mitra, M. R. Petraglia, and O. V. Shentov, "DFT calculation via subband decomposition," *Proc. Fifth European Signal Processing Conference (EUSIPCO)*, Barcelona, Spain, pp. 501-504, September 1990.

[6] J. S. Lim and A. V. Oppenheim, *Advanced Topics in Signal Processing*, Prentice-Hall, Inc., Englewood Cliffs, NJ, 1988.

[7] J. W. Woods, *Subband Image Coding*, Kluwer Academic Publishers, Boston, MA, 1991.

Time-Varying Image Processing and
Moving Object Recognition, 3 – V. Cappellini (Ed.)

A constant–geometry multidimensional FFT

R. Bernardini, G. Cortelazzo, G.A. Mian

Dipartimento di Elettronica e Informatica, Universitá di Padova
via Gradenigo 6/A, 35131 Padova, Italy

1. Introduction

Phase–correlation is a well–known technique for estimating the displacement between two pictures relative to the same scene [1]. The displacement can then be used in order to determine the velocity of the moving object [2].

The idea behind this method is very simple. Let $L_1(\mathbf{f})$ represent the Fourier transform of the image of a scene with a single object moving on a uniform background, taken at time t_1. The Fourier transform of an image relative to this scene, taken at time t_2 is $L_2(\mathbf{f}) = L_1(\mathbf{f})e^{-j2\pi \mathbf{f}^T \mathbf{v}\Delta t}$ where $\mathbf{v}$ is the object's velocity and $\Delta t = t_2 - t_1$.

Quantity

$$c(\mathbf{x}) = \mathcal{F}^{-1}\left[\frac{L_1(\mathbf{f})L_2^*(\mathbf{f})}{|L_1(\mathbf{f})L_2^*(\mathbf{f})|}\right] \tag{1}$$

called phase correlation surface in this case is

$$c(\mathbf{x}) = \delta(\mathbf{x} - \mathbf{v}\Delta t) \tag{2}$$

Therefore the velocity can be determined by localizing the maximum of function $c(\mathbf{x})$ (and by scaling its coordinates by Δt).

Equality (2) strictly holds only if the illumination is constant, there is a single moving object in the scene, and if the background is uniform. The method however is rather robust. It can be proven to be intrinsecally indipendent from illumination variation of type $l_2(\mathbf{x}) = \alpha + \beta l_1(\mathbf{x} - \mathbf{v}\Delta t)$ with $\alpha, \beta \geq 0$. The maximum of (1) is localized at the correct displacement and it remains relatively easy to detect, even though the background is not uniform, as long as the background intensity function is inferior to the object's one. The spurious information determined by the possible (partial) entrance of other objects into the scene can be limited by the use of suitable windows.

Such characteristics makes phase–correlation applicable also to television applications [2], [3]. This context clearly requires real–time velocity estimates, a goal achievable only by means of dedicated hardware. The hardwarization would be extremely simplified by FFT algorithms having all stages with identical geometrical configuration [4]. In the one–dimensional case the sequential FFT algorithm, due to Singleton [5], is known to have such a interesting characteristic.

This work presents an original algorithm for the sequential computation of the M–D FFT, which also enjoys constant geometry.

Motion estimation by means of phase–correlation, is only one of the many application calling for the use of "on–chip" FFT's in sequence processing, other applications are spectral estimation [6] and fast convolution [7]. The usefulness of the proposed "constant–geometry" algorithm in these context is apparent.

2. Background and problem statement

2.1. Definitions

Let N be a non–singular integer matrix, lattice $\Lambda(\mathbf{N})$ is defined as [7], [8]

$$\Lambda(\mathbf{N}) \triangleq \{\mathbf{m} : \mathbf{m} = \mathbf{N}\mathbf{k}; \mathbf{k} \in Z^M\} \tag{3}$$

Multidimensional signal $x(\mathbf{p})$, $\mathbf{p} \in Z^M$ is periodic on lattice $\Lambda(\mathbf{N})$ if

$$x(\mathbf{p}) = x(\mathbf{p} + \mathbf{N}\mathbf{m}); \qquad \forall \mathbf{m}, \mathbf{p} \in Z^M \tag{4}$$

Signal $x(\mathbf{p})$ is completely defined by its $|\det(\mathbf{N})|$ values on any of its periods I_N. The Fourier transforms of $x(\mathbf{p})$ is defined as [7], [8]

$$X(\boldsymbol{\lambda}) = \sum_{\mathbf{p} \in I_N} x(\mathbf{p}) \exp(-i2\pi \boldsymbol{\lambda}^T \mathbf{N}^{-1}\mathbf{p}) \tag{5}$$

$$x(\mathbf{p}) = \frac{1}{|\det \mathbf{N}|} \sum_{\boldsymbol{\lambda} \in I_{N^T}} X(\boldsymbol{\lambda}) \exp(i2\pi \boldsymbol{\lambda}^T \mathbf{N}^{-1}\mathbf{p}) \tag{6}$$

and it is periodic on $\Lambda\left(\mathbf{N}^T\right)$, i.e., $X(\boldsymbol{\lambda}) = X(\boldsymbol{\lambda} + \mathbf{N}^T\mathbf{k})$, $\forall \boldsymbol{\lambda}, \mathbf{k} \in Z^M$. Let I_{N^T} denote a period of $X(\boldsymbol{\lambda})$.

If $\mathbf{N} = \mathbf{P}_1\mathbf{Q}_1$, with $\mathbf{P}_1$ and $\mathbf{Q}_1$ integer matrixes, $X(\boldsymbol{\lambda})$ can be computed by the multidimensional extension of the algorithm of Cooley and Tukey [9] which corresponds to

$$X\left(\boldsymbol{\mu}_1 + \mathbf{Q}_1^T\boldsymbol{\lambda}_1\right) = \sum_{\mathbf{p}_1 \in I_{P_1}} \exp\left(-i2\pi \boldsymbol{\lambda}_1^T \mathbf{P}_1^{-1}\mathbf{p}_1\right) \exp\left(-i2\pi \boldsymbol{\mu}_1^T \mathbf{N}^{-1}\mathbf{p}_1\right)$$

$$\times \sum_{\mathbf{q} \in I_{Q_1}} \exp\left(-i2\pi \boldsymbol{\mu}_1^T \mathbf{Q}_1^{-1}\mathbf{q}_1\right) x(\mathbf{P}_1\mathbf{q}_1 + \mathbf{p}_1) \tag{7}$$

If $\mathbf{N} = \prod_{i=1}^{\nu} \mathbf{P}_i$ expression (7) after ν iterations becomes

$$X \left(\boldsymbol{\lambda}_\nu + \mathbf{P}_\nu^T \boldsymbol{\lambda}_{\nu-1} + \mathbf{P}_\nu^T \mathbf{P}_{\nu-1}^T \boldsymbol{\lambda}_{\nu-2} + \cdots + \mathbf{P}_\nu^T \mathbf{P}_{\nu-1}^T \cdots \mathbf{P}_2^T \boldsymbol{\lambda}_1 \right)$$

$$= \sum_{\mathbf{p}_1 \in I_{P_1}} \exp\left(-i2\pi \boldsymbol{\lambda}_1{}^T \mathbf{P}_1^{-1} \mathbf{p}_1 \right) \mathrm{Tw}_{\nu-1}$$

$$\times \sum_{\mathbf{p}_2 \in I_{P_2}} \exp\left(-i2\pi \boldsymbol{\lambda}_2{}^T \mathbf{P}_2^{-1} \mathbf{p}_2 \right) \mathrm{Tw}_{\nu-2}$$

$$\cdots$$

$$\times \sum_{\mathbf{p}_\nu \in I_{P_\nu}} \exp\left(-i2\pi \boldsymbol{\lambda}_\nu{}^T \mathbf{P}_\nu^{-1} \mathbf{p}_\nu \right) x \left(\mathbf{p}_1 + \mathbf{P}_1 \mathbf{p}_2 + \mathbf{P}_1 \mathbf{P}_2 \mathbf{p}_3 + \cdots + \mathbf{P}_1 \mathbf{P}_2 \cdots \mathbf{P}_{\nu-1} \mathbf{p}_\nu \right)$$

$$\tag{8}$$

Symbols Tw_i represent the twiddle factor of the i-th iteration and $\boldsymbol{\lambda}_i \in I_{P_i}$, $i = 1, 2, \ldots, \nu$. The twiddle factors do not play any role in the development of the sequential multidimensional Cooley–Tukey algorithm, therefore in the following they will just be denoted as Tw.

According to (8) the computation of the DFT corresponds to a sequence of DFT followed by the corresponding twiddle–factor multiplication. Since the output-data of the i-th DFT, denoted as $\chi_i(\mathbf{p}_1, \mathbf{p}_2, \ldots, \mathbf{p}_{\nu-i}, \boldsymbol{\lambda}_{\nu-i+1}, \ldots, \boldsymbol{\lambda}_\nu)$ has $\nu - i$ spatial indexes $(\mathbf{p}_1, \mathbf{p}_2, \ldots, \mathbf{p}_{\nu-i})$ and i frequency indexes $(\boldsymbol{\lambda}_{\nu-i+1}, \ldots, \boldsymbol{\lambda}_\nu)$, the MD Cooley–Tukey algorithm can be described by the following operations

$$\chi_0(\mathbf{p}_1, \mathbf{p}_2, \ldots, \mathbf{p}_\nu) = x(\mathbf{p}_1 + \mathbf{P}_1 \mathbf{p}_2 + \cdots + \mathbf{P}_1 \ldots \mathbf{P}_{\nu-1} \mathbf{p}_\nu)$$

$$\chi_1(\mathbf{p}_1, \mathbf{p}_2, \ldots, \boldsymbol{\lambda}_\nu) = \mathrm{Tw}\ \mathrm{DFT}_{\mathbf{p}_\nu \to \boldsymbol{\lambda}_\nu} \chi_0(\mathbf{p}_1, \mathbf{p}_2, \ldots, \mathbf{p}_\nu)$$

$$\chi_2(\mathbf{p}_1, \mathbf{p}_2, \ldots, \boldsymbol{\lambda}_\nu) = \mathrm{Tw}\ \mathrm{DFT}_{\mathbf{p}_{\nu-1} \to \boldsymbol{\lambda}_{i-1}} \chi_1(\mathbf{p}_1, \mathbf{p}_2, \ldots, \boldsymbol{\lambda}_\nu)$$

$$\vdots$$

$$\chi_\nu(\boldsymbol{\lambda}_1, \boldsymbol{\lambda}_2, \ldots, \boldsymbol{\lambda}_\nu) = \mathrm{Tw}\ \mathrm{DFT}_{\mathbf{p}_1 \to \boldsymbol{\lambda}_1} \chi_{\nu-1}(\mathbf{p}_1, \boldsymbol{\lambda}_2, \ldots, \boldsymbol{\lambda}_\nu)$$

$$X(\boldsymbol{\lambda}_\nu + \mathbf{P}_\nu^T \boldsymbol{\lambda}_{\nu-1} + \cdots + \mathbf{P}_\nu^T \ldots \mathbf{P}_2^T \boldsymbol{\lambda}_1) = \chi_\nu(\boldsymbol{\lambda}_1, \boldsymbol{\lambda}_2, \ldots, \boldsymbol{\lambda}_\nu)$$

$$\tag{9}$$

Symbol $\mathrm{Tw}\ \mathrm{DFT}_{\mathbf{p}_i \to \boldsymbol{\lambda}_i}$ correspond to the execution of the DFT with respect to $\mathbf{p}_i$ and to the multiplication by the corresponding twiddle-factor.

2.2. The one–dimensional sequential FFT

Extensive presentations of the Singleton algorithm can be found in popular signal processing textbooks [10]. This subsection points out some fundamental characteristics of the algorithm, retained, in some way, by its multidimensional extension. The one-dimensional algorithm uses two files for input storage and two files for output storage. The sequential/parallel version of the algorithm reads a pair of data from the first input file (it switches to the second, when the first file is finished) and writes the two corresponding output data, one per output file. On the converse, the parallel/sequential version, reads a datum from each input file, and writes a pair of data to the output file. If the number of input samples is $N = p^{\nu}$, where p is an integer, every stage of the algorithm has the same structure. Because of such a distinctive feature, the one-dimensional sequential FFT, is often referred to as constant-geometry FFT.

3. A parallel/sequential M-D Cooley-Tukey FFT

This section presents a sequential access algorithm for computing the iterations of (9) which reads the input data of the i-th iteration from $|\det(\mathbf{P}_{\nu-i})|$ files of $|\det(\mathbf{N})|/|\det(\mathbf{P}_{\nu-i})|$ data (organized in a single file) and writes the output data of the i-th iteration into a file of $|\det(\mathbf{N})|$ locations.

Let us denote that $y(\mathbf{n})$ is the address of $x(\mathbf{n})$ as

$$x(\mathbf{n}) \rightarrow \text{MEM}[y(\mathbf{n})] \tag{10}$$

The desired input data ordering formally corresponds to an input addressing rule $S(\mathbf{p})$, denoted as

$$x(\mathbf{p}) \rightarrow \text{MEM}[S(\mathbf{p})] \tag{11}$$

such that the input data read at the i-th iteration are organized as

$$\chi_i(\mathbf{p}_1, \mathbf{p}_2, \mathbf{p}_{\nu-i}, \boldsymbol{\lambda}_{\nu-i+1}, \ldots, \boldsymbol{\lambda}_{\nu}) \rightarrow \text{MEM}[n_{i+1} + \tfrac{|\det(\mathbf{N})|}{|\det(\mathbf{P})|_{\nu-i}} f_{\nu-i}(\mathbf{p}_{\nu-i})] \tag{12}$$

where $f_{\nu-i}$ is a one–to–one and onto mapping $I_{P_{\nu-i}} \rightarrow \{0, 1, \ldots, |\det(\mathbf{P}_{\nu-i})| - 1\}$, and n_i is the number of the n-th butterfly of the i-th stage, i.e., $n_i \in \{0, 1, \ldots, |\det(\mathbf{N})|/|\det(\mathbf{P}_{\nu-i})|\}$. The i-th iteration output of the P/S algorithm is written according to rule

$$\chi_i(\mathbf{p}_1, \mathbf{p}_2, \ldots, \mathbf{p}_{\nu-i}, \boldsymbol{\lambda}_{\nu-i+1}, \ldots, \boldsymbol{\lambda}_{\nu}) \rightarrow \text{MEM}[n_i| \det(\mathbf{P}_{\nu-i+1})| + g_{\nu-i+1}(\boldsymbol{\lambda}_{\nu-i+1})] \tag{13}$$

where $g_{\nu-i+1}$ is is a one-to-one and onto mapping, $I_{P_{\nu-i+1}} \rightarrow \{0, 1, \ldots, |\det(\mathbf{P}_{\nu-i+1})| - 1\}$. The determination of the P/S algorithm requires an input-ordering rule

$$x(\mathbf{p}) \rightarrow \text{MEM}[S_{\text{P/S}}(\mathbf{p})] \tag{14}$$

allowing for iteration (12) at the first stage. It is also necessary to determine the output ordering rule

$$X(\boldsymbol{\lambda}) \rightarrow \text{MEM}[R_{\text{P/S}}(\boldsymbol{\lambda})] \tag{15}$$

Both tasks can be accomplished by means of an iterative definition of quantity n_1.

The input data to be read at the first iteration must be put into $|\det(\mathbf{P}_\nu)|$ files according to (12), i.e.,

$$\chi_0(\mathbf{p}_1, \mathbf{p}_2, \ldots, \mathbf{p}_\nu) \to \text{MEM}[n_1(\mathbf{p}_1, \mathbf{p}_2, \ldots, \mathbf{p}_{\nu-1}) + \frac{|\det(\mathbf{N})|}{|\det(\mathbf{P}_\nu)|} f_\nu(\mathbf{p}_\nu)] \qquad (16)$$

From (13) the first stage output is written as

$$\chi_1(\mathbf{p}_1, \mathbf{p}_2, \ldots, \mathbf{p}_{\nu-1}, \boldsymbol{\lambda}_\nu) \to \text{MEM}[n_1(\mathbf{p}_1, \mathbf{p}_2, \ldots, \mathbf{p}_{\nu-1})|\det(\mathbf{P}_\nu)| + g_\nu(\boldsymbol{\lambda}_\nu)] \qquad (17)$$

As data (17) are the second iteration input, they must also be ordered according to (12), i.e.,

$$\chi_1(\mathbf{p}_1, \mathbf{p}_2, \ldots, \mathbf{p}_{\nu-1}, \boldsymbol{\lambda}_\nu) \to \text{MEM}[n_2(\mathbf{p}_1, \mathbf{p}_2, \ldots, \mathbf{p}_{\nu-2}, \boldsymbol{\lambda}_\nu) + \frac{|\det(\mathbf{N})|}{|\det(\mathbf{P}_{\nu-1})|} f_{\nu-1}(\mathbf{p}_{\nu-1})]$$
$$(18)$$

If in (17) one uses position

$$n_1(\mathbf{p}_1, \mathbf{p}_2, \ldots, \mathbf{p}_{\nu-1}) = \frac{|\det(\mathbf{N})|}{|\det(\mathbf{P}_{\nu-1})||\det(\mathbf{P}_\nu)|} f_{\nu-1}(\mathbf{p}_{\nu-1}) + n_1^1(\mathbf{p}_1, \mathbf{p}_2, \ldots, \mathbf{p}_{\nu-2}),$$
$$(19)$$

where quantity n_1^1 will be subsequently determined, the argument of (17) becomes of type (18), with

$$n_2(\mathbf{p}_1, \mathbf{p}_2, \ldots, \mathbf{p}_{\nu-2}, \boldsymbol{\lambda}_\nu) = |\det(\mathbf{P}_\nu)| n_1^1(\mathbf{p}_1, \mathbf{p}_2, \ldots, \mathbf{p}_{\nu-2}) + g_\nu(\boldsymbol{\lambda}_\nu) \qquad (20)$$

By iterating the above procedure at the subsequent stages, one can define quantity n_1^m in terms of subsequent stage quantity n_1^{m+1} as

$$n_1^m = n_1^{m+1} + \frac{|\det(\mathbf{N})|}{\prod_{k=m}^\nu |\det(\mathbf{P}_k)|} f_m(\mathbf{p}_m) \qquad (21)$$

Quantity n_1 can be determined by means of (21), resulting

$$n_1(\mathbf{p}_1, \mathbf{p}_2, \ldots, \mathbf{p}_{\nu-1}) = \sum_{m=1}^{\nu-1} \prod_{k=1}^{m-1} |\det(\mathbf{P}_k)| f_m(\mathbf{p}_m) \qquad (22)$$

From (16) and (22) the wanted input ordering $S_{P/S}(\mathbf{p})$ is

$$S_{P/S}(\mathbf{p}) = \sum_{m=1}^\nu \prod_{k=1}^{m-1} |\det(\mathbf{P}_k)| f_m(\mathbf{p}_m) \qquad (23)$$

It is worth remembering that the input data of the P/S FFT are progressively ordered and not scrambled in bit-reversal fashion. By similar arguments intermediate stage $\chi_i, i = 1, 2, \ldots, \nu$, can be seen to be organized as

$$\chi_i(\mathbf{p}_1, \mathbf{p}_2, \mathbf{p}_{\nu-i}, \boldsymbol{\lambda}_{\nu-i+1}, \ldots, \boldsymbol{\lambda}_\nu)$$

$$\rightarrow \mathrm{MEM}[\textstyle\sum_{m=\nu-i+1}^{\nu} \prod_{k=\nu-i+1}^{m-1} |\det(\mathbf{P}_k)| g_m(\boldsymbol{\lambda}_m) \tag{24}$$

$$+ \prod_{k=\nu-i+1}^{m-1} |\det(\mathbf{P}_k)| \textstyle\sum_{m=1}^{\nu-i} \prod_{l=1}^{m-1} |\det(\mathbf{P}_l)| f_m(\mathbf{p}_m)$$

From (24) with $i = \nu$, one obtains

$$R_{\mathrm{P/S}}(\boldsymbol{\lambda}) = \sum_{m=1}^{\nu} \prod_{k=1}^{m-1} |\det(\mathbf{P}_k)| g_k(\boldsymbol{\lambda}_k) \tag{25}$$

4. Constant–geometry characteristic

The constant geometry feature refers to the fact that in the Singleton's algorithm (if input record length is the power of an integer) all the stages have identical geometrical configuration.

In the case of arrays with $\mathbf{N} = \mathbf{P}^\nu$ (and $\mathbf{P}$ generic integer matrix) the following two considerations verify at each stage:

- $\mathbf{P}_i = \mathbf{P}$ (and $|\det(\mathbf{P}_i)| = |\det(\mathbf{P})|$)

- mappings g_i can be chosen identical to that of the first stage.

The two conditions above used in (12) and (13) eliminate the stage–iteration index from the reading and writing expression of the MD S/P algorithm.

Acknowledgement

Work carried out with the partial contribution of C.N.R. Progetto Finalizzato Tele-comunicazioni, contract n. 92.00972.PF71

References

[1] J. Pearson, D. Hines, S. Golosman, and D. Kuglin, "Video rate image correlation Processor," in *Proc. SPIE (IOCC 1977)*, pp. 197–204, SPIE, 1977.

[2] G. Thomas and B. Hons, "Television motion measurements for DATV and other applications." BBC Technical Reports 1987/11.

[3] C. Reventlow, K. Muller, C. Stoppers, and J.Reimers, "Phasecorrelation chipset for TV/HDTV real time motion estimation," Nov 1992.

[4] J. M. C.S. Joshi and R. Steinvorth, "A video rate two dimensional FFT processor," in *ICASSP*, pp. 774–777, IEEE, 1980.

[5] R. C. Singleton, "A metod for computing the Fast Fourier Transform with auxiliary memory and limited high-speed storage," *IEEE Transaction on Audio Electroacoustic*, vol. 17, pp. 91–97, June 1967.

[6] G. M. Cortelazzo, G. Mian, and R. Rinaldo, "Spectral estimation of video signals," *Multidimensional Systems and Signal Proccesing*, vol. 2, pp. 131–160, Mar. 1992.

[7] D. Dudgeon and R. Mersereau, *Multidimensional Digital Signal Processing*. New Jersey: Englewood Cliffs, 1984.

[8] E. Dubois, "The sampling and reconstruction of time-varing imagery," *Proc. of the IEEE*, vol. 73, pp. 502–522, Apr. 1985.

[9] R. Mersereau and T. Speake, "A unified treatment of Cooley–Tukey algorithms for the evaluation of the multidimensional DFT," *IEEE Transaction on Acoustics, Speech and Signal Processing*, vol. ASSP-29, pp. 1011–1017, Oct. 1981.

[10] A. V. Oppenheim and R. W. Schafer, *Discrete-time Signal Processing*. Prentice Hall, 1989.

Time-Varying Image Processing and
Moving Object Recognition, 3 – V. Cappellini (Ed.)
© 1994 Elsevier Science B.V. All rights reserved.

Estimation of the Measurement Covariance Matrix in a Kalman Filter

Oddbjørn Bergem

Norwegian Defence Research Establishment and University of Trondheim

P.b. 115, N–3191 Horten, NORWAY

A method is presented to dynamically estimate the measurement covariance matrix in a Kalman filter. The variance is estimated using the relative uncertainty of the matches in the search region, resulting in a measurement covariance matrix which vary according to the underlying data and model.

1   INTRODUCTION

The Kalman filter has been known since the beginning of the 1960s, and we see today an increasing number of applications inside the computer vision field which use the filter for parameter estimation, prediction, and sensor fusion. Originally, the theory arose because of the inadequacy of the Wiener–Kolmogorov theory for coping with certain applications in which non–stationarity of the signal and/or noise was intrinsic to the problem. The assumption that the underlying signal and noise processes are stationary was crucial to the Wiener–Kolmogorov theory, and the Kalman filter that did not require this stationarity assumption filled a gap in the theory at the right time[1]. Today the Kalman filter is used also for stationary processes, and truly the stationary process is a particular type of non–stationary processes. However, it seems to be a trend to use the Kalman filter and *assume stationary processes* even if the signal or noise processes are non–stationary. Particularly, this is the case for applications where the measurements are results from a matching or correlation between sensor data and a model. Typically this situation arises in tracking, object velocity estimation, positioning systems, and in general in time–varying image processing. In such situations the certainty of the match may vary according to the underlying data and model, and the measurement process noise will not be constant.

2   OVERVIEW OF THE KALMAN FILTER

The Kalman filter is a least square estimation technique using a Bayesian approach[2]. The filter is used to track the state of stochastic dynamic systems being observed with noisy sensors. Essentially, the filter is based on three separate probabilistic models. The first model describes the evolution over (discrete) time by the equation

$$\mathbf{x}(k + 1) = \mathbf{F}(k)\mathbf{x}(k) + \mathbf{G}(k)\mathbf{u}(k) + \mathbf{v}(k) \qquad (2.1)$$

where $\mathbf{x}(k)$ is the state at time k, $\mathbf{u}(k)$ is the controlled input signal entering the system with gain matrix $\mathbf{G}$, and $\mathbf{v}(k)$ is a sequence of zero–mean white Gaussian process noise with covariance matrix $\mathbf{Q}(k)$. $\mathbf{F}$ is the transition matrix of the system.

The second model describes the measurements, and relates the measurement vector $\mathbf{z}$ to the current state through a measurement matrix $\mathbf{H}$:

$$\mathbf{z}(k) = \mathbf{H}(k)\mathbf{x}(k) + \mathbf{w}(k) \qquad (2.2)$$

Here $\mathbf{w}(k)$ is assumed to be a sequence of zero–mean white Gaussian noise with covariance

$$\mathbf{E}\left[\mathbf{w}(k)\mathbf{w}^T(j)\right] = \mathbf{R}(k)\delta_{kj} \qquad (2.3)$$

where δ_{kj} is the Kronecker delta. The third model describes the knowledge about the system state and its covariance before the first measurement is taken. Usually the initial system state is assumed to be a normally distributed random variable with a known mean and a given covariance matrix:

$$\mathbf{x}(0) \sim \mathbf{N}\left[\hat{\mathbf{x}}(0|0), \mathbf{P}(0|0)\right] \qquad (2.4)$$

where $\mathbf{P}$ is the associated conditional state error covariance matrix and the vector $\hat{\mathbf{x}}(a|b)$ is the conditional state estimate. The one step prediction stage can be calculated using the formula

$$\hat{\mathbf{x}}(k + 1|k) = \mathbf{F}(k)\hat{\mathbf{x}}(k|k) + \mathbf{G}(k)\mathbf{u}(k) \qquad (2.5)$$

where the prediction is calculated by applying the transition matrix $\mathbf{F}(k)$ to the previous state estimate and adding the control input. The system state estimate may be calculated by the formula

$$\hat{\mathbf{x}}(k + 1|k + 1) = \mathbf{F}(k)\hat{\mathbf{x}}(k|k) + \mathbf{W}(k + 1)\left[\mathbf{z}(k + 1) - \mathbf{H}(k + 1)\hat{\mathbf{x}}(k + 1|k)\right] \qquad (2.6)$$

This shows that the new estimate, using the measurement $\mathbf{z}(k+1)$, is formed by extrapolating the estimated vector, and then adding a correction term. This correction term, called the innovation (or measurement residual), is formed by subtracting the predicted measurement vector from the new observation vector $\mathbf{z}(k+1)$. The gain matrix $\mathbf{W}$ contains information of how much the innovation should affect the new state estimate. In a Kalman filter, the gain is related recursively to the covariance matrix $\mathbf{P}(k+1|k)$ by the formula

$$\mathbf{W}(k + 1) = \mathbf{P}(k + 1|k)\mathbf{H}^T(k + 1)\mathbf{S}^{-1}(k + 1) \qquad (2.7)$$

where $S^{-1}(k + 1)$ is given by the inverse of

$$S(k + 1) = H(k + 1)P(k + 1|k)H^T(k + 1) + R(k + 1) \qquad (2.8)$$

The state prediction covariance matrix is

$$P(k + 1|k) = F(k)P(k|k)F^T(k) + Q(k) \qquad (2.9)$$

and the updated state covariance matrix is

$$P(k + 1|k + 1) = P(k + 1|k) - W(k + 1)S(k + 1)W^T(k + 1) \qquad (2.10)$$

For a deeper description of the Kalman filter refer to [2].

3   THE MEASUREMENT UNCERTAINTY

Essentially, the main task of the Kalman filter is to adjust the gain matrix W to select between the state deduced from the measurement and the state predicted by the system dynamics encoded in F. To do this, the filter looks at the covariance matrices (equation 2.7). The better the measurement is, the more the measurement is taken into account. The information of the exactness of the measurement is put into the covariance matrix R, which represents the covariance of a Gaussian white noise process added to the measurements. For many sensors, it is easy to compute the variance in the measurements and code this information in R. For applications where the measurements are results from a matching process between the data and a model, it is less clear what the measurement variance is. At least three different alternatives may be used:

A) Assume that the measurement covariance is independent of the actual data and model for the given time step, and use an average value for the measurement covariance based on earlier experiments or post analysis of the matches. The result is a measurement covariance matrix which is constant and independent of the time step k.

B) Assume that the measurement covariance is dependent of the goodness of the match. This means that a short distance (given a selected metric) between the data and the model results in a small variance, and a long distance gives a high variance. The measurement covariance matrix will then be a function of the best match, but the variance will not be influenced by other matches in the search region.

C) Assume that the measurement covariance is a function of the relative goodness of the match. If a distinct match is found this results in a small variance, but if several matches are found the variance is increased.

To illustrate the alternatives above, assume that the measurement problem at hand is to find the position of a box in a picture, i.e. traditional template matching. The measurement process must return the position of the box and the measurement covariance associated with the position. Alternative A) corresponds to the situation where the covariance returned is independent of whether a box is found, whether the box eventually found looks more like a circle, whether there actually were 5 boxes in the picture and so on. Alternative B) will return a covariance which is dependent of the goodness of the match between the best box found and the picture. The problem with this method is that a situation where there are many box–like objects in the picture may result in a small covariance, even if the the actual position is uncertain. For alternative C), the covariance is determined based on how god the best match is compared to the other matches. If there are several good matches, the covariance will be increased, but if there is just one match, the covariance will be low even if the match itself not is perfect. One problem with this method, is the possibility to have a low position covariance even if there is no box in the picture. However, this problem does not occur if it is assumed that the box really exists in the picture, and the probability for this assumption may be controlled by sizing the search area according to the prediction covariance matrix S and the innovation as proposed in [2]. In section 5, experiments are carried out where the problem is to match a depth profile against a depth map of the sea floor. If the sea floor is flat, a good match may be obtained resulting in a low variance when using alternative B), but the position associated with this match is highly uncertain because there are many possible positions where the profile matches the sea floor.

If the match is evaluated over the total search area for alternative C), we get a surface relating the match to the actual position of the measurement. Given this matching strength surface, the covariance matrix may be estimated for almost any matching function by using second order moments as showed in the next section.

4   ESTIMATION OF THE MEASUREMENT COVARIANCE MATRIX

To compare the actual data with the reference map for a position (x, y) corresponding to alternative C) in the previous section, we define a matching strength function $f_v(x, y)$. The matching strength function $f_v(x, y)$ is a function that returns a positive number describing the goodness of the match between the data and the model for every position (x,y) inside the search area V, and which returns zero otherwise. The function must define a total ordering of the matches where the best match is represented by the highest value. The matching strength function may be a correlation type of function, an LMS type of function, or any function which describes the goodness of the match in a "reasonable" manner. For details on the restrictions of $f_v(x, y)$ see [5].

Suppose $\mathbf{X}^T = [X_1, X_2]$ is a random vector described by the joint probability density function $f(x_1, x_2) = f(\mathbf{x})$. The expected values of the random vector $\mathbf{X}$ is contained in the vector of means $\bar{\mathbf{x}} = E(\mathbf{X})$, and the covariances are contained in the matrix $\mathbf{C} = E(\mathbf{X} - \bar{\mathbf{x}})(\mathbf{X} - \bar{\mathbf{x}})^T$. Specifically,

$$C = E(X - \bar{x})(X - \bar{x})^T = E\left[\begin{bmatrix} X_1 - \bar{x}_1 \\ X_2 - \bar{x}_2 \end{bmatrix} [X_1 - \bar{x}_1, \ X_2 - \bar{x}_2]\right]$$

$$= \begin{bmatrix} E(X_1 - \bar{x}_1)^2 & E(X_1 - \bar{x}_1)(X_2 - \bar{x}_2) \\ E(X_1 - \bar{x}_1)(X_2 - \bar{x}_2) & E(X_2 - \bar{x}_2) \end{bmatrix} = \begin{bmatrix} \sigma_{20} & \sigma_{11} \\ \sigma_{11} & \sigma_{02} \end{bmatrix} \qquad (4.1)$$

where

$$\sigma_{pq} = \int_{-\infty}^{\infty} \int_{-\infty}^{\infty} f(x_1, x_2)(x_1 - \bar{x}_1)^p (x_2 - \bar{x}_2)^q dx_1 dx_2 \qquad (4.2)$$

The mean is given by

$$\bar{x} = E(X) = \begin{bmatrix} E(X_1) \\ E(X_2) \end{bmatrix} = \begin{bmatrix} \bar{x}_1 \\ \bar{x}_2 \end{bmatrix} = \begin{bmatrix} \int_{-\infty}^{\infty} \int_{-\infty}^{\infty} x_1 f(x_1, x_2) dx_1 dx_2 \\ \int_{-\infty}^{\infty} \int_{-\infty}^{\infty} x_2 f(x_1, x_2) dx_1 dx_2 \end{bmatrix} \qquad (4.3)$$

Equation 4.1 contains the covariance matrix for the joint probability density function $f(x_1, x_2) = f(\mathbf{x})$.

To compute the covariance matrix for the matching strength surface for a given time step k, the matching strength function $f_v(x_1^k, x_2^k)$ must be normalized to represent a joint probability density function. This can be done by dividing $f_v(x_1^k, x_2^k)$ by the total probability mass, i.e. the volume of the matching strength function:

$$f_N(x_1^k, x_2^k) = \frac{f_v(x_1^k, x_2^k)}{\int\int_V f_v(x_1^k, x_2^k) dx_1^k dx_2^k} \qquad (4.4)$$

The dynamic measurement covariance matrix (DMCM) is then computed for a given time step k by the equation

$$\mathbf{R}(k) = \begin{bmatrix} \sigma'_{20} & \sigma'_{11} \\ \sigma'_{11} & \sigma'_{02} \end{bmatrix} \qquad (4.5)$$

where

$$\sigma'_{pq} = \int\int\limits_{V} f_N(x_1^k, x_2^k)(x_1^k - \bar{x}^k_1)^p (x_2^k - \bar{x}^k_2)^q dx_1^k dx_2^k \qquad (4.6)$$

As seen from the equations above, it is clear that the DMCM will vary with the time step k because the matching function vary with k. However, the covariance is not a stochastic function of k, but rather a deterministic function of the position and size of the validation gate and of the actual data. As a consequence a dependency is introduced between the states and the measurements, which means that the gain matrix no longer is precomputeable, but the Kalman equations are still valid.

5   EXPERIMENTAL RESULTS

The technique has been tested on a positioning system for an autonomous underwater vehicle (AUV), where a multibeam sonar is used to extract the sea floor profile. This profile is matched against a detailed depth map, and the position of the vehicle is estimated. The experiments described here is based on data from a 4 km^2 area in Oslofjorden, recorded with a multibeam sonar with 60 beams. The reference map used had a resolution of 10 meter and an accuracy of approximately 1%. To illustrate the importance of the measurement variances, a track which consists of 270 measurements starting at position (1000,800) is selected (figure 5.1). This track is run both by using the DMCM suggested in section 4, and by using different constant measurement covariance matrices corresponding to alternative A). Experiments using alternative B) give for this task very poor results (which is not surprising), and for a description of these results and a more exhaustive description of other tracks and experiments the reader is referred to [3][4][5].

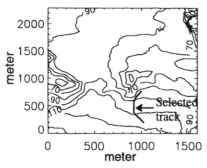

Figure 5.1 Contour map and selected track.

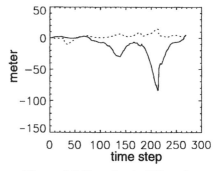

Figure 5.2 Error in x(solid) and y. R matrix varying

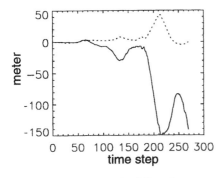

Figure 5.3 Error in x(solid) and y.
R values equal 200.

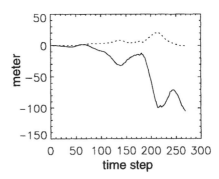

Figure 5.4 Error in x(solid) and y.
R values equal 2000.

Figure 5.2 shows the error in the estimated x and y position when using the DMCM*. Figure 5.3 and 5.4 show the error in the estimated position when a constant value of 200 and 2000 respectively are used for the measurement variances. A value of approximately 200 meter is the average variance in the x and y estimates based on a post analysis of 15000 measurements in the area, but for this track a higher variance gives better results. Table 5.1 shows the standard deviation and maximum error for the selected track for different values of the measurement variances. As seen in the table the best constant value for this track is 2000 meter, but even then the error is much higher than using the DMCM.

R VALUES	SD X	SD Y	MAX X	MAX Y
DMCM	13.7	3.7	83.6	14.7
200	37.5	6.7	151.6	42.3
500	34.0	5.6	123.6	31.1
1000	31.6	5.4	111.7	26.6
1500	30.5	5.2	103.8	23.3
2000	30.4	5.0	99.0	21.9
2500	31.4	5.1	110.8	20.9

Table 5.1   Standard deviation and maximum error for the selected track.

In figure 5.5 and 5.6 the values for the x variance and y variance in the DMCM is plotted as a function of the time–step k. From time–step 0 to 80 the vessel is traversing an area with a steep slope, and the matching function returns an almost unique answer resulting in low variances. From time–step 100 to 200 the underlying sea floor gets more flat, and the variances increase. The vessel makes a turn at time–step 170 which is difficult to detect from the profile due to the flatness of the sea floor. As a result the error increases, but looking at figure 5.5 a few good measurements are taken around time–step 215 causing the error to decrease when the DMCM is used.

*For the actual positioning system on the AUV a speed senor also gives input to the system, causing the standard deviation to be less than 3 meter.

26

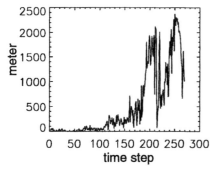

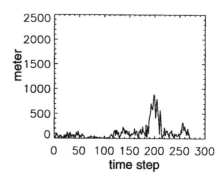

Figure 5.5 R_x value. Figure 5.6 R_y value.

When using the constant values for the measurement variances, a small constant value causes the filter to believe to much in the measurement and therefore increase the error, while a high constant value causes the filter to discard the good measurements.

6   CONCLUSION

In this paper we have presented a technique that incorporates the uncertainty in the matching process when using a Kalman filter for navigation or tracking purposes. The technique estimates the measurement variance in the search region, resulting in a measurement covariance matrix varying dynamically according to the underlying model and data. Experiments indicate important improvements compared to a Kalman filter where a constant measurement covariance matrix is used. The improvements are mainly a consequence of using all accessible information about the match between model and data to give a correct assessment of the measurement error.

REFERENCES

[1]   Anderson, B.D.O. and Moore, John B. "Optimal filtering". *Prentice–Hall*, 1979.

[2]   Bar–Shalom, Y. and Fortmann, E. "Tracking and Data Association". *Academic Press*, 1988.

[3]   Bergem, O et al. "Using Match Uncertainty in the Kalman Filter for a Sonar Based Positioning System". Proc. of *8th Scandinavian Conference on Image Analysis*, Tromsø, Norway, 1993.

[4]   Bergem, O. "A Multibeam Sonar Based Positioning System for an AUV". Accepted to *8th Int. Symposium on Unmanned Untethered Submersible Technology*, New Hampshire, USA, 1993.

[5]   Bergem, O. "A Sonar Based Positioning System for an Autonomous Underwater Vehicle". *Ph.D. thesis, University of Trondheim,* Norway, to be published.

Time-Varying Image Processing and
Moving Object Recognition, 3 – V. Cappellini (Ed.)

Real-Time Computation of Statistical Moments on Binary Images Using Block Representation

Iraklis M. Spiliotis and Basil G. Mertzios

Automatic Control Systems Laboratory, Department of Electrical Engineering, Democritus University of Thrace, Xanthi 67100, HELLAS.
Fax: +30-541-26947,
e-mail: iraklis%xanthi@ics.forth.gr or mertzios%xanthi@ics.forth.gr

This paper presents a new approach and an algorithm for binary image representation, which is applied for the fast and efficient computation of moments in binary images. In the terminology of this paper the binary image representation scheme is called block representation, since it represents the image as a set of nonoverlapping rectangular areas. The fast computation of moments in block represented images, is achieved exploiting the rectangular structure of the blocks.

1. INTRODUCTION

This paper presents a new advantageous representation for binary images, which in the sequel will be called *block representation*. In the block representation process the whole binary image is decomposed in a set of rectangular areas with object level. The block representation exploits the fact that many compact areas of a given binary image have the same value.

Various sets of two-dimensional (2-D) statistical moments constitute a well-known image analysis and pattern recognition tool [1]-[5]. In pattern recognition applications a small set of the lower order moments is used to discriminate among different patterns. The most common moments are the geometrical moments, the central moments, the normalized central moments and the moments invariants [1]. Other sets of moments are the Zernike moments and the Legendre moments which are based on the theory of orthogonal polynomials [4], [6] and the complex moments [5].

One main difficulty concerning the use of moments as features in pattern recognition applications is the implied high computational time. In the proposed approach, which is based on block represented images, the computational complexity of the moments calculation is in the general case independent of the size of the image and depends only on the image content and on the number of the required moments.

2. BLOCK REPRESENTATION

A bilevel digital image is represented by a 2-D array. Without loss of generality, we suppose that the object pixels are assigned to level 1 and background pixels to level 0. Due to this kind of representation, there are rectangular areas of object value 1 with edges parallel to the image axes. At the extreme case one pixel is the minimum rectangular area of the image. These rectangulars will be called *blocks* in the terminology of this paper.

Consider a set that contains as members all the nonoverlapping blocks of a specific binary image, in such a way that no other block can be extracted from the image (or equivalently each pixel with object level belongs to only one block). This set represents the image without loss of information. It is always feasible to represent a binary image with such a set, of all the nonoverlapping blocks with object level. We call this representation of the binary image, *block representation*. Figure 1 illustrates an image of the character d and the blocks.

The block representation concept leads to a simple and fast algorithm, which requires just one pass of the image and simple bookkeeping process. Consider a binary image $f(x,y)$, $x=0,1, .. ,N_1-1$, $y=0,1, .. , N_2-1$. The block extraction process requires a pass from each line y of the image. In this pass all object level intervals are extracted and compared with the previous extracted blocks.

In the following, a uniquely determined block representation algorithm, which is based on the row by row processing, is given.

Block representation algorithm
Step 1: For each line y of the image f,
Step 2: Find the object level intervals in line y
Step 3: Compare intervals and blocks that have pixels in line $y-1$,
Step 4: If an interval does not matches with any block, this is the beginning of a new block.
Step 5: If a block matches with an interval, the end of the block is in the line y.

As a result of the application of the above algorithm, we obtain a set of all the rectangular areas with level 1, that form the object. The block extraction process is implemented easily with low computational complexity, since it is a pixel checking process without numerical operations. In Figure 2, the application of the block representation algorithm in a binary image, is illustrated.

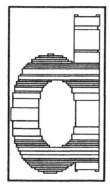

Figure 1. Image of the character d and the blocks.

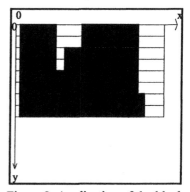

Figure 2. Application of the block representation algorithm.

It is possible to implement various processing and analysis algorithms with block represented binary images, since the most of these algorithms operate on image regions (blocks). The most important characteristic of the block representation is that a perception of image parts greater than a pixel, is provided to the machine. Therefore, all the operations on the pixels belonging to a block are substituted by a simple operation on the block; thus the block representation is advantageous for image modelling.

The block representation may be seen as a physical way for the representation of binary images. Each block is represented with four integers, the coordinates of the upper left and down right corner in vertical and horizontal axes. The block representation is an information lossless representation. It is noticeable, that the total amount of information required for the block representation of an image remains unchanged, under the scale of the image. We conclude that the block representation of a binary image is dependent only on the image content and is independent of the image size.

3. GEOMETRICAL MOMENTS

Consider a binary digital image $f(x,y)$, with N_1 pixels in horizontal axis and N_2 pixels in vertical axis. The 2-D geometrical moments of order (p,q) of the image are defined by the relation:

$$m_{pq} = \sum_{x=0}^{N_1-1} \sum_{y=0}^{N_2-1} x^p y^q f(x,y), \qquad p,q = 0,1,2,... \tag{1}$$

If the image is block represented, it is represented as the set of the nonoverlapping blocks that constitute the image, as follows:

$$f(x,y) = \{ b_1, b_2, ..., b_k \} \tag{2}$$

where k is the number of the blocks. Since the background level is 0, only the pixels with level 1 are taken into account for the computation of the moments. Thus the 2-D geometrical moments of order (p,q) of the image $f(x,y)$ are defined by the relation:

$$m_{pq} = \sum_x \sum_y x^p y^q \qquad \forall x,y: \ f(x,y) = 1 \tag{3}$$

Since the image pixels with level 1 belong to the image blocks, equation (3) is rewritten as:

$$m_{pq} = \sum_{i=1}^{k} \sum_{x=x_{1,b_i}}^{x_{2,b_i}} \sum_{y=y_{1,b_i}}^{y_{2,b_i}} x^p y^q \tag{4}$$

where x_{1,b_i}, x_{2,b_i} and y_{1,b_i}, y_{2,b_i} are the coordinates of the i-th block with respect to the horizontal axis and to the vertical axis, respectively.

3.1 Computation of the geometrical moments of one block

Consider the block b, with x_{1b}, x_{2b} coordinates with respect to the horizontal axis and y_{1b}, y_{2b} with respect to the vertical axis. Then the 2-D geometrical moments m_{pq}^b for the block b are given by

$$m_{pq}^b = \sum_{x=x_{1b}}^{x_{2b}} \sum_{y=y_{1b}}^{y_{2b}} x^p y^q = x_{1b}^{\ p} \sum_{y=y_{1b}}^{y_{2b}} y^q + (x_{1b}+1)^p \sum_{y=y_{1b}}^{y_{2b}} y^q + \ldots + x_{2b}^{\ p} \sum_{y=y_{1b}}^{y_{2b}} y^q =$$

$$= \sum_{x=x_{1b}}^{x_{2b}} x^p \sum_{y=y_{1b}}^{y_{2b}} y^q \qquad (5)$$

Using the rectangular form appeared within the block, the computation effort for equation (4) is reduced from $O(N^2)$ to $O(N)$ in equation (5). For the computation of (5), it is enough to calculate the summations of the powers of x and y.

Moreover, taking into account that the sum $1^m + 2^m + \ldots + n^m$, is provided for every $m \in Z$, by the following analytical formulae:

$$\sum_{i=1}^{n} i = \frac{n(n+1)}{2}, \quad \sum_{i=1}^{n} i^2 = \frac{n(n+1)(n+2)}{6}, \quad \sum_{i=1}^{n} i^3 = \frac{n^2(n+1)^2}{4}$$

$$\qquad (6)$$

$$\binom{m+1}{1}\sum_{i=1}^{n} i + \binom{m+1}{2}\sum_{i=1}^{n} i^2 + \ldots + \binom{m+1}{m}\sum_{i=1}^{n} i^m = (n+1)^{m+1} - (n+1)$$

the sum of the powers of x is computed using (6) and the following formula:

$$\sum_{x=x_{1b}}^{x_{2b}} x^p = \sum_{x=1}^{x_{2b}} x^p - \sum_{x=1}^{x_{1b}-1} x^p \qquad (7)$$

The sum of the powers of y in (5) is computed in a similar manner. Fast computation of the 2-D geometrical moments of one block, according to (5), is achieved with the above simple and analytical formulae (6) and (7).

3.2 Computation of the geometrical moments of the whole image

According to the equation (5), the 2-D geometrical moments of the whole image are computed as the summation of the 2-D geometrical moments of all the individual blocks of the binary image.

4. COMPUTATIONAL COMPLEXITY

The following analysis refers to the computational complexity of the geometrical moments. The other sets of moments can be computed in a similar manner. Consider that a binary image contains one rectangular block with level 1. For simplicity and without loss of generality, suppose a square block with MxM points. In the sequel we estimate the computational complexity required for the geometrical moments computation up to the order $(L-1,L-1)$.

The direct computation from equation (1) of one geometrical moment of that block requires M^2 power computations, M^2 multiplications and M^2 additions. For the computation of L^2 moments, it results that L^2M^2 power computations, L^2M^2 multiplications and L^2M^2 additions are required.

Consider equation (5), which exploits the rectangular form appearing within the block. For the computation of the sum of the powers of x, LM power calculations and LM additions are required. The same number of operations are required for the computation of the sum of the powers of y. Therefore, for the computation of L^2 geometrical moments from equation (5), $2LM$ power calculations, L^2 multiplications and $2LM$ additions, are required for one block.

Now, consider the analytical formula (6). The binomial coefficients are appeared in the computation formula for any specific geometrical moment of every block of the image; therefore the computational effort is reduced by the number of the blocks. Moreover the factorials that have to be computed, for the determination of the binomial coefficients, require least effort, i.e. one multiplication for the calculation of $m!$ in terms of $(m-1)!$. Therefore the complexity for the computation of the binomial coefficients of (6) is minor.

The sum x^i, $i = 1,2,...,p-1$, that have been computed previously and their values are stored, are used for the computation of the sum of x^p. Thus the computation of the sum of x^p from equations (6) and (7) requires 2 power calculations, p multiplications and p additions. The computation of the sum of y^q requires 2 power calculations, q multiplications and q additions. For L^2 moments, $4L$ power calculations, $3L^2-L$ multiplications and L^2-L additions, are required, for the computation of all the L^2 square moments.

Table 1 demonstrates the above results. The complexity is reduced from 2-D form from (1) to 1-D by the use of block representation and equation (5). Moreover, the complexity is independent to the size with the use of the analytical formulae (6) and (7). The required number of power calculations, of multiplications and of additions for the computation of the geometrical moments up to the order (4,4) of a block with MxM pixels, where M varies from 1 to 100, is illustrated in Figure 3.

Lemma 1
Assuming that the complexity of raising a number to a power is the same as one multiplication, the computation of (5) and (6) requires L^2+2LM and $3L^2+2L$ multiplications respectively. Comparing the above number of multiplications it is concluded that (5) has less computational complexity than (6) when

$$L^2 + 2LM \leq 3L^2 + 2L \Rightarrow M \leq L+1 \tag{8}$$

However, in typical pattern recognition applications, the higher order moments are not used since are very sensitive to noise; usually they are calculated up to the order (4,4). Usually, in images with a high entropy value, like images of text, where a significant number of small blocks appears, the time for moments computation is reduced by a factor among 10 and 50, using block representation. In images with large areas of object level, like images of industrial parts, aircrafts, ships e.t.a. the factor of time reduction is much greater. Consider the aircraft image, illustrated in Figure 4. The image size is 287x207 points. Using the algorithm described in Section II, 195 blocks extracted from the image. The total number of points with object level is 19922. The geometrical moments of that image have been computed up to the order (4,4), using the different methods described in this paper. For the direct computation from (1), 4.77 sec are required. Using the block represented image and equation (5), 42 msec are required. The use of the analytical formulae (6) and (7) increases the computation time to 55 msec, since in most of the extracted blocks one edge has width 1 or 2 points. Using the criterion, provided by Lemma 1 for $L=5$, the computational time is decreased to 15 msec; that implies a factor of time reduction equals to 318 times less, in comparison with the time required for the computation by (1). The above results are shown in Table 2. From Table 2, it results that an average rate greater than 40 frames/sec is achieved for the block representation and the computation of the moments. Therefore, even with the software implementation, the method is characterized as real-time.

Table 1.

The required number of operations for the computation of geometrical moments up to the order $(L-1,L-1)$, of one block with MxM pixels.

Operations number	Direct computation from equation (1)	Computation from equation (5)	Computation from equations (6), (7)
L^2M^2	power calculations	$2LM$	$4L$
L^2M^2	multiplications	L^2	$3L^2-2L$
L^2M^2	additions	$2LM$	L^2-2L

5. CONCLUSIONS

In this paper the block representation idea and the associated algorithm are presented. Block representation is a useful binary image representation that allows the development of more efficient algorithms for various image processing and analysis tasks. Owing to the nature of the digital image, only rectangular areas with the same level are present. Block representation uses these rectangular similarities and offers advantages in image handling and computational cost. Also the block representation provides a perception about image pieces greater than a pixel. Two dimensional moments is a classical image analysis tool and the use of block represented binary images decrease dramatically the computation effort. The complexity of the algorithm for the computation of moments in block represented images, is independent of the image size. The use of the block representation permits the real-time computation of moments. Other image processing and analysis tasks can be implemented in block represented images. At this time this is a research subject.

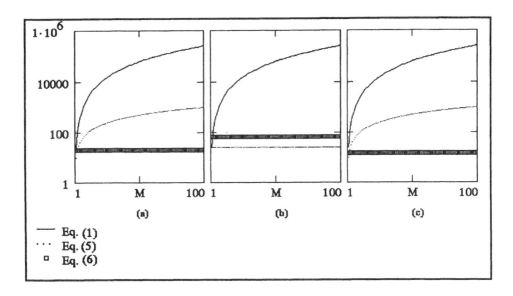

Figure 3. Number of operations for the geometrical moments computation, of a MxM block, from equations (1), (5) and (6) with L=5 and M=1,2, ... , 100. (a) Number of power calculations. (b) Number of multiplications. (c) Number of additions.

Figure 4. Aircraft image.

Table 2.

Computation of the geometrical moments up to the order (4,4) of the aircraft image, using different methods. The use of block representation, results to a great factor of time reduction.

Computation of the geometrical moments of the aircraft image	Time in seconds	Reduction factor with respect to (1)
from equation (1)	4.771	
from equation (5)	0.042	114
from equation (6)	0.055	87
using the Lemma 1	0.015	318

REFERENCES

1   M.-K. Hu, "Visual pattern recognition by moment invariants", *IRE Trans. Inform. Theory*, vol. IT-8, pp. 179-187, Feb. 1962.

2   G.L. Cash and M. Hatamian, "Optical character recognition by the method of moments", *Computer Vision, Graphics and Image Processing*, vol. 39, pp. 291-310, 1987.

3   K. Tsirikolias and B.G. Mertzios, "Statistical pattern recognition using efficient 2-D moments with applications to character recognition", *Pattern Recognition*. To appear.

4   M.R. Teague, "Image analysis via the general theory of moments", *J. Opt. Soc. Amer.*, vol. 70, pp. 920-930, Aug. 1980.

5   Y.S. Abu-Mostafa and D. Psaltis, "Image normalization by complex moments", *IEEE Trans. Pattern Anal., Machine Intel.*, vol. PAMI-7, No. 1, pp. 46-55, Jan. 1985.

6   C.-H. Teh and R.T. Chin, "On image analysis by the methods of moments", *IEEE Trans. Pattern Anal., Machine Intel.*, vol. 10, No. 4, pp. 496-513, 1980.

7   A. Papoulis, *Propability, Random Variables, and Stochastic Processes*, McGraw-Hill, New York, 1965.

Time-Varying Image Processing and
Moving Object Recognition, 3 – V. Cappellini (Ed.)
© 1994 Elsevier Science B.V. All rights reserved.

Reducing segmentation errors through iterative region merging°

Silvana DELLEPIANE

DIBE Department of Biophysical and Electronic Engineering
 University of Genoa
via Opera Pia 11A,   I-16145 Genova - ITALY

The errors incurred in using classical thresholding and region-growing methods as well as in region merging are compared. As a result, a novel approach is proposed that exploits the capabilities of the region-growing scheme (able to ensure a null undersegmentation error for small regions) by integrating it with the region-merging scheme that makes use of global region features and relations. The use of the distance between the gray-level averages allows errors to be predicted, hence the adaptivity to the noise level or a multilevel result can be achieved.

One of the major advantages of the proposed approach is the possibility of changing threshold values during processing without affecting the error risk. The method is then independent of parameter and threshold values.

1.INTRODUCTION

A digital image acquired from a real-world situation depicts a configuration of real objects, O_i's, also named "world elements". An efficient region-segmentation algorithm aims at localizing a set of image elements (primitives) that well correspond to world elements.

Obviously, the acquisition and transmission processes, together with object properties, set severe limits on the resolution details in both the intensity and topological spaces that can be achieved by the "optimal" segmentation algorithm. Moreover, the optimal segmentation process strongly depends on the subsequent utilization of the acquisition signal, hence it cannot be objectively defined. Finally, the noise level is usually unknown, so it cannot be of any help during the segmentation process.

°This work was partially supported by the European project AIM-COVIRA [1].

The aim of this paper is to define a relationship between the segmentation error and the noise level. The measure of the degree of intensity resolution will be taken into account to describe the errors incurred in using classical thresholding and region-growing methods as well as in region merging. The paper reports a comparison of perfomances, and shows that an iterative merging of region pairs results in a sensible reduction in the error rate.

The relationship between intensity resolution and noise level also allows one to adaptively change algorithm parameters and thresholds according to the noise level in the image portion under analysis, or according to the required segmentation resolution detail.

The paper draws some conclusions about the possibility of utilizing the region-merging approach to reduce segmentation errors. The practical implementation of a novel region-merging method is described in detail in [2], which also reports results obtained on real images.

2. IMAGE MODEL , SEGMENTATION ERRORS, AND SIGNAL-TO-NOISE RATIO

The simplest image model (piece-wise constant) assumes that a source object produces a homogeneous response to an acquisition signal. Moreover, acquisition and transmission noises are globally described by independent additive components, identically distributed, and characterized by Gaussian density functions with zero mean and variance equal to σ^2_n.

Under these assumptions, we consider two object classes, O_A and O_B, characterized by the probability density functions $f_{gl/_A}(gl)$ and $f_{gl/_B}(gl)$ and the a-priori probabilities $P(A)$ and $P(B)$, respectively. The possibility of separating two distinct objects on the basis of their intensity distributions only is physically limited and follows the traditional rules of statistics.

The error-minimization criterion we adopted takes into account the fact that oversegmentation errors are easy to reduce (if they are not too many) through a manual or automatic region-merging process. On the contrary, undersegmentation errors are more difficult to overcome. In this context, the proposed approach aims at reducing the segmentation process to a correct sequence of region-merging steps. Avoiding undersegmentation errors is then equivalent to avoiding incorrect region mergings.

To find a relationship between noise level and errors, we consider the signal-to-noise ratio between two distinct objects, defined as:

$$SNR = 10 \cdot Log \frac{\delta_{AB}^2}{\sigma_n^2} \qquad (1)$$

where δ_{AB} is the absolute distance between the two average values (gl_A and gl_B) of the gray-level distributions of the two objects O_A and O_B, and can be expressed as:

$$\delta_{AB} = \delta = |gl_A - gl_B|. \qquad (2)$$

This parameter defines a relationship between the gray-level averages of the two objects, and is very useful for the segmentation process, but it does not provide any information about the separability of the two objects.

3. HISTOGRAM THRESHOLDING ERRORS

Let a set of regions X_i's be the iconic representation of the object class O_A; the relationship between the model and the region histograms $h(X_i)$ is given by:

$$E\left\{\frac{h(X_i)}{dim X^{IM}}\right\} = P(A) \cdot f_{gl/_A}(gl) \tag{3}$$

where $dim X^{IM}$ is the number of pixels in the whole image.

An analysis of the histogram behaviour allows one to make some hypotheses about the best threshold.

The gray-level distance is a useful measure to decide if a pixel or a group of pixels belong to an object, whose average gray level is known. Moreover, from the analysis of the image gray-level histogram, distance measures related to the maxima may be derived, in order to subdivide the feature range into various decision regions. However, the minimum possible error (if the optimum threshold has been chosen) depends not only on the distance between modes but also on the distribution variance.

Therefore, when one analyzes the histogram for the purpose of its segmentation, or when one compares two region gray-level distributions in order to vote for a possible merging, one should use a more appropriate metric that globally evaluates the distribution distance, taking also variance into account.

As an example, the Bhattacharyya distance is an easy and suitable feature to evaluate distribution separability. Its formula, in the form specific for the simplified case of Gaussian distributions [3], defines the distance value b_{AB} between the distributions of the two objects, O_A and O_B, as:

$$b_{AB} = \frac{1}{8} \cdot \frac{|gl_A - gl_B|^2}{\dfrac{\sigma^2_A + \sigma^2_B}{2}} + \frac{1}{2} \cdot \ln \frac{(\sigma^2_A + \sigma^2_B)}{2\sigma_A \sigma_B}. \tag{4}$$

This value is very easy to compute, as it is a combination of the average gray levels and the variances of the two sets. When σ_A and σ_B are equal to σ_n, the distance value reduces to:

$$b_{AB} = \frac{1}{8} \cdot \frac{\delta^2}{\sigma^2_n} \tag{5}$$

From this distance value, an upper bound to the related error can be determined by using the formula for the Bhattacharyya bound:

$$\varepsilon_b = \sqrt{P(A) \cdot P(B)} \cdot e^{-b_{AB}} \tag{6}$$

38

which reduces to

$$\varepsilon_b = \frac{1}{2} \cdot e^{-\frac{\delta^2}{8\sigma_n^2}} \qquad (7)$$

when $P(A)$ and $P(B)$ are equal to 1/2.

This value, corresponding to the percentage error on the global image size, is the maximum error that may be incurred in classifying the points belonging to two objects on the basis of the decision threshold selected by applying the maximum-a-posteriori criterion.

By eq. (7), we obtain a relation between the upper bound to the errors on a first- order statistical segmentation and the signal-to-noise ratio. This function, displayed in Fig.1 allows us to define the achievable intensity resolution detail *minδ* as a function of noise, which is generally unknown for real images. Given a fixed value *εmax* of an acceptable error, we cannot ensure that it will be possible to separate, on the basis of the only global intensity distribution, contiguous objects that, in the intensity space, are at a distance from each other that is smaller than a *minδ* value, defined as:

$$min\,\delta = 2.82 \cdot \sigma_n \sqrt{-ln\,2 \cdot \varepsilon_{max}} \qquad (8)$$

(obtained by reversing equation 7).

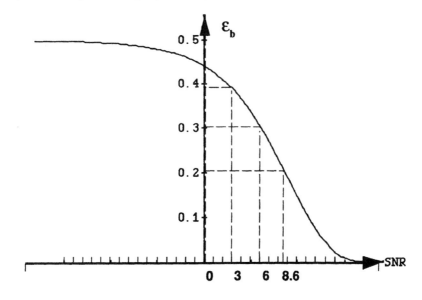

Figure 1. Upper bound to the Bayes error vs. SNR.

This formula includes the noise standard deviation σ_n as a parameter. Then, the achievable intensity resolution detail $min\delta$ can be predicted on the basis of a predicted σ_n value, or can be computed for various noise levels.

For instance, when the tolerated error ε_{max} is taken equal to 20% (corresponding to an SNR equal to 8.6), the $min\delta$ value is equal to $2.7\sigma_n$. This means that, if the SNR between the two objects is higher than 8.6, the error will be smaller than ε_{max}.

Using topological information amounts to adding a new dimension to the feature space. Therefore, the segmentation method proposed in this paper takes into account both intensity separation and region connectivity by performing a first numerical processing (which should prevent initial errors), followed by iterative merging steps. In this way, punctual, local and global information is simultaneously considered.

4. REGION-GROWING ERRORS

To evaluate the errors that may be incurred when applying a classical region-growing method, an upper bound based on the parameters δ and σ_n can be set. The maximum error may be incurred when a region X_i is grown exactly between two objects. This means that the average level, gl_{X_i}, takes the value:

$$gl_{X_i} = \frac{gl_A + gl_B}{2} \tag{9}$$

Such a region may represent an undersegmentation error; in the worst case, its pixels belong to both objects.

The probability of finding pixels belonging to two classes in the neighbourhood of gl_{X_i} depends on the separability δ of the two objects and on their probability distributions. This allows us to estimate the size of an error as the sum of the probabilities of finding pixels belonging to the first and second objects in the range centered in gl_{X_i}, with a width W_i equal to the gray-level dynamics of X_i.

Possible undersegmentation errors on the two objects can then be expressed as:

$$\varepsilon_{u,A} = \frac{dim\{p_{X_i} \in A\}}{dim X^{IM}} \tag{10}$$

$$\varepsilon_{u,B} = \frac{dim\{p_{X_i} \in B\}}{dim X^{IM}} \tag{11}$$

which correspond to the errors incurred when region X_i is associated with object O_B in the former equation, and with object O_A in the latter.

An upper bound to each of the above errors is given by:

$$\varepsilon_u = \varepsilon_{u,A} = \frac{dim\{p_{Xi} \in A\}}{dim\, X^{IM}} \leq P(A) \cdot \int_{\overline{p_{Xi}}-\frac{W_i}{2}}^{\overline{p_{Xi}}+\frac{W_i}{2}} f_{gl/A}(gl)\, dgl \approx P(A) \cdot W_i \cdot f_{gl/A}(gl_{Xi}) \qquad (12)$$

that is,

$$\varepsilon_u = \frac{1}{2} \cdot \left[W_i \Big/ \sqrt{2\pi} \cdot \sigma \right] \cdot e^{-\frac{\delta^2}{8\sigma^2}} \qquad (13)$$

Then the relation $\varepsilon_u \leq \varepsilon_b$, holds for every $W_i \leq \sqrt{2\pi} \cdot \sigma$. Such a condition is realistic in the case of small W_i values, which ensure no undersegmentation error.

To exploit this interesting property, the proposed approach makes use of a region-growing algorithm for the initialization of the merging method. This pre-segmentation step is characterized by small regions, i.e., with very small W_i values.

5. ITERATIVE REGION-MERGING APPROACH

As the pre-segmentation step produces too large a region set, iterative mergings reduce the number of regions, thus the set gains some significance. Then the segmentation problem becomes one of finding the correct merging sequence.

Region-merging steps are performed on the basis of region similarities. Such steps aim to reduce possible errors by avoiding risky mergings, and obtain reliability values for the resulting regions.

The segmentation process can now be formulated as a hypothesis test problem. In the pairwise merging scheme selected, the null hypothesis H_0 is defined as:

hypothesis H_0: region X_i and region X_j belong to the same population **M**

The population mean value is μ and its variance is σ^2.
The opposite hypothesis H_1 is defined as:

hypothesis H_1: region X_i and region X_j DO NOT belong to the same population **M**

In this case, the means of the two populations are μ_i and μ_j, respectively, with the same variance σ^2.

The null hypothesis test is applied to candidate region pairs. If it holds, a merging step is performed. A lower complexity is achieved by considering only the difference in region gray-level average. More precisely, the average gray-level distance between the two regions is compared with a threshold value k_m that subdivides the feature space of the average distance into acceptance and rejection regions.

According to the error-minimization criterion mentioned before, the error probability reduces to:

$$P(\varepsilon) = P(H_0 \text{ accepted} \mid H_1 \text{ true}) = P(\varepsilon \mid H_1 \text{ true}) \tag{14}$$

We denote by Δm_{H1} the random variable $\Delta m \mid H_1 \text{ true}$, that is, the distance of gray-level averages under hypothesis H1. Its probability density function is:

$$N_{H1} \equiv N(\Delta\mu, \sigma_{\Delta m}^2) \tag{15}$$

where

$$\Delta\mu = |\mu_i - \mu_j| \quad \text{and} \quad \sigma_{\Delta m}^2 \equiv \frac{n_i + n_j}{n_i \cdot n_j} \cdot \sigma^2 \tag{16}$$

The probability error can then be espressed by using the N_{H1} function, as follows:

$$P(\varepsilon) = P\left(\varepsilon \mid H_1 \text{ true}\right) = P(|\Delta m| \geq k_m \mid H_1 \text{ true}) = P\left(\left|\frac{\Delta m}{\sigma_{\Delta m}}\right| \geq w_m \mid H_1 \text{ true}\right)$$

$$\tag{17}$$

$$P(\varepsilon) \cong 2k_m \cdot N_{H1}(0) \cong 2w_m \cdot N(\frac{\Delta\mu}{\sigma_{\Delta m}})$$

where

$$k_m = w_m \cdot \sigma_{\Delta m} \tag{18}$$

Then the error turns out to be a function of the selected threshold w_m, of the average gray-level distance $\Delta\mu$, and of the variance $\sigma_{\Delta m}^2$. From equation (17), we deduce that minimizing the error corresponds to minimizing $\sigma_{\Delta m}^2$, or, equivalently, the factor $\frac{n_i + n_j}{n_i \cdot n_j}$.

Such a factor is minimized when new and larger regions are grown as a result of the merging steps.

This can lead to very important conclusions. First of all, if the selected threshold is maintained during the processing, the error is reduced when larger regions are grown. Similarly, we can adopt a dynamic threshold value that relaxes the hypothesis test as mergings steps are executed. This will allow us to maintain a certain probability error during the whole processing, by using a small threshold value at the beginning (thus avoiding incorrect mergings) and by increasing this value when merging steps result in larger regions, with more statistical significance and fewer error risks.

This novel approach allows several merging iterations to be perfomed during each cycle, for different threshold values and different merging levels. The picture graph generated by the successive merging steps associates each intermediate level with a reliability factor that defines the qualities of the regions obtained, and that is characterized by a specific level of detail.

6. CONCLUSIONS

As suggested by the evaluation of the errors due to a purely statistical segmentation approach, the integration of local and global analyses with the statistical one allows a sensible reduction in the error rate. To this end, the proposed approach exploits the capabilities of the region-growing scheme (able to ensure a null undersegmentation error for small regions) by integrating it with the region-merging scheme that makes use of global region features and relations. The use of the distance between the gray-level averages allows errors to be predicted, hence the adaptivity to the noise level or a multilevel result can be achieved.

In conclusion, one of the major advantages of the proposed approach is its independence of parameter and threshold values.

REFERENCES

1. COVIRA - Project A2003 of the AIM programme of the European Community, Consortium Partners:
- Philips Medical Systems, Best (NL) and Madrid (E); Corporate Research, Hamburg (D) (prime contractor) - Siemens AG, Erlangen (D) and Munich (D) - IBM UK Scientific Centre, Winchester (UK) - Gregorio Maranon General Hospital, madrid (E) - University of Tuebingen, Neuroradiology and Theoretical Astrophysics (D) - German Cancer Research Centre, Heidelberg (D) - University of Leuven, Neurosurgery, Radiology and Electrical Engineering (B) - University of Utrecht, Neurosurgery and Computer Vision (NL) - Royal Marsden Hospital/Institute of Cancer Research, Sutton (UK) - National Hospital for Neurology and Neurosurgery, London (UK) - Foundation of Research and Technology, Crete (GR) - University of Sheffield (UK) - University of Genoa (I) - University of Aachen (D) - University of Hamburg (D) - Federal Institute of Technology, Zurich (CH)
2. S. Dellepiane, Multilevel segmentation through iterative region merging, 7 ICIAP, Bari (Italy), Sept. 93, (in press).
3. C. W. Therrien, "Decision, Estimation and Classification", John Wiley & Sons, 1989.

B
PATTERN RECOGNITION

Time-Varying Image Processing and
Moving Object Recognition, 3 – V. Cappellini (Ed.)
© 1994 Elsevier Science B.V. All rights reserved.

Hypergraph Based Feature Matching in a Sequence of Range Images

Bikash Sabata[a], and J.K. Aggarwal[b]

[a] Computer Science Department, Wayne State University, Detroit, MI 48202

[b] Computer & Vision Research Center
Department of Electrical & Computer Engineering
The University of Texas at Austin, Austin, Texas 78712

Abstract

The key issue in motion estimation and tracking an object over a sequence of images is establishing correspondence between the features of the object in the different images of the sequence. This paper considers the problem of establishing correspondences between surfaces in a sequence of range images. We present a novel procedure for finding correspondence and show the results on real range image sequences. A hypergraph search procedure forms the basis for the algorithm that computes the correspondence between surfaces. The solution uses geometrical and topological information derived from the scenes to direct the search procedure. Two scenes are modeled as hypergraphs and the hyperedges are matched using a sub-graph isomorphism algorithm. Further, we present a sub-hypergraph isomorphism procedure to establish the correspondences between the surface patches and demonstrate the algorithm on different types of real range image sequences. We present results that show that the algorithm is robust and performs well in presence of occlusions and incorrect segmentations.

1 Introduction

Matching of object features is the key to motion estimation and tracking an object over a sequence of images. In this paper we deal with the tracking of objects in a sequence of range images to estimate the motion of the camera (range sensor) in the environment. Range images sense the surface of the objects, so it is natural to use surface segments and boundaries between the surfaces as the features of interest; this translates the tracking of objects into finding a match between the surface segments in a pair of range images of the scene. This paper considers the finding of correspondences between surfaces in a sequence of range images. Finding correspondence or a match between features is not isolated to object tracking, but is also central to other computer vision tasks including navigation, object recognition, target tracking, and map building. We present a novel procedure for establishing correspondence and show the results on real range image sequences.

A hypergraph search procedure forms the basis for the algorithm that computes the correspondence between surfaces. The solution uses geometrical and topological information derived from the scenes to direct the search procedure. In general, the input to the matching algorithm is the output from a segmentation algorithm that partitions the image into surface segments. The performance of the matching depends greatly on the results of the segmentation algorithms. Incorrect segmentation causes poor estimation of the surface parameters and affects the performance of the matching algorithm. Occlusion of features and appearance of new features further complicates the problem. We address these issues and obtain a solution that is robust and able to handle occlusions of surfaces, noise in data, and incorrect segmentation from a segmentation algorithm. In the present implementation, we assume that the images have planar, cylindrical and conical surfaces; however, the procedure is general enough to be extended to other surface classes.

The question of finding correspondences between features has been studied extensively (see [1, 5, 6, 7]) but, most of these approaches deal with matching a scene to a model of the object. The fundamental difference between model-to-scene matching and scene-to-scene matching is that in the former, the model description of the object is complete, and to that we match the incomplete description of the object obtained from the scene. However, in the case of scene-to-scene matching, both descriptions of the object are incomplete and we must find a match between two incomplete descriptions. By incomplete, we mean that all the features are not present in the description of the object because of occlusions and sensor errors. This difference makes it impossible to use the strategies obtained for object recognition in the domain of object tracking; new strategies based on the constraints of the problem have to be designed.

Fundamental to our strategy to match features over a sequence of range images is a hypergraph representation of the scenes. The two scenes are modeled as hypergraphs and the hyperedges are matched using a sub-graph isomorphism algorithm. To reduce the complexity of the matching task, heuristics derived from the topological and the geometrical information available from the scene are used to direct the search. The hierarchical representation of hypergraphs not only reduces the search space significantly, but also facilitates the encoding of the topological and geometrical information. Hyperedges are formed by grouping the surface features, which reduces the search space. Using a priori knowledge arising out of the physical constraints of laser scanning, a fast matching algorithm is designed.

The paper is organized as follows. In section 2 we introduce the hypergraph structure and present a procedure of polynomial complexity to generate the hypergraphs. Next, in section 3 we give the matching strategy and a sub-hypergraph isomorphism procedure to establish the correspondences between the surface patches. We also discuss the complexity issues and show how the search space is reduced. In section 4 the algorithm is demonstrated on different types of real range image sequences. We present results that show that the algorithm is robust and performs well in presence of occlusions and incorrect segmentations. Finally we conclude in section 5.

2   Hypergraph Representation

Representation of the available information in a suitable form is the key to the solution of matching. The output of the segmentation module is a low level description of the geometrical and the topological information. Each pixel in the range image is assigned a label and to each label a surface parameter vector is associated. Such representation is not conducive to reasoning about the topology of the scene. A representation that facilitates reasoning about the geometry and the topology of the scene is required.

The attributes computed by the low level processing module are represented using a hypergraph data structure. The representation encodes the topological and geometrical information extracted from the range image of the scene. Hypergraphs are generalizations of graphs. The arc (or edge) is generalized as a hyperedge, where a set of vertices forms the hyperedge, instead of just two vertices forming the arc. The group of vertices forming the hyperedge may share some common property. Although hypergraphs have been used in vision and robotics applications [10, 11], a new definition of the hyperedge and a novel method for constructing the hypergraphs that makes them a powerful tool for vision applications is presented here.

Attributed hypergraphs are a concise way of representing objects such that both quantitative and qualitative information are encoded in the representation. The formal definition is as follows:

Definition 2.1 *The Hypergraph [2] is defined as an ordered pair* $H = (X, E)$ *where* $X = \{x_1, x_2, \cdots, x_n\}$ *is a finite set of attributed vertices and* $E = \{e_1, e_2, \cdots, e_m\}$ *are the hyperedges of the hypergraph. The set E is a family of subsets of X (i.e. each* e_i *is a subset of X) such that (1)* $e_i \neq \emptyset, i = 1, \cdots, m$ *and (2)* $\bigcup_{i=1}^{m} e_i = X$.

A graph is a hypergraph whose hyperedges have cardinality of two.

To arrive at the hypergraph representation, the scene is first represented as an attributed graph. Each surface patch in the range image forms an attributed vertex. The attribute values are the surface property values. For each pair of surfaces that are connected, an attributed arc is formed. The attributes of the arc describe the interfacing edge and the relative geometrical information between the two surfaces. Groups of the attributed vertices (surface patches) form an hyperedge, and with each hyperedge we associate an attributed graph that describes the topology of the component attributed vertices (surface patches).

2.1 Hyperedge Construction

The set of vertices that form the hyperedge should represent a topologically significant feature in the graph so that the matching task is guided by the topology of the scene. Cliques in the graph are significant features that are rich in information.

Definition 2.2 *A clique of a graph G is a maximal complete subgraph of G, i.e., a complete subgraph of G not properly contained in any other subgraph of G.*

Physically, the cliques represent groups of surfaces that are adjacent to each other. Since a clique provides a larger attribute set and many geometrical properties, the probability of a false positive match (between two cliques) is reduced significantly. Each clique forms a hyperedge in the hypergraph and the attributed graph describing the clique is the associated attribute of the hyperedge. Figure 1 illustrates the formation of a hypergraph from a scene.

The complexity of computing the cliques in a graph is exponential, so the advantage gained by matching using the hypergraph would be lost because of the cost of constructing the hypergraph. However, the physics of the range imaging process restricts the size of the cliques in the scenes that are observed. The laser scanning process results in a depth map which describes the depth z in terms of the image plane coordinates (u, v). The depth map description is a geographic map of the depth values on a plane, and for a planar map it has been proved that

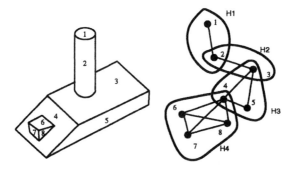

Figure 1: *An object and its corresponding Hypergraph representation*

four colors are sufficient to color the map [3]. What the result signifies is that there cannot exist more that four regions that touch each other. If the map is represented as a graph, where the vertices are the regions and an arc exits between two vertices corresponding to adjacent regions, then there cannot exist a clique with more than four vertices in the graph. This result is stated as a lemma:

Lemma 2.1 *For a range image described as a depth map, $z = f(u, v)$, there cannot exist a group of more than four surface segments that are adjacent to each other.*

Once the upper bound on the size of the cliques is known, the complexity of computing the cliques becomes $O(n^3)$. (A tighter bound with a lower polynomial order can be obtained.) The strategy of the algorithm to detect the cliques is to consider one vertex at a time and find all the cliques that the vertex can form with its neighbors. Once all the cliques have been found, the vertex is removed from the graph and not considered again. Since the cardinality of the cliques can only be two, three or four, the algorithm first creates all the groups of two vertices, which includes the current vertex. From the set of groups of two vertices, groups of three vertices are formed. The groups of four vertices are then formed from the group of three vertices. These groups are the cliques of the graph with the current vertex as a member of each clique.

The cliques are the hyperedges of the hypergraph representing the scene. Each clique forms a hyperedge and the attributed graph associated with the clique is associated with the corresponding hyperedge. The hyperedges in the scene hypergraph represent the regions of "high activity" in the scene. These regions are rich in geometrical information; therefore, the probability of making a false positive match between hyperedges is reduced significantly. The topological information is also available implicitly because of the nature of the representation. The topological information "steers" the matching algorithm in the proper direction so that the solution is obtained in the minimum number of steps.

3 The Matching Procedure

This section presents the matching procedure used to derive the surface correspondences in a sequence of range images. If the scene is represented as a graph, then matching of features, in general, is equivalent to subgraph isomorphism, where the first graph is matched to a subgraph

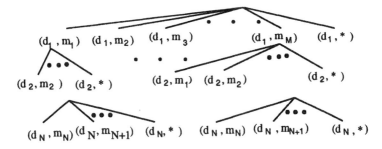

Figure 2: *The interpretation tree used in the matching procedure.*

of the second. However, scene-to-scene matching differs from this in that the first graph is not entirely a subgraph of the second graph, but a subgraph of the first graph matches a subgraph of the second graph. This additional variation makes the problem of scene-to-scene matching more complicated than the scene-to-model matching.

The matching procedure uses the information encoded in the hypergraph representaiton to result in the final pairings. The heart of the matching procedure is a *directed tree search* algorithm that tests various hypotheses and rejects the impossible ones. The interpretation that gives the largest match is selected as the solution. Constrained tree search algorithms have been used in many applications [6, 7, 8]. (Most of the constraint search algorithms are equivalent to the tree search algorithm [8].) Data pairings are formed by a depth-first search of an *interpretation tree* (figure 2). Each node of the tree represents a possible pairing. The first data (surface patch) is taken from the first scene and paired with each of the data in the second scene. These form the nodes in the first level of the tree. To account for missing surface segments due to occlusions, the data is also paired with a *wild card* $\star$. Subsequent levels of the tree correspond to pairings of other vertices. Each branch of the tree represents a partial matching of the scenes. The constraints are used to prune the search tree and thus reduce the search space.

A variation to the constrained tree search is presented, in which the search is directed based on the current hypothesis. The directed search, coupled with the termination conditions, reduces the search space. The key idea is to use the topological constraints of the scene to determine the next most likely match, and to accept or reject the matches based on the geometrical constraints.

The features used in the matching process are surface segments. The segmentation module segments the range image into surface segments and the surface parameters of each region are computed. The interfacing edges between the surface segments are detected and their properties are computed. The properties of the edge segments used are: (1) the edge type (straight line or curved), (2) the edge length, and (3) the depth discontinuity. The depth discontinuity across the edge implies that one surface may be occluding (partially or completely) another surface. The information about occlusion is incorporated in the attribute list of the surface patches.

The constraints used are similar to the *unary* and *binary* constraints developed by Grimson and Lozano-Perez [4]. The only unary constraint used is the surface type classification (planar, cylindrical, conical, etc.). Other properties used in model based object recognition, such as area, perimeter, and compactness, are very sensitive to occlusion, and since occlusion may occur in either of the range images, these properties cannot be used as constraints. The binary constraints describe the relative properties between pairs of surface segments. The properties used are: (1) connectivity, (2) the angle between the surface patches, (3) the range of distances between the two surface patches, (4) the range of the components of the vector spanning the two surface patches, and (3) the properties of the interfacing edge.

Each constraint is measured and tested against a predetermined threshold. For surface segments that have an occluding edge, the neighbors information is not complete (a neighbor may be hidden) and the connectivity information may be inaccurate. Therefore, for such cases only a *weak arc* is formed in the attributed graph of the scene. A match based on a weak arc is subject to confirmation or rejection based on further evidence.

Matching between the two hypergraphs representing the scenes is achieved by computing the match between the component hyperedges. The algorithm 3.1 gives the hypergraph matching procedure.

Algorithm 3.1 Hypergraph matching algorithm

begin
   1. ORDER*(hedge_list$_1$, vertex_list$_1$, Hgraph$_1$)*
   2. ORDER*(hedge_list$_2$, vertex_list$_2$, Hgraph$_2$)*
   3. INITIALIZE*(CH, CV, Hgraph$_1$, Hgraph$_2$)*
   4. level = 0
   5. FIND_MATCH*(CH, CV)*
end

The algorithm starts by ordering the hyperedges and the vertices in the scene hypergraphs (steps 1 and 2 of the algorithm). The order determines the branches taken in the interpretation tree. The ordering is done by selecting the first hyperedge H_1 in the hypergraph. The selection criterion can be (1) the size of the hyperedge (cardinality), or (2) the hyperedge with the vertex corresponding to the largest surface segment (area), or (3) the hyperedge containing the vertex with the highest degree. The ordered list of hyperedges is then formed by sorting the hyperedges in the order of the distance to the first selected hyperedge H_1. The distance is defined by the shortest distance between the component vertices of the hyperedges. If two hyperedges have vertices in common, then the distance is zero. When distances between two pairs of hyperedges are the same, then the selection criterion that was used to select the first hyperedge is used to break the tie. The order of the vertices is determined by ordering the vertices in the hyperedges in terms of the area or degree of the vertex. Hyperedges are matched by matching the attributed graphs representing the hyperedges. The unary and the binary constraints are checked to evaluate the match between the hyperedges. Once the hyperedge-match has been established, the second set of hyperedges are selected. The procedure goes down the list of all the vertices in the hypergraphs in a predetermined order. Once a match for a vertex or hyperedge is found, that vertex or hyperedge is marked as *matched*. The marked vertices and hyperedges are not considered in the future hypotheses.

The matching information is encoded in a *compatibility matrix*. The matching between the hyperedges is represented using the *hyperedge compatibility matrix CH*, where the (i, j) entry represents the match (or possibility of a match) between the ith hyperedge in the first hypergraph with the jth hyperedge in the second hypergraph. Similarly the *vertex compatibility matrix CV* represents pairings of vertices. The (i, j) entry gives the match (or possible match) between the

i^{th} vertex of the first hypergraph with the jth vertex in the second hypergraph. The entries of CV correspond to the nodes of the intrerpretation tree.

The next step of the matching procedure (step 3) initializes the compatibility matrices (INITIALIZE). Each entry of the hyperedge compatibility matrix is initialized to 1 and the entries of the vertex compatibility matrix are initialized by considering the unary constraints. Entries corresponding to compatible pairs are set to 1 and the rest are set to 0. A depth first search of the interpretation tree is done in a recursive function (FIND_MATCH) and the *level* gives the depth of the recursion. The variable *level* is the level at which the current hypothesized match is in the interpretation tree. Algorithm 3.2 gives the recursive function FIND_MATCH.

Algorithm 3.2 FIND_MATCH

begin
    1. *while(level $\geq$ 0) do*
        1.1 *if(¬*Find_Level*(CH, level, row, col)) then*
            1.1.1 *level := level − 1*
            1.1.2. *return*
        1.2. $CH_{bak} \longleftarrow CH$; $CV_{bak} \longleftarrow CV$
        1.3. Rm_Hedge_Incompatibilities*(CH, row, col)*
        1.4. Extract_Submtx*(CV_{sub}, CV, row, col)*
        1.5. Rm_Submtx_Incompatibilities*(CV_{sub}, CV)*
        1.6. Get_Match_Set*(CV_{sub}, track_match)*
        1.7. Rm_Vertex_Incompatibilities*(CV_{sub}, CV)*
        1.8. *if(*Is_Solution*(CV)) then*
            1.8.1. Update_Best*(CV)*
        1.9. *else*
            1.9.1. *level := level + 1*
            1.9.2. Find_Match*(CH, CV)*
        1.10. $CH \longleftarrow CH_{bak}$; $CV \longleftarrow CV_{bak}$
        1.11. *if(*All_Sets_Matched*(track_match)) then*
            1.11.1 Remove_Entry*(CH, level)*
    od
end

The matching algorithm 3.2 recursively searches the tree in a depth first manner. The function is called at the root of the interpretation tree (*level* = 0). The root of the tree corresponds to the first entry in the vertex compatibility matrix. The first non-zero entry in the hyperedge compatibility matrix provides with the first hypothesis i.e., the hyperedges H_1 and H'_1 match. The hyperedge pairings are investigated by going down the interpretation tree. The *level* increases as the algorithm moves forward in the tree, and level decreases when the algorithm backtracks. When all the possible branches are exhausted, the *level* becomes less than 0. The search continues until there is a possibility of obtaining another match (the *while* loop in step *1*). A hypothesis is generated by selecting an entry from the hyperedge compatibility matrix. Step *1.1* selects the *level^{th}* non-zero entry from the CH matrix. If no non-zero *level^{th}* entry exists in CH then the search has exhausted all the possible positive matches along the current branch, so the algorithm backtracks by decreasing the *level* (steps *1.1.1* and *1.1.2*). If, however, the *level^{th}* entry exists ((row, col) entry in the CH matrix) then the search progresses in the forward direction. The search involves updating the compatibility matrices; therefore, to enable backtracking, the matrices are backed up in step *1.2*. At each stage, when a match is obtained it is recorded in a *matched list* of all matches.

The hypothesis corresponding to the (row, col) entry of the CH matrix is, "hyperedge H_{row} in the first hypergraph corresponds to hyperedge H'_{col} in the second hypergraph." Each hyperedge has an unique match; therefore, in step *1.3* the entries corresponding to matches between H_{row} and other hyperedges, and those corresponding to matches between H'_{col} and any other hyperedges, are set to 0. The test of hyperedge matching is done by testing the component attributed graphs. The submatrix in CV, representing the component vertices of the two hyperedges H_{row} and H'_{col}, is extracted to compute the vertex matches (step *1.4*). The entries in the submatrix CV_{sub} are tested for compatibility with all the previous matches (in the *matched list*), i.e., all the pairings corresponding to the nodes of the interpretation tree in the same branch as the current hypothesis but above the current node are tested for compatibility with the pairings in the submatrix CV_{sub}. All incompatibilities in CV_{sub} are removed.

The submatrix CV_{sub} represents many sets of valid pairings. Each branch of the subtree rooted at the current hypothesis node is a valid pairing set. The algorithm investigates each set of pairings by keeping track of the sets it has already tested (*track_match*), i.e., the branches of the tree, fanning out of the current node, that have been tested are marked. Based on the sets already investigated, the next set for matching is extracted in step *1.6*. The pairing set is a set of hypothesized vertex pairings. All entries in the CV matrix that are incompatible with the vertex pairings are removed in step *1.7*.

At this point, if the algorithm obtains a solution (a pairing for each vertex in the graphs) or discovers that a match better than one obtained earlier is not feasible, then the algorithm backtracks to investigate other branches and obtain a better solution if possible. Step *1.8* checks for this condition and, if a solution is obtained, the current best match is updated. If it is determined that the algorithm does not have to backtrack, then the *level* is incremented and the search progresses in the forward direction by making a recursive call to FIND_MATCH. The algorithm investigates all the branches after the current node and returns to the current node on the way back. If all the possible pairings for the component vertices have been exhausted (step *1.11*), then it implies that the best possible match with the current hypothesis has been obtained. The next hypothesis is investigated by removing the current entry in the restored CH matrix. The search continues and next time step *1.1* is executed to determine the $level^{th}$ entry in the CH matrix, the new hypothesis is selected. If, however, it is determined in step *1.11* that all the possible pairings for the component vertices have not been exhausted, then the entry in the restored CH is not removed. This results in step *1.1* selecting the same hypothesis again, and step *1.6* ensures that the next pairing set is investigated by the algorithm.

Occlusions can cause vertices in the hypergraph to disappear. This may cause hyperedges to split up and form new hyperedge sets. This phenomenon is accounted for in the matching algorithm, i.e., a provision has been made to dynamically change the hypergraph. By detecting the occluding edges, the locations in the hypergraph where occlusions may occur and modify the hypergraph are identified. For this reason, surfaces that are connected by an interfacing occluding edge form a weak arc in the graph representation. The algorithm will break the weak arc to form a new hypergraph when if breaking the weak arc will give a better match.

When the algorithm determines that the best possible match that can be obtained by moving down the current branch is smaller than a threshold value, then it backtracks without going any further. Termination of the matching procedure occurs if the fraction of surface segments matched exceeds a threshold. Once a match has been determined (i.e., the search procedure has reached the leaf node of the tree), the number of positive pairings (i.e., non-wild card pairings) is computed. If this number is less than the threshold fraction, then the procedure backtracks and searches other branches. At every stage, the best possible match is compared with the current best match. If the best possible match is smaller than the current match, then the search along that branch is abandoned and the next branch is investigated.

4   Results

The algorithm was tried successfully on different types of range image sequences. In this section we present an example of a range image sequence and describe how the matching algorithm computes the surface correspondences.

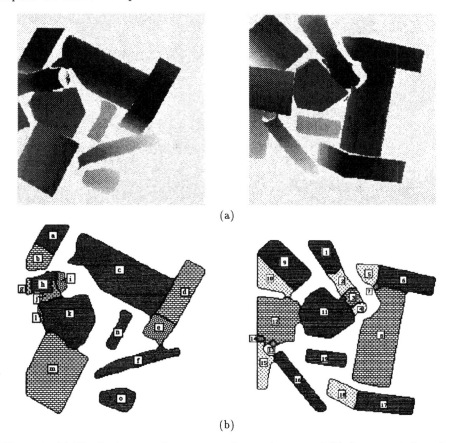

(a)

(b)

Figure 3: *(a) The depth maps of a sequence of range images and (b) the segmented results.*

Figures 3-4 illustrate the algorithm on the example. Figure 3.a shows the depth maps of two frames in the sequence of range images. The scenes consist of a jumble of different kinds of objects. The camera is moved to obtain the second frame of the sequence. The segmented results are shown in figure 3.b. It may be noticed that this example has many instances of incorrect segmentations and occlusions. The first step of the algorithm generates the attributed graph of the scene and computes the cliques in the graph. The hypergraphs generated are shown in the figure 4. In the figure the arcs of the attributed graph are shown in the hyperedges. If there exists an occluding edge between two surfaces then the arc in the attributed graph is *weak* and shown in the figure 4 with dotted lines.

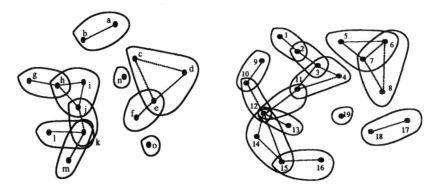

Figure 4: *The generated hypergraphs of the range images.*

The first hyperedge pair hypothesized to match is $\{h, i, j\}$ in the first scene matches $\{3, 4, 11\}$. The vertex with the highest degree h is considered as the first vertex. The unary constraints leave only one option i.e., $(h, 3)$ as the first node in the interpretation tree. However, the next two vertices i and j do not match any vertex so they are matched with the wild card $\star$. Note that in the final match that is obtained the pairing $(h, 3)$ is an incorrect pairing. The algorithm backtracks and finds the correct match even though we start with an incorrect match. The second hyperedge considered for match is $\{h, g\}$ because it is connected to the first hyperedge at h. Since the current hypothesis is $(h, 3)$, the next hyperedge match considered is between $\{h, g\}$ and $\{3, 2\}$. The unary constraints are satisfied between the pair $(g, 2)$ so the binary constraints of angle, distance and the spanning vector are tested. All the constraints are satisfied so the match pair is accepted in the current hypothesis. The next hyperedge considered now is $\{j, k\}$ as it is connected to the first hyperedge. The match between $\{j, k\}$ and $\{11, 12\}$ is tried and the pairing $(k, 11)$ satisfies all the constraints, but the connectivity is not satisfied (k is not connected to h while 11 is connected to 3). At this point we use the fact that the arc between 3 and 11 is a *weak* one so it can be broken and all the constraints are satisfied.

The procedure continues till a complete match (i.e., all the vertices are accounted for) is obtained. The match size is evaluated and if a better match can be obtained, the procedure backtracks to improve the results. The final matching results are:

I	a	b	c	d	e	f	g	h	i	j	k
II	6	5	8	17	18	16	1	2	3	$\star$	11
I	l	m	n	o	$\star$	$\star$	$\star$	$\star$	$\star$	$\star$	
II	$\star$	12	19	15	4	7	9	10	13	14	

It may be observed that in the example shown there are many errors in segmentation (for eg. surfaces j, 7, 14, 13, etc.) and there are surfaces that get occluded in one of the scenes (for eg. 4 and l); notwithstanding, the algorithm performs well and the correspondences are evaluated.

5   Conclusion

Computing motion and tracking an object over a sequence of range images involves establishing correspondence between the features of the object in different images in the sequence. The question of finding correspondence in a sequence of range images is very different from finding correspondence between a model and an object description. The fundamental difference lies in the fact that the model description of the object is complete, while in case of a sequence of range images, both descriptions of the scene are incomplete. The lack of information forces one to impose only weak constraints and allow for larger tolerances.

We presented a new framework and procedure to compute the correspondences between surface segments in a sequence of range images. Fundamental to our framework is the hypergraph representation of the range images. The hierarchical representation of hypergraphs not only reduces the search space significantly, but also facilitates the encoding of the topological and geometrical information. In addition to the topological and geometrical information obtained from the scene we also use a priori knowledge of the scene obtained from the physics of the laser scanning process used to produce the range images. Each piece of information used reduces the complexity of the matching procedure by pruning the search space. The solution is robust and accounts for errors in segmentation, occlusions of surfaces, and noise in the data. By using the topological information to guide the search procedure, the average case complexity of the algorithm is reduced significantly.

Acknowledgments

We would like to thank Debi Paxton for proof reading the paper. This research was supported in part by Army Research Office, Contract *DAAL-03-91-G-0050* and Air Force Office of Sponsored Research, Contract *F49620-92-C-0027*.

References

[1] F. Arman and J.K. Aggarwal. Model–based object recognition in dense range images – a review. *ACM Computing Surveys*, 25(1):5–43, March 1993.

[2] C. Berge. *Graphs and Hypergraphs*. North–Holland Publishing Company, Amsterdam, 1983.

[3] N. Deo. *Graph Theory with Applications to Engineering and Computer Science*. Prentice Hall, Englewood Cliffs, New Jersey, 1974.

[4] W. E. L. Grimson and T. Lozeno-Perez. Localizing overlapping parts by searching the interpretation tree. *IEEE Transactions on Pattern Analysis and Machine Intelligence*,

9(4):469–482, Apr 1987.

[5] W.E.L. Grimson. The combinatorics of object recognition in cluttered environments using constrained search. *Artificial Intelligence*, 44:121–165, 1990.

[6] W.E.L. Grimson. The combinatorics of heuristic search termination for object recognition in cluttered environments. *IEEE Transactions on Pattern Analysis and Machine Intelligence*, 13(9):920–935, Sept. 1991.

[7] W.E.L. Grimson and D.P. Huttenlocher. On the verification of hypothesized matches in model-based recognition. *IEEE Transactions on Pattern Analysis and Machine Intelligence*, 13(12):1201–1213, Dec. 1991.

[8] V. Kumar. Algorithms for constraint satisfaction problems: A survey. *AI Magazine*, pages 32 – 44, Spring 1992.

[9] B. Sabata, F. Arman, and J. K. Aggarwal. Segmentation of 3–d range images using pyramid data structures. *Computer Vision, Graphics, and Image Processing: Image Inderstanding*, 57(3):373–387, May 1993.

[10] A.K.C Wong. Knowledge representation for robot vision and path planning using attributed graphs and hypergraphs. In A.K.C. Wong and A. Pugh, editors. *Machine Intelligence and Knowledge Engineering for Robotic Applications*, pages 113–143. Springer Verlag, Berlin Heidelberg, 1987.

[11] A.K.C. Wong, S.W. Lu, and M. Rioux. Recognition and shape synthesis of 3–d objects based on attributed hypergraphs. *IEEE Transactions on Pattern Analysis and Machine Intelligence*, 11(3):279 – 290, March 1989.

*Time-Varying Image Processing and
Moving Object Recognition, 3* – V. Cappellini (Ed.)

A Geometrical Correlation Function for Shape Recognition

*A. Loui * and A.N. Venetsanopoulos +*

*Bell Communications Research, 331 Newman Springs Road, Red Bank, New Jersey 07701, U.S.A.

+Department of Electrical and Computer Engineering, University of Toronto, Toronto, Ontario, Canada M5S 1A4

A new descriptor for the representation of two-dimensional continuous or discrete signals is introduced in this paper. The proposed shape descriptor, which we call the *geometrical correlation function* (GCF), is based on the idea of morphological covariance [1]. It is shown that the family of GCFs associated with different orientations of a particular shape is both translation, scale, and rotation invariant. In addition, geometrical properties, such as the area and perimeter of the shape can be derived from the GCF family. The GCF family can be computed using the associated *morphological correlator* which is composed of m parallel computation units and a feature-function selection unit, where a small subset of the GCF family is selected for classification. Hence, the GCF family can be used efficiently for shape-recognition. In addition, it is shown that with a suitable criterion for selecting the feature function, promising results for successful classification are obtained.

1. INTRODUCTION

The effectiveness of shape recognition is important to many applications such as motion analysis in video coding, and machine parts identification in automated or flexible manufacturing. In many of these applications, the task of shape recognition may also include identification of position, orientation and velocity of moving objects.

There are basically two major steps involved in shape recognition, namely feature extraction and shape classification. The selection of meaningful features has been traditionally based on the structures used by human beings in interpreting pictorial information [2]. Given that a shape descriptor has been chosen, the first step is then to represent the unknown shape using this shape descriptor. This shape descriptor should be efficient in its use of computation, as well as faithful in providing a usable representation of the original shape. Though there exist many quantitative measures of shape, there is yet, in general, no agreement on a minimum set of shape descriptors to adequately quantify object form. Once a feature set is established, the second step in the recognition process is the classification of the unknown shape based on the creation of a suitable measure. This usually involves the application of some criterion such as Minimum Mean Square Error (MMSE).

Recently, the use of morphological techniques [3,4] for image analysis has become very popular due to the simple and fast implementations of many morphological operations [5]. In this paper, a 2-D shape descriptor, namely the *geometrical correlation function* (GCF), is described and its applications to shape recognition are demonstrated. In section 2, the definition and properties of the GCF are presented. Section 3 then describes a shape-recognition system, which utilizes the proposed shape descriptor and employs an associated *morphological correlator*. Finally, in section 4, the performance of the proposed scheme is analyzed using real image data obtained from a conveyor-belt system.

2. GEOMETRICAL CORRELATION FUNCTIONS (GCFs)

In this section, the mathematical definition and properties of the *geometrical correla-tion function*, which is closely related to the morphological covariance as defined in [1], are given. The morphological covariance descends from the theory of random functions of order two, and it provides a measure on the correlation properties of 2-D signals.

Specifically, the morphological covariance is defined as the measure of the eroded set by the structuring element B, which consists of a pair of points separated by a distance equivalent to the amount of spatial shift h, at an angle ϕ.

Let X be a deterministic compact set in $\mathbf{R}^n$, and B^ϕ be a structuring element which is composed of two single points separated by a distance h at an angle ϕ relative to $0°$. The morphological covariance $C(h)$ is defined as the measure of the set $X \ominus B^\phi$, which is X eroded by B^ϕ. If we consider the translation B_x^ϕ of B by x at an angle ϕ, the point x belongs to the eroded set $X \ominus B^\phi$ if and only if x and $x+h \in X$. Hence,

$$X \ominus B^\phi = X \cap X_{-h}^\phi .$$

(1)

The covariance $C(h)$ is given by

$$C(h) = Mes \ [X \ominus B^\phi]$$

(2)

where Mes is defined as the digital area of binary object X.

2.1 Definitions

For a 2-D closed subset X of the Euclidean space $\mathbf{E} = \mathbf{R}^2$, the GCF is defined as

$$K_\phi(h) \triangleq \frac{Mes \ [X \ominus B_h^\phi]}{Mes \ [Y]}$$

(3)

where B_h^ϕ is a two-point structuring element, and Y is a pre-defined standard shape (e.g. a square of size 200×200 pixels), or Y can be chosen to be equal to X. For example, the GCF can be restricted to two particular directions, either vertical or horizontal. For this situation, the vertical GCF is defined as follows:

$$K_{\pi/2}(h) = \frac{Mes \ [X \ominus k_1(h)]}{Mes \ [Y]}$$

(4)

where

$$k_1(h) = (1 * \cdots * 1)^T \ ;$$

(5)

and the corresponding horizontal GCF is defined as

$$K_0(h) = \frac{Mes\ [X\ \ominus k_2(h)]}{Mes\ [Y\ ]}$$ (6)

where

$$k_2(h) = (1 * \cdots * 1)\ ,$$ (7)

i.e., $k_1(0) = (1)^T$, $k_2(0) = (1)$, $k_1(1) = (1\ 1)^T$, and $k_2(1) = (1\ 1)$, etc.

2.2 Properties

In this section we investigate some of the basic properties of the GCF as a shape descriptor. These include possible invariance under translation, rotation and scale change. As for any other shape descriptor, these characteristics are important properties to use when applying GCF for shape description.

Property 1: The GCF is an even positive-valued function.

This is quite obvious since the GCF evaluated at direction ϕ and at direction $\phi \pm \pi$ are the same.

Property 2: The maximum value of the GCF occurs at the origin, i.e., $K(0) \geq K(h)$.

This is the direct consequence of property 1.

Property 3: The area of the shape can be derived from the GCF directly as

$$Are\ [X] = Mes\ [X] = K_\phi(0)Mes\ [Y]$$ (8)

Property 4: The perimeter of the shape can be derived from its GCFs as [6]

$$Per\ [X] = -\frac{Mes\ (X)}{2} \int_0^{2\pi} K_\phi'(0)d\phi$$ (9)

Property 5: The Fourier transform of the family of GCFs is given by

$$\Gamma(w) = F\ [K(n)] = \sum_{n \in Z^2} K(n)e^{-jwn}$$ (10)

where $n = \{n_1, n_2\}$ such that $n_1 = h\cos\phi$ and $n_2 = h\sin\phi$, and $w = \{w_1, w_2\}$ is the corresponding vector in the frequency domain.

$\Gamma(w)$ can be considered as the *area spectral density* [6] of the binary image.

Property 6: The GCF is invariant under signal translation.

This follows since computing the GCF is a morphological operation.

Property 7: In general, the GCF is not invariant under signal scale change. However with some normalization, the GCF can be made scale-invariant.

Property 8: The GCF alone is not invariant under signal rotation. However, the GCF family is invariant under signal rotation.

Property 9: The GCF of a signal is not in general unique to that signal.

This is because the GCF contains only second-order information of a 2-D signal, and hence, the original signal cannot in general be recovered solely from the GCF.

3. SHAPE-RECOGNITION BASED ON THE GCF

In this section, a practical shape-recognition system based on the GCF is presented. A simplified block diagram of the proposed shape-recognition system is shown in Figure 1. The major task of this system is to analyze sensed information and to take appropriate action depending on the outcome of the classifier. In this paper, we assume that the sensed information is an image which corresponds to the 2-D projection of a real 3-D object.

According to Figure 1, the first step corresponds to the acquisition of the input signal. After acquisition, the incoming signal is digitized through an A/D converter and then applied to the preprocessor. The preprocessor then enchances and restores the digitized signal. The next step is the major task of feature extraction. This is accomplished by the *morphological correlator* in which the GCFs are computed. However, if we want the system to be size invariant, the object or shape to be described must be pre-scaled before the GCFs are computed. To accomplish this, the 2-D signal that represents the unknown object is linearly and isotropically scaled along the two mutually perpendicular coordinate axes, so that the resulting measure is equal to a pre-defined standard measure.

The GCFs that we obtain from the *morphological correlator* can provide us with information regarding geometrical properties of the unknown object. For example, as noted, both the area (equation (8)) and perimeter (equation (9)) can be deduced from the GCF. In addition, the orientation of the unknown shape can be estimated from the angle associated with the individual GCF. This can be done by comparing the angles associated with the corresponding reference object's GCF and those of the rotated and translated object's GCF. For example, a square rotated by 45° will result in a GCF (evaluated at 0°) that looks like the GCF for the reference square evaluated at 45°. Hence, the difference between the angles associated with the two GCFs of the same shape gives an indication of the orientation of the unknown object. As indicated in Figure 1, this additional knowledge can be obtained using an auxiliary module which is connected to the data base of the system.

According to the block diagram of Figure 1, the next step is to determine whether the system is in a training mode. If this is the case, the appropriate attributes of the reference object are stored in the data base. If not, the output of the *morphological correlator* is passed on to the classifier. The classifier then decides which object is present based on the attributes of the unknown relative to those of the reference set. During the classification step, the descriptor values of the unknown object are compared and matched with the reference values, with classification made according to some decision rules. Additional information such as area, perimeter, and orientation if available, can be used in the classification stage. Finally, the outcome of the classifier is passed on to the last module which performs an action such as sending a feedback signal to a robot or activating some other device.

The *morphological correlator* is the heart of the proposed object-identification system. A block diagram of the morphological correlator is shown in Figure 2. The correlator basically consists of m parallel computation units and a feature-selection unit. The m parallel units compute the GCF of the unknown 2-D image signal for m different directions. The feature-selection unit selects a small subset, m_f where $m_f \ll m$, among the m functions based on some pre-defined criterion.

One feature subset is based on the maximum and/or minimum area under a member of the family of GCFs of a given object. The area under the GCF provides an indication of the geometrical similarity of an object, i.e., the smaller the area, the bigger the change in correlation as the object is moved. This in turn implies that the dimension of the object in one direction is relatively smaller. Other criteria, such as the slope at the origin of the GCF and/or the number of slope changes, can also be used. While the area under the GCF provides a global view of the geometry of the object, the number of slope changes gives an indication of its local geometry. For example, successive peaks imply that some part of the object is periodic.

4. EXPERIMENTAL RESULTS

In this section, the performance of the GCF as a shape descriptor is examined. A criterion based on the area under a single GCF curve is used. The number of computation units, $m = 8$ and the number of feature-functions, $m_f = 1$. The classifier used is a simple deterministic minimum-distance classifier, which in this case computes the sum of the squared distances, $d(u,r)$ between the reference GCF, $K_{\phi}^r(h)$ and the unknown GCF, $K_{\phi}^u(h)$, i.e.

$$d(u,r) = \sum_{h=0}^{N-1} w(h)[K_{\phi}^u(h) - K_{\phi}^r(h)]^2 \tag{11}$$

where $w(h)$ is a weighting factor for the h component of the GCF, and N is the maximum extent of the GCF. In our results, $w(h)$ is set to 1. The criterion for selecting the respective feature functions is based on the area under the GCF curve, i.e., the quantity $A[K_{\phi}]$, where

$$A[K_{\phi}] = \sum_{h=0}^{N-1} K_{\phi}(h) \quad . \tag{12}$$

Hence, the maximum feature function, $K_{\phi_{max}}$ is defined as

$$K_{\phi_{max}} = \{K_{\phi}: A[K_{\phi}] \text{ is maximun } \}, \ 0° \le \phi \le 180° \quad . \tag{13}$$

Similarly, the minimum feature function, $K_{\phi_{min}}$ is defined as

$$K_{\phi_{min}} = \{K_{\phi}: A[K_{\phi}] \text{ is minimum } \}, \ 0° \le \phi \le 180° \quad . \tag{14}$$

Preliminary experimental results using synthetic images were reported in [7]. From [7], it was observed that the performance of the minimum feature function is better than the maximum feature function. In this paper, we evaluate the performance of the GCF using real object images obtained from an experimental conveyor belt system at the Computer-Integrated-Manufacturing Laboratory of the University of Toronto. The test object was placed on a pallet of a moving conveyor belt in an arbitrary orientation. Figure 3 shows images of some sample test objects placed on top of a pallet. Note also that no special lighting was arranged. This implies that different workpieces will exhibit different reflections, depending on their surface finish.

The processing and analysis of the video data were carried out using the SUN 3/260 workstation at the Signal Processing Laboratory. Note that quality of the digitized images is degraded by both the transfer process from video camera to VCR as well as that from VCR to computer memory through the digitizer. The stationary images of Figures 3 are used as a reference set for object identification. The performance of the minimum-distance classifier, using the MAT obtained from the morphological correlator, is illustrated in Tables 1 to 3. Tables 1, 2 and 3 are organized in such a way that comparisons (the distance $d(u,r)$) between an unknown object and all the reference objects are listed in a column format. When all these columns are combined , a confusion matrix is formed.

Table 1 shows the matching results using only the maximum feature functions; while table 2 shows the results using only the minimum feature functions. Table 3 tabulates the results using both the maximum and minimum feature functions at the classification stage. That is, the value corresponding to the minimum of the confusion matrices in Tables 1 and 2, is used. The results show that all objects are identified correctly in all three cases. However, in order to evaluate the performance of an individual criterion, we propose to use a probability of misclassification measure, which will give us an idea how the different criteria compare to each other. We define the probability of misclassification, P_m, as

$$P_m = \frac{\text{sum of all diagonal elements of the confusion matrix}}{\text{sum of all the elements of the confusion matrix}} \qquad (15)$$

Based on this measure, P_m is found to be 0.00301, 0.00274, and 0.00112 for the case in Tables 1, 2 and 3 respectively. Hence, in this case, the criterion based on a minimum-area GCF is more robust than that based on maximum area GCF. Further, the combined criterion of Table 3 gives the smallest probability of misclassification, which is approximately 3 times smaller than the criterion based on maximum area GCF. In general, observations in the real image situation agree well with those in the synthetic situation reported in [7], in the sense that the criterion based on minimum area GCF seems to perform better than the corresponding maximum criterion. One conjecture that can be made from the experimental results is that the minimum feature function is less sensitive to rotation than the maximum feature function and thus, it is inherently more stable and reliable than other members of the same GCF family.

5. SUMMARY

In this paper, a new 2-D shape descriptor is introduced and its performance is evaluated using different real test objects. This shape descriptor, which we have called the *geometrical correlation function* (GCF), is based on the idea of morphological covariance and is related to 2nd order properties of the object. Experimental results show that the algorithm works well for real image objects despite the presence of various kinds of noise, non-linearity, and interference. The performance of the estimation scheme may be improved by increasing 1) the resolution of the table-lookup operations associated with the algorithm, and 2) the number of GCFs used. The potential application of the GCF representation scheme is not limited to object identification. Since the GCF contains information about the spatial distribution of features of the object, it is possible to extend the GCF idea to the problem of motion parameter estimation.

REFERENCES

[1] J. Serra, *Image Analysis and Mathematical Morphology*, Academic Press, 1982.

[2] M. Levine, *Vision in Man and Machine*, McGraw-Hill Inc., 1985.

[3] R.M. Haralick, S.R. Sternberg and X. Zhuang, "Image analysis using mathematical morphology," *IEEE Trans. on PAMI*, Vol. PAMI-9, No. 4, pp. 532-550, July 1987.

[4] P. Maragos and R.W. Schafer, "Morphological filters - part I: their set-theoretic analysis and relations to linear shift-invariant filters," *IEEE Trans. on ASSP*, Vol. ASSP-35, No. 8, pp. 1153-1169, Aug 1987.

[5] A. Loui, A.N. Venetsanopoulos, K.C. Smith, "Flexible architectures for morphological image processing and analysis," *IEEE Trans. on Circuits and Systems for Video*

Technology, Vol. 2, No. 1, pp. 72-83, March 1992.

[6] A. C. P. Loui, "A morphological approach to moving-object recognition with applications to machine vision,", *Ph.D Thesis*, Dept. Elec. Eng., University of Toronto, Toronto, Canada, 1990.

[7] A. Loui, A.N. Venetsanopoulos, K.C. Smith, "Morphological autocorrelation transform: A new representation and classification scheme for two-dimensional images," *IEEE Trans. on Image Processing*, Vol. 1, No. 3, pp. 337-354, July 1992.

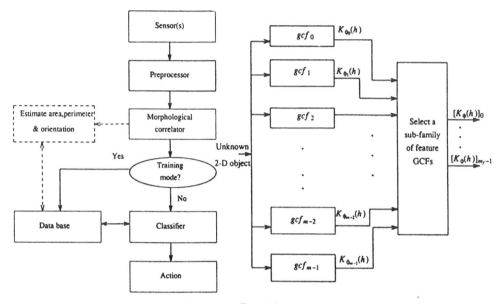

Figure 1 Block diagram of the shape-recognition system.   Figure 2 Block diagram of the morphological correlator.

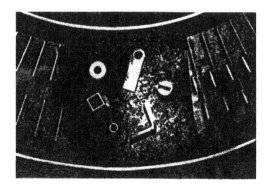

Figure 3 Samples of mechanical test objects used in the experiment.

Euclidean distances from a randomly oriented real object to reference object set						
Shape	square	square1	bracket	annulus	disk	triangle
square	0.094	0.099	0.346	16.486	7.710	25.596
square1	0.094	0.099	0.346	16.486	7.710	25.596
bracket	1.797	1.817	0.285	21.426	12.343	30.898
annulus	14.795	14.748	18.588	0.063	2.755	1.781
disk	6.488	6.415	9.923	2.830	0.009	7.854
triangle	19.466	19.414	23.707	0.549	4.768	0.424
Decision Correct?	yes	yes	yes	yes	yes	yes

Table 1   A confusion matrix for 6 real machine-parts using only the maximum-feature-function criterion.

Euclidean distances from a randomly oriented real object to reference object set						
Shape	square	square1	bracket	annulus	disk	triangle
square	0.044	0.058	12.899	14.160	6.430	25.679
square1	0.044	0.058	12.899	14.160	6.430	25.679
bracket	11.542	11.398	0.009	0.564	1.575	4.594
annulus	10.345	10.209	0.749	0.331	1.735	3.887
disk	3.718	3.625	3.170	4.294	0.266	12.761
triangle	25.406	25.221	4.957	3.064	11.058	0.040
Decision Correct?	yes	yes	yes	yes	yes	yes

Table 2   A confusion matrix for 6 real machine-parts using only the minimum-feature-function criterion.

Euclidean distances from a randomly oriented real object to reference object set						
Shape	square	square1	bracket	annulus	disk	triangle
square	0.044	0.058	0.346	14.160	6.430	25.596
square1	0.044	0.058	0.346	14.160	6.430	25.596
bracket	1.797	1.817	0.009	0.564	1.575	4.594
annulus	10.345	10.209	0.749	0.063	1.735	1.781
disk	3.718	3.625	3.170	2.830	0.009	7.854
triangle	19.466	19.414	4.957	0.549	4.768	0.040
Decision Correct?	yes	yes	yes	yes	yes	yes

Table 3   A confusion matrix for 6 real machine-parts using the minimum MMSE obtained from the maximum and minimum feature criteria.

*Time-Varying Image Processing and
Moving Object Recognition, 3* – V. Cappellini (Ed.)
© 1994 Elsevier Science B.V. All rights reserved.

Spotting Recognition of Human Gestures from Motion Images

K. Takahashi S. Seki and R. Oka

Tsukuba Research Center, Real World Computing Partnership,
Tsukuba-Mitsui bldg. 16F, 1-6-1 Takezono, Tsukuba-shi, Ibaraki 305, Japan
Email: takahasi@trc.rwcp.or.jp

A spotting algorithm to recognize the meanings of human gestures from motion images
is proposed. Our algorithm consists of two principles. One uses spatio-temporal vector
fields composed of three dimensional edges for recognition. These are also used to represent gesture models (we call them "standard sequence patterns"). The other recognizes
gestures using a spotting algorithm. The spotting algorithm removes the need for temporal segmentation of gesture duration and introduces frame-wise recognition. We carried
out some experiments into gesture recognition and the results have confirmed that our
algorithm is robust with various clothing textures and backgrounds.

1. INTRODUCTION

Human beings receive many motion images through their eyes and use these to govern
how they act. If we could give computers the ability to manage motion images, some of
the tasks of human beings could be done by computer. This provides the opportunity for
a new type of human interface.

In this research field, many research projects have been proposed which treat vision
as a dynamic process and acquire visual features from multiple images [1-4]. But as far
as we know, few methods have been reported to understand the meanings of motion.
One such research is a human action recognition method based on the Hidden Markov
Model[5]. In this research, they observed the way in which tennis players swing their
rackets and carried out some experiments to recognize the different swings. In the field of
gesture recognition from motion images, there are many technological breakthroughs to
be overcome such as extracting features, temporal segmentation of gesture duration, and
flexible matching between a model and input images.

In this paper, we present a novel algorithm to recognize gestures from motion images.
Our main principles are:

- use of spatio-temporal vector fields composed of three dimensional edges

- spotting recognition by Continuous Dynamic Programming (CDP) which realizes
  segmentation free and frame-wise recognition.

We do not use geometric model but instead employ a simple description of the feature
sequence to represent the gesture. We revealed that the features obtained by spatial-
reduction, temporal-averaging operation, and saturation operation of several independent

66

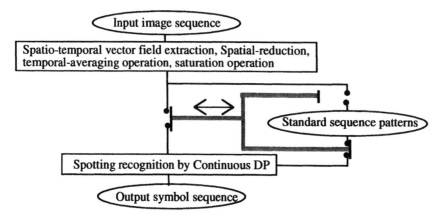

Figure 1: System structure

scalar fields, each of which corresponds to a partial derivative of motion images, have enough robustness and stability for gesture recognition.

The spotting recognition by CDP works frame-wise without segmentation of gesture duration. The frame-wise computation scheme is also suitable both to evaluate and respond to gestures.

2. SYSTEM STRUCTURE

Figure 1 shows the overall structure of our spotting recognition system. Our system consists of two main parts. One extracts spatio-temporal vector fields from input image sequences and makes standard sequence patterns that correspond to gestures. This process must be done before gesture recognition. The other recognizes gestures using spotting recognition by Continuous Dynamic Programming.

3. FEATURE EXTRACTION

Many kinds of features can be extracted from motion image sequences. Our system utilizes horizontal, vertical, and temporal edges to recognize gestures. This is because these features make the system robust and have sufficient resolution for motion image understanding, only by means of spatial-reduction, temporal-averaging operation, and saturation operation. Edge detection is relatively easy so it is suitable for use in a real-time recognition system.

Human gestures are observed as sequences of images through a camera, and we could derive edge feature images (spatio-temporal vector field) from these sequences by using the following steps. First, we represent input images:

$$\{f(x, y, t) \mid 0 \leq x \leq L, 0 \leq y \leq L, 0 \leq t < \infty\}, \tag{1}$$

where x and y denote the coordinates of the position in an image and t denotes time.

$f(x, y, t)$ denotes the gray value of point(x, y) in the image at time t. Next, the spatio-temporal vector $V(x, y, t)$ which consists of three kinds of edges is described as follows:

$$V(x, y, t) = (v_1, v_2, v_3) = (\frac{\partial f(x, y, t)}{\partial x}, \frac{\partial f(x, y, t)}{\partial y}, \frac{\partial f(x, y, t)}{\partial t}). \tag{2}$$

Elements v_1, v_2, and v_3 correspond to the horizontal edge, vertical edge, and temporal edge, respectively. However x, y, and t are actually discrete, each vector element can be calculated as follows:

$$\begin{cases} v_1 &= \sum_{i,j=1}^{3} w_{ij} f(x+1, y-2+i, t-3+j) - \sum_{i,j=1}^{3} w_{ij} f(x-1, y-2+i, t-3+j) \\ v_2 &= \sum_{i,j=1}^{3} w_{ij} f(x-2+i, y+1, t-3+j) - \sum_{i,j=1}^{3} w_{ij} f(x-2+i, y-1, t-3+j) \\ v_3 &= \sum_{i,j=1}^{3} w_{ij} f(x-2+i, y-2+j, t) - \sum_{i,j=1}^{3} w_{ij} f(x-2+i, y-2+j, t-2) \end{cases}, (3)$$

where its weight values w_{ij} are:

$$W = (w_{ij}) = \begin{pmatrix} \frac{1}{\sqrt{3}} & \frac{1}{\sqrt{2}} & \frac{1}{\sqrt{3}} \\ \frac{1}{\sqrt{2}} & 1 & \frac{1}{\sqrt{2}} \\ \frac{1}{\sqrt{3}} & \frac{1}{\sqrt{2}} & \frac{1}{\sqrt{3}} \end{pmatrix}. \tag{4}$$

Finally, in order to make them robust as features, spatial-reduction, temporal-averaging operation, and saturation operation are adapted to each vector element. Spatial-reduction compresses an $x - y$ plane and reduces grids along the x and y axes from L to $N(\leq L)$. Temporal-averaging operation calculates an average value on each grid from the $(t-K+1)$-th frame to the t-th frame and makes it the new value of the t-th frame. Saturation operation saturates each element value with a logarithmic function to emphasize the existence of edges. These operations are described in the following formula:

$$u_\omega(l, m, t) = log(1 + | \sum_{\substack{0 < \alpha, \beta \leq h \\ 0 \leq k < K}} v_\omega ([l \cdot h] - \alpha, [m \cdot h] - \beta, t - k) |), \tag{5}$$

where α and β are both integers, $h = L/N$, $\omega = 1, 2, 3$, $1 \leq l \leq N$, $1 \leq m \leq N$, and $[\ ]$ denotes Gauss's symbol. Edge feature image sequences are described as:

$$U = \{u_\omega(l, m, t) \mid 1 \leq \omega \leq 3, 1 \leq l \leq N, 1 \leq m \leq N, 1 \leq t < \infty\}. \tag{6}$$

This feature image sequence is also a spatio-temporal vector field and we call this the "feature vector field."

4. STANDARD SEQUENCE PATTERN

A standard sequence pattern is a gesture model and is represented by a feature vector field with a beginning and end point.

A series of frames, which correspond to one gesture, is determined by segmenting an image sequence in advance. Then, a feature vector field with a beginning and end point is acquired by applying the procedure described in Section 3 to such an image sequence. This is the standard sequence pattern corresponding to the input gesture. One standard sequence pattern corresponds to one gesture and the frame length of each standard sequence pattern may be different.

5. SPOTTING RECOGNITION

In this section, we show our matching algorithm between an input image sequence and a set of standard sequence patterns stored in advance. Each input image frame observed by a device like a CCD camera is transformed into the feature vector field shown in formula 6 by the procedure mentioned in Section 3 as soon as it is input into the computer. In the matching stage, assuming that the input frame observed at each time t corresponds to the last frame of the standard sequence pattern, the best correspondence between input frames and the standard sequence pattern is searched for along the time axis. The value calculated along the best matching path is used to generate a result at that time. This matching method is called "spotting recognition" in speech recognition, and the "Continuous Dynamic Programming method(CDP)[6,7]" is one of the most famous of these methods. We adopt CDP to match feature vector fields, i.e., to recognize gestures.

The feature vector field at time t is described as:

$$u(t) = \{u_\omega(l, m, t) \mid 1 \leq \omega \leq 3, 1 \leq l \leq N, 1 \leq m \leq N\}. \tag{7}$$

A standard sequence pattern is described as:

$$Z = \{z(\tau) \mid 1 \leq \tau \leq T\}, \tag{8}$$

where

$$z(\tau) = \{z_\omega(l, m, \tau) \mid 1 \leq \omega \leq 3, 1 \leq l \leq N, 1 \leq m \leq N\}. \tag{9}$$

Here, we define the distance between a frame in the feature vector field and one in the standard sequence pattern. The distance $d(t, \tau)$ between $u(t)$ and $z(\tau)$ is defined as:

$$d(t, \tau) = \|u(t) - z(\tau)\| = \sum_{1 \leq l,\, m \leq N} \sum_\omega |\, u_\omega(l, m, t) - z_\omega(l, m, \tau) \,| \,, \tag{10}$$

where $\omega \in \{1, 2, 3\}$, $\omega \in \{1, 2\}$, or $\omega \in \{3\}$. The value of ω can be changed according to which features we pay attention to. Then, the accumulated distance $S(t, \tau)$ between u and z at (t, τ) is defined as follows:

Initial condition:

$$S(-1, \tau) = S(0, \tau) = \infty \quad (1 \leq \tau \leq T). \tag{11}$$

Formulas($t \geq 1$):

$$S(t, 1) = 3 \cdot d(t, 1) \tag{12}$$

$$S(t, 2) = \min \begin{cases} S(t - 2, 1) + 2 \cdot d(t - 1, 2) + d(t, 2) \\ S(t - 1, 1) + 3 \cdot d(t, 2) \\ S(t, 1) + 3 \cdot d(t, 2) \end{cases} \tag{13}$$

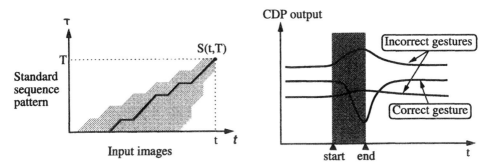

Figure 2 : Continuous Dynamic Programming   **Figure 3: CDP output peculiarity**

$$S(t,\tau) = \min \begin{cases} S(t-2,\tau-1) + 2\cdot d(t-1,\tau) + d(t,\tau) \\ S(t-1,\tau-1) + 3\cdot d(t,\tau) \\ S(t-1,\tau-2) + 3\cdot d(t,\tau-1) + 3\cdot d(t,\tau) \end{cases} \cdot \tag{14}$$
$$(3 \le \tau \le T)$$

Assuming $\tau = T$ (in other words, the gesture finishes at time t), $S(t,T)$ means the accumulated distance along a best matching path between the input image sequence and the standard sequence pattern. Figure 2 illustrates the correspondence of frames between an input image sequence and a standard sequence pattern by CDP. The dotted region is CDP's search area and the thick line passing through the center of the dotted region is the best matching path. The value of $S(t,T)$ can be normalized by the sum of weight $3\cdot T$ as follows:

$$A(t) = \frac{1}{3\cdot T}\, S(t,T). \tag{15}$$

This normalized value $A(t)$ is the CDP output value at time t and corresponds to the unlikelihood of finishing the gesture at that time. This normalization absorbs the effect of the difference of frame length among each standard sequence pattern.

If I standard sequence patterns exist, I CDP output values, $A(t)$'s, also exist at each time. Let $A_\ell(t)$ be CDP output values corresponding to category ℓ at time t, where $\ell = 1, 2, ..., I$. Then, the category number which means the subject's intention is defined as follows:

$$\ell^*(t) = \begin{cases} \mathrm{Arg}\{\min_{1\le\ell\le I}(A_\ell(t) - h_\ell)\} & \text{if } \exists\ell \text{ so that } A_\ell(t) \le h_\ell \\ \text{null} & \text{otherwise} \end{cases}, \tag{16}$$

where Arg means a function which returns its argument, h_ℓ denotes thresholds defined for each gesture to satisfy its "if" condition part only when the subject performs its gesture, and null denotes empty. The CDP output peculiarity is illustrated in Figure 3. This figure indicates to us that the output of CDP corresponding to a correct gesture doesn't become a minimum while the subject is doing the gesture, but does when the subject finishes the gesture.

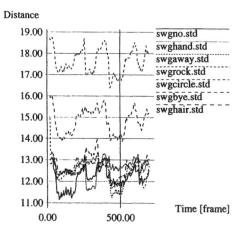

Figure 4: A gesture "circle"   **Figure 5: Example of CDP output**

6. EXPERIMENTS

We have implemented our algorithm and have carried out experiments to examine the effects of variations in clothes and backgrounds with different textures, and the effectivity of edge features.

6.1. Experimental Procedure

The experiments were conducted indoors and only one subject was tested. The CCD camera was fixed so that the background of the image basically did not move. The subject was always near the center of the image and filled the same proportion of the image. In this experiment, the input image size is 64×64 (L=64), the spatial-reduced image size N is 16 and the temporal-averaging length K is 3.

The gestures used in these experiments were (1) no, (2) handclap, (3) away, (4) rock-paper-scissors, (5) circle (making circle with hands over one's head), (6) bye, (7) scratch (scratch one's head). A snapshot of the "circle" gesture is illustrated in Figure 4. The subject did all gestures at a natural speed. The input image sequences to make standard sequence patterns were all taken under the same conditions of clothing and background. Each standard sequence pattern was made using the typical motion of its gesture. An input image sequence to be recognized is composed of these seven gestures in this order.

In order to examine the effect of different clothes and background when making standard sequence patterns and when recognizing them, we set up four situations: (s1) same clothes, same background, (s2) different clothes, same background, (s3) same clothes, different background, and (s4) different clothes, different background.

In order to examine the effectivity of three kinds of edge features, we set up 3 situations: (f1) use u_1 and u_2, (f2) use u_3 only, and (f3) use u_1, u_2, and u_3.

In each condition (12 conditions in total), we applied some kinds of input image sequence and calculated the recognition rate.

Table 1
recognition rate (%)

	edge feature		
	f1	f2	f3
s1	86	100	100
s2	76	95	90
s3	36	100	93
s4	62	100	83

Table 2
recognition rate (%)

s1	64
s2	67
s3	7
s4	5

6.2. Experimental Results

The recognition results were shown in Table 1. Figure 5 depicts an example of $A_l(t)$. This table shows us that u_3 only; u_1, u_2, and u_3; and u_1 and u_2 are efficient, in this order, in terms of edge features. In terms of effect by clothing and background, the case using u_3 only and that of u_1, u_2, and u_3 weren't affected by clothing and background.

6.3. Experiment to Confirm Edge Feature Superiority

In our proposal, we obtained the features by spatial-reduction, temporal-averaging operation, and saturation operation of several independent scalar field, each of which corresponds to a partial derivative. This was done because we believed that this procedure would produce robust features for motion recognition. Here, in order to confirm this assumption, we applied gray images instead of a feature vector field to CDP's input data.

Table 2 shows the recognition results using gray image information. This shows us that the recognition rate when using gray images as the CDP input is lower than when using edge features. Especially, in the case of different backgrounds, this method based on gray image information very rarely recognizes gestures correctly. From these results, we can conclude that the edge feature we proposed is more effective for motion recognition than gray image information.

7. CONSIDERATION

We consider the efficiency of our method based on the above results.

First, we got to know that our method isn't affected by different clothing textures and backgrounds when using u_3 for recognition. We think that this is because u_3 reflects parts where its pixel value changes along the time axis and minimizes the effect of different clothes or a different background. On the other hand, when u_1 and u_2 are used, even though the occluding contours of a subject are really matched with the one of a standard sequence pattern, the edges of the textures of clothes and backgrounds constitute noise, and this makes the recognition rate low.

Feature u_3 was the best feature overall in our experiments. But when the clothes and background are the same when making standard sequence patterns, that uses u_1, u_2, and u_3 is also perfect. We think that this is because u_1 and u_2 which represent shape information, and u_3 which represents motion information, are mixed adequately. From this consideration, it is not adequate to merely find the sum of u_1, u_2, and u_3. But, after using u_3 to extract moving objects and motion parameters, the system recognizes their shapes using u_1 and u_2. This might make it possible to realize high-precision motion recognition.

8. CONCLUSION

We proposed a gesture recognition method based on a spotting algorithm. Our main principles are:

- use of spatio-temporal vector fields composed of three dimensional edges to recognize gestures and to represent gesture models,

- spotting recognition by Continuous Dynamic Programming which realizes segmentation free and frame-wise recognition.

We carried out some experiments to study the influence of clothing and background and the usefulness of edge features. Our experiments showed that our method is robust against variations in clothing and background.

From here, we intend to consider a method using three kinds of edges in order to remove two kinds of dependence, that is, individuality and the location of the person in the image.

ACKNOWLEDGEMENTS

The authors would like to thank Dr. Junichi Shimada, the director of Real World Computing Partnership, for his encouragement and support of this work.

REFERENCES

1. H.H.Baker, R.C.Bolles: "Generalizing Epipolar-Plane Image Analysis on The Spatiotemporal Surface", Proc. CVPR, pp.2-9, 1988.
2. R.C.Bolles, H.H.Baker, D.H.Marimont: "Epipolar-Plane Image Analysis:An Approach to Determining Structure from Motion", International Journal of Computer Vision, 1, pp.7-55, 1987.
3. A.P.Pentland: "Visually Guided Graphics", International AI Symposium 92 Nagoya Proceedings, pp.37-44, 1992.
4. M.A.Turk,A.P.Pentland: "Face Recognition Using Eigenfaces", Proc. CVPR, pp.586-590, 1991.
5. J.Yamato, J.Ohya, K.Ishii: "Recognizing Human Action in Time-Sequential Images Using Hidden Markov Model", Proc. CVPR, pp.379-385, 1992.
6. R.Oka: "Continuous Word Recognition with Continuous DP", Report of the Acoustic Society of Japan, S78, 20, 1978 [in Japanese].
7. S.Hayamizu and R.Oka: "Experiment and Discussion of Continuous Word Recognition with Continuous DP", The Transactions of IEICE(D), J67-D, pp.677-684, 1984 [in Japanese].

C
IMAGE RESTORATION

*Time-Varying Image Processing and
Moving Object Recognition, 3* – V. Cappellini (Ed.)
© 1994 Elsevier Science B.V. All rights reserved.

Motion-compensated filtering of noisy image sequences[*]

Richard P. Kleihorst, Jan Biemond and Reginald L. Lagendijk

Delft University of Technology,
Department of Electrical Engineering,
Information Theory Group,
P.O. Box 5031, 2600 GA Delft,
The Netherlands.

A description is given of the activities in the field of image sequence noise filtering in general, and in particular the research activities in our own institution. In order to classify the different image sequence filtering methods we make the rather important distinction between filters that use explicit motion estimation and compensation and those that do not. The filter structure itself is usually nonlinear to cope with non-stationarities in the temporal signal. We will distinguish the class of order statistic filters and the signal adaptive filters. Finally, we discuss some present and future developments, which are mainly application oriented.

1. INTRODUCTION

Image sequences are consecutive (digital) recordings of a time varying scene [1]. Often, image sequences are corrupted by noise because of imperfections of the scanner or recording device. To improve the appreciation, coding and/or analysis of the sequences, the noise has to be reduced by applying noise filtering strategies. The research for filtering strategies was not directly application oriented until recently when approaches dealing with improvement of medical sequences, film material and video recordings have been reported.

A frequently used observation model of a noisy image sequence is given by:

$$g(i, j, k) = f(i, j, k) + n(i, j, k). \tag{1}$$

Here, g is the observation of the original signal f corrupted by noise n. The indices for vertical, horizontal and temporal directions are i, j and k, respectively. In most publications, the noise is thought as as being white and independent of f.

Because of motion in general and object motion in particular and due to scene changes, the temporal component of the image sequence signal shows a heavy non-stationary behavior as illustrated in Figure 1. As it concerns image data, the signals are spatially non-stationary as well. A lot of research is devoted to the specific goal of designing image sequence noise filters that can handle these non-stationarities and prevent the blurring of spatio-temporal edges.

In this paper, we will give a description of the activities that appeared within this field of research during the past years. Our main objective is to analyze the different

---

[*]This work was supported in part by NATO under Grant 0103/88

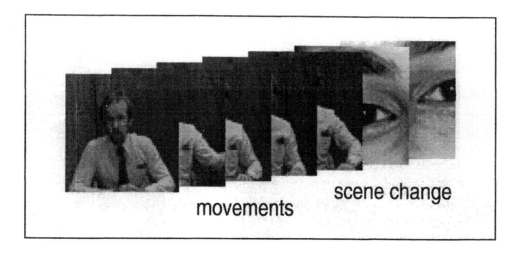

Figure 1. Image sequences are consecutive recordings of a time varying scene. The signal along the temporal direction is non-stationary because of object motion and scene changes. The signal in the spatial direction is non-stationary because it concerns image data.

"flows of interests" and categorize them accordingly. We will restrict ourselves to the most accessible and referenced publications that appeared. To give a personal note to this paper we will emphasize the activities of our research group in the image sequence filtering area.

The outline of the paper is as follows. In Section 2 we will briefly discuss motion estimation and compensation. The use of explicit motion estimation in the filtering algorithms appears to be a suitable classification method. In Section 3 filtering methods and applications without explicit motion compensation are presented. In Section 4 and 5 we focus on motion compensated filtering techniques using regular and noise robust motion compensation as a preprocessing step. In Section 6 we will discuss some future developments which are mainly application oriented. Finally, in Section 7 we draw some conclusions. The Reference section contains many entrances to gather additional information.

2. CLASSIFICATION OF METHODS

As noise filtering has a low pass characteristic, high frequencies are reduced which causes blurring of edges, both spatially and temporally. To avoid blurring of temporal "edges" a filtering strategy will at least have to rely on some form of *implicit* motion detection or estimation.

To classify the different image sequence filtering methods we make use of the rather important distinction whether the filter uses *explicit* motion estimation and compensation or not to account for most of the non-stationarities along the temporal trajectories.

Motion estimation involves the estimation of the motion vectors $(\Delta i, \Delta j)^T$, that relate similar pixels or image patches in subsequent frames, from the noisy observations. This is done by solving:

$$\hat{\Delta}i, \hat{\Delta}j \leftarrow \min_{\Delta'i,\Delta'j} \text{Crit}\left\{g(i,j,k) - g(i + \Delta'i, j + \Delta'j, k - 1)\right\}, \tag{2}$$

where $(\hat{\Delta}i, \hat{\Delta}j)^T$ is the resulting estimate, and $\text{Crit}\{\}$ is an appropriately chosen criterion function. For instance, the maximum absolute difference or square error [2]. To overcome the ill-posedness of the problem, and to avoid an exhaustive search, use is made of the correlation between neighboring motion vectors. This has resulted in several well-known algorithms such as block-matching and pel-recursive motion estimators [3, 4].

Motion compensated noise filtering is performed by orienting the temporal part of the data window along the estimated motion trajectories instead of using a data window with a fixed orientation along the temporal axis.

During the developments of the first filtering techniques in the early 80's, motion compensation was already suggested as a useful tool that would improve the results of the filter [5]. However, in practical applications and subsequent publications it was not yet used, mainly because of the complexity and also because of the noise sensitivity of the motion estimator. Instead, nonlinear filters, capable of dealing with non-stationary signals were investigated. These comprise the first category of filters discussed in Section 3.

Along the line of nonlinear filters, motion compensation again emerged as a preprocessing step. This was mainly because of research results in the area of motion compensated sequence coding. However, these motion estimators were not able to give accurate estimates in the presence of noise. Section 4 discusses this second category.

Finally, robust motion estimation in the presence of noise was investigated and used as a preprocessing step in the third category of filtering techniques. This category of filters is therefore able to tackle moderate to severe noise levels, and is the subject of Section 5.

3. NON MOTION-COMPENSATED TECHNIQUES

With the exception of the trivial method of temporal averaging [5], the proposed methods in this class all relied on nonlinear filters to cope with the non-stationarities in the temporal signals. The class of nonlinear approaches can be divided in two major groups: the signal adaptive and the order statistic filters.

For the signal adaptive class the following filter, based on a Kalman structure, was substantially exploited:

$$\hat{f}_a(i,j,k) = \hat{f}_b(i,j,k) + C(i,j,k)[g(i,j,k) - \hat{f}_b(i,j,k)], \tag{3}$$

with $\hat{f}_b$ the estimate before updating and $\hat{f}_a$ the final estimate after updating. The estimate before updating was chosen as:

$$\hat{f}_b(i,j,k) = \hat{f}_a(i,j,k-1), \tag{4}$$

which means a recursion in the temporal direction. This was a great advantage as it requires only 1 frame memory. The relative simple structure of the filter is illustrated in Figure 2.

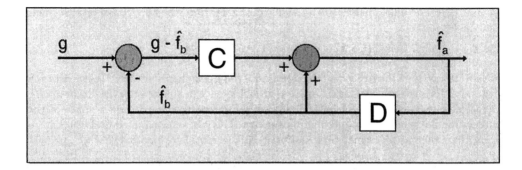

Figure 2. Among the nonlinear approaches without motion compensation this recursive signal adaptive filter was often exploited. The actual differences in implementation were in the control parameter C.

The actual differences between the several proposals based on Equation (3) lie in the derivation of the control parameter C. The value of this parameter must be a function of the prediction error: $[g - \hat{f}_b]$, therefore able to "switch-off" the filter on the occurrence of motion signaled by temporal edges. Several methodologies were proposed. Among those are 1), a threshold function switching C to 0 or 1 [6]; 2), a specially designed monotone increasing function of the prediction error [7, 8]; 3), a global optimal gain coefficient [9]; and 4), a local optimal gain coefficient [10, 11].

On another front it was recognized that order statistic filters are able to process non-stationary signals without blurring the signal too much. They have the following structure:

$$\hat{f}(i, j, k) = \sum_{l=1}^{m} c_l(i, j, k) g_{(l)}(i, j, k), \tag{5}$$

where $g_{(l)}(i, j, k)$ are the ordered observations from a window of size m centered at position i, j, k, and c_l are the filter weights. Note that this group of filters needs an ordering stage, is more memory consuming than the data adaptive class, and also has a greater time lag. However, the ordering action enables to perform a very basic signal segmentation. Outliers are automatically grouped with the largest or smallest observations. The median filter, with only $c_{m/2+1} = 1$ and all other weights set to zero, is a well known member of the group that takes advantage of this effect.

The median filter is often used in image sequence noise filtering. The differences between the several proposals are mainly in the size and orientation of the operating window. The proposed filtering algorithms range from a temporal median filter [12] proposed for real time television purposes, to compound median filters with a 3D support [13, 14] where the results of several median filters with different orientations are combined.

Order statistic filters can be used as estimators of the local statistics, the local mean μ and local deviation σ. This is done by adjusting the weights c_l accordingly. With these

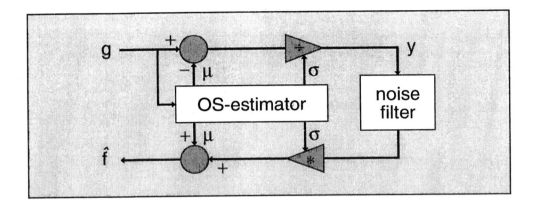

Figure 3. An order statistic estimator is used to estimate the local mean μ and deviation σ from the observation. They are used in creating a normalized signal y which can be filtered by a regular noise filter.

estimates the non-stationary temporal signal can first be normalized as follows:

$$y(i,j,k) = \frac{g(i,j,k) - \mu(i,j,k)}{\sigma(i,j,k)}, \tag{6}$$

and next filtered by a regular filter as shown in Figure 3. A method following this approach was proposed by our group in a temporal [15, 16] and a spatio-temporal implementation [17]. An advantage of this technique is that the noise filter used can be application dependent and even linear, but the same structure can still be used.

4. TECHNIQUES USING REGULAR MOTION ESTIMATION

As the algorithms evolved, motion estimation became recognized as an important preprocessing step to solve most of the problems with the non-stationary temporal signals. At that time, motion estimation algorithms were "borrowed" from the field of motion compensated coding research. These motion estimators were definitely not able to estimate motion in the presence of severe noise, because that had not been one of their design considerations.

To filter slightly by noise corrupted sequences, many new algorithms involving motion compensation were proposed. A lot of those "new" algorithms can be viewed as a concatenation of a motion compensator and an existing filter as described in the previous section. Similar filters as the nonlinear filter from Equation (3) were used in a motion compensated form in [9, 16, 18–20]. In these cases, however, the control parameter C could be controlled directly by the registration error of the motion estimator. The motion estimation and compensation was cleverly performed within the recursion loop (Figure 4). Note that previously filtered frames are used for motion estimation.

Also in the field of order statistic filters, motion compensation was used as a preprocessing step. Some published results are: 1) the motion compensated median filters by:

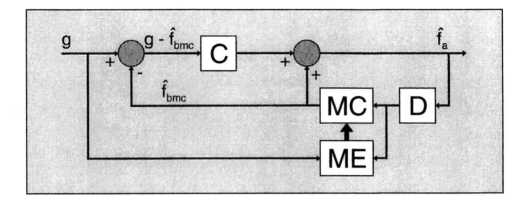

Figure 4. The signal adaptive nonlinear filter from Figure 2 can be extended with a motion estimation and compensating section operating on the existing memory in the loop.

[5, 21, 22]; and 2), a motion compensated version of our normalizing method from Figure 3 [23, 24].

The arrival of cheaper memory stimulated an increase in the use of FIR filters. These filters are preferred over the IIR type (as shown in Figure 2 and Figure 4), because they avoid the accumulation of errors connected with the recursive temporal filters, which present themselves as disturbing "comet tails" in the filtered sequences. The FIR filters ranged from the motion compensated order statistic filters as discussed earlier, to motion compensated polynomial (or Volterra) filters [25]. Also, a FIR filter can be injected into the structure of (4). In this case $\hat{f}_b$ is replaced by the output of a FIR filter. The motion compensated FIR filters used in this way were temporal and spatio-temporal estimates of the local mean [21, 26].

Despite the promising improvement in results for slightly by noise corrupted sequences, most methods failed when operating on severely corrupted image sequences. This was because of the break-down of the pel-recursive motion estimators by the noise as reported in [5, 9, 16, 18, 19] and because of "noise matching" by the block matching algorithms used in [22, 26, 27]. The problem of matching to the noise is that the temporal compensated signal becomes colored. This will cause the final result (after temporal filtering) to contain annoying "spots" [23, 24].

5. TECHNIQUES USING NOISE ROBUST MOTION ESTIMATION

The improvement in results of motion compensated schemes and the problem of ill-behavior of the motion estimator in the presence of moderate to severe noise triggered the efforts in the development of noise robust motion estimators within the context of image sequence filtering.

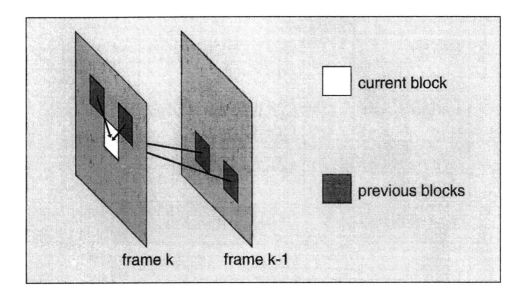

Figure 5. In the 3-D recursive block matching algorithm the results from previous blocks are used as candidate vectors for the current block. The most promising vector is selected by a criterion involving tripple correlation.

Some investigators modified an existing motion estimator to be noise robust, such as the algorithm by Fogel [28], used in [29–32]. Other methods were proposed that used consistent motion estimation algorithms derived for coding applications. One of these motion estimation algorithms is the multilevel block matching algorithm by Bierling [3], which is used in [32–34] and appears to be quite noise robust.

For a motion estimation algorithm to be noise robust, it has to produce consistent vector fields. This can be stimulated in a block matching algorithm in two ways: first by using a limited set of candidate vectors, for instance recursively chosen from the results of previous matches such as shown in Figure 5. Second, by using a noise robust selection criterion.

Our own research in this area has led to a recursive block matching algorithm with a criterion based on tripple correlation [35, 36]:

$$\text{Crit}\{g, \Delta'i, \Delta'j\} = \left[\sum_{i,j,\in S} g(i + \Delta'i, j + \Delta'j, k - 1)\, g(i, j, k)\, g(i - \Delta'i, j - \Delta'j, k + 1) \right]^{-1}. \quad (7)$$

Here, S denotes the support of the block. Note that the displacement vector is extended to the next frame, ignoring object acceleration, but also enforcing temporal smoothness. It was shown in [35] that this algorithm could estimate useful vector fields from image sequences corrupted with 0dB Gaussian noise. This noise robust motion estimation method functioned in the filtering approach presented in [36].

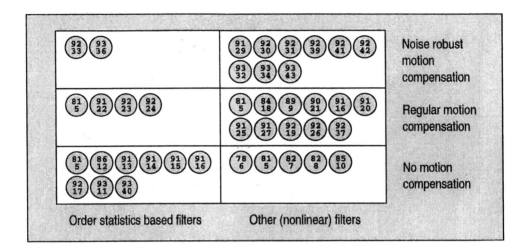

Figure 6. This figure shows an overview of the different approaches to image sequence noise filtering. Each circle represents an approach; the upper number represents the year of publication, the lower number its entry in the reference section.

6. FUTURE DEVELOPMENTS

By looking at recent literature, it is not difficult to predict where the future image sequence noise filtering methods are heading to. Three main directions can be discerned: applications to specific problems, simultaneous motion estimation and filtering, and combined deblurring and filtering.

Regarding the specific applications, much attention has been, and will be devoted to removal of noise other than Gaussian, e.g. the removal of speckle noise [33]. Another application area is the degradation reduction of film material [37]. Current emphasis is also placed on the improvement of medical sequences such as the removal of quantum mottle or Poisson noise from Röntgen recordings resulting from digital angiography [27, 38–40] and the enhancement of MRI heart recordings [20].

With the second direction, simultaneous motion estimation and filtering, we mean an integration of the filtering and motion estimation algorithm. This can be a method where the motion estimation is performed on previously filtered frames [34], or algorithms where the motion and noise process are closely combined in and solved from the observation model such as a Markov random vector and image field approach [41–43].

The third main direction is aimed at the combined removal of noise and spatial distortions from image sequences. In this way a distortion operator is included in the observation equation and solved for. Some promising publications in this field are [31, 42–45].

7. CONCLUSIONS

We have seen that the first image sequence noise filtering methods did not include motion compensation for several reasons. Some methods, mostly those based on order statistics, perform reasonably well without motion compensation. The first motion compensated methods suffered from ill-behavior of the motion estimator due to the noise. Recently, noise robust motion estimators have been used that allow motion compensated filtering of sequences corrupted with a wide range of noise levels. It is important to notice that the structure of the noise filters themselves did not change much over the last decade. Most developments have been in the motion estimator and its combination with the noise filter.

Figure 6 shows that order statistic based methods where used during the entire timespan and that the other (nonlinear) methods have evolved during the years from non-compensated and compensated to methods with noise robust estimation and compensation. The entries for this figure are the proposals from our reference list. Some papers include more methods and accordingly have multiple entries in the figure. It can be clearly seen that the use of order statistics based filters is more frequent without motion compensation and with regular motion compensation. The use of order statistic based noise filters is infrequent compared to the use of other nonlinear approaches in the class of noise robust motion estimation.

The world of image sequence noise filtering has grown to a mature state. The combination of noise robust motion estimation, compensation and nonlinear filtering is able to reduce the noise sufficiently without blurring the original signal too much. The many papers that have appeared discussing specific applications of noise filtering and removing "other than normal" noise corruptions and even removing distortions reinforce this claim. A more detailed overview including a comparison in filter performance will appear in [46].

REFERENCES

1. M. I. Sezan and R. L. Lagendijk, *Motion analysis and image sequence processing.* Kluwer Academic Publishers, 1993.

2. G. de Haan, *Motion estimation and compensation.* Ph.d. dissertation, Delft University of Technology, Dept of EE,, Delft, the Netherlands, Sept. 1992.

3. M. Bierling, "Displacement estimation by hierarchical block matching," in *Proc. SPIE Conf. Visual Commun. and Image Processing,* (Cambridge, Massachusetts, USA), Nov. 9–11, 1988.

4. J. Biemond, L. Looijenga, D. E. Boekee, and R. H. J. M. Plompen, "A pel-recursive Wiener-based displacement estimation algorithm," *Signal Processing,* vol. 13, pp. 399–412, Dec. 1987.

5. T. S. Huang and Y. P. Hsu, "Image sequence enhancement," in *Image sequence analysis* (T. S. Huang, ed.), ch. 4, pp. 289–309, Springer-Verlag, 1981.

6. R. H. McMann, "Digital noise reducer for encoded NTSC signals," *SMPTE journal,* vol. 87, pp. 129–133, Mar. 1978.

7. T. J. Dennis, "A spatio/temporal filter for television picture noise reduction," in *Proc. Int. IEE Conf. on electronic image processing,* (York, U.K.), pp. 243–249, July 1982.

8. D. I. Crawford, "Spatio/temporal prefiltering for a video conference coder," in *Proc. Int. IEE Conf. on electronic image processing*, (York, U.K.), pp. 236–242, July 1982.

9. A. K. Katsaggelos, J. N. Driessen, S. N. Efstratiadis, and R. L. Lagendijk, "Temporal motion-compensated filtering of image sequences," in *Proc. SPIE Conf. Visual Commun. and Image Processing*, vol. 1199, (Boston, Massachusetts, USA), pp. 61–70, SPIE, 1989.

10. D. Martinez and J. S. Lim, "Implicit motion compensated noise reduction of motion video scenes," in *IEEE Proc. ICASSP85*, (Tampa, Florida, USA), pp. 375–378, 1985.

11. R. P. Kleihorst, R. L. Lagendijk, and J. Biemond, "An order-statistics supported non-linear filter with application to image sequence filtering," in *Proc. IEEE Winter Workshop on Nonlinear Digital Signal Processing*, (Tampere, Finland), pp. 1.2–2.1 to 1.2–2.5, Jan. 17–20, 1993.

12. S. S. H. Naqvi, N. C. Gallagher, and E. J. Coyle, "An application of median filters to digital television," in *IEEE Proc. Int. Conf. Acoust., Speech, and Signal Proc.*, (Tokyo, Japan.), pp. 2451–2454, 1986.

13. G. R. Arce, "Multistage order statistic filters for image sequence processing," *IEEE Trans. Signal Processing*, vol. 39, pp. 1147–1163, May 1991.

14. M. B. Alp and Y. Neuvo, "3-dimensional median filters for image sequence processing," in *IEEE Proc. Int. Conf. Acoust., Speech, and Signal Proc.*, vol. 4, (Toronto, Canada), pp. 2917–2920, May 14–17 1991.

15. R. P. Kleihorst, R. L. Lagendijk, and J. Biemond, "Nonlinear filtering of image sequences using order statistics," in *Proc. 12th Symposium on information theory in the Benelux*, (Veldhoven, The Netherlands), pp. 49–55, May 23–24 1991.

16. A. K. Katsaggelos, R. P. Kleihorst, S. N. Efstratiadis, and R. L. Lagendijk, "Adaptive image sequence noise filtering methods," in *Proc. SPIE Conf. Visual Commun. and Image Processing*, vol. 1606, (Boston, Massachusetts, USA), pp. 716–727, Nov. 10–13, 1991.

17. R. P. Kleihorst, "Nonlinear filtering of image sequences using order statistics," Chartered Designer's report, Delft University of Technology, Delft, The Netherlands, Jan. 1992.

18. E. Dubois and S. Sabri, "Noise reduction in image sequences using motion compensated temporal filtering.," *IEEE Trans. on COMM.*, vol. 32, pp. 826–831, July 1984.

19. E. Dubois, "Motion compensated filtering of time-varying images," *Multidimensional Systems and Signal Processing*, vol. 3, pp. 211–239, May 1992.

20. T. A. Reinen, "Noise reduction in heart movies by motion compensated filtering," in *Proc. SPIE Conf. Visual Commun. and Image Processing*, vol. 1606, pp. 755–763, 1991.

21. D. S. Kalivas and A. A. Sawchuck, "Motion compensated enhancement of noisy image sequences," in *IEEE Proc. ICASSP90*, pp. 2121–2124, 1990.

22. C. T. Lee, B. S. Jeng, R. H. Ju, H. C. Huang, K. S. Kan, J. S. Huang, and T. S. Liu, "Postprocessing of video sequence using motion dependent median filters," in *Proc. SPIE Conf. Visual Commun. and Image Processing*, vol. 1606, (Boston, Massachussets, USA), pp. 728–734, SPIE, Nov. 1991.

23. R. P. Kleihorst, G. de Haan, R. L. Lagendijk, and J. Biemond, "Noise filtering of image sequences with double compensation for motion," in *13th Symposium on information*

theory in the Benelux, (Enschede, The Netherlands), pp. 197–204, June 1–2 1992.

24. R. P. Kleihorst, G. de Haan, R. L. Lagendijk, and J. Biemond, "Motion compensated noise filtering of image sequences," in *Proc. Eusipco '92*, (Brussels, Belgium), pp. 1385–1388, Aug. 24–27, 1992.

25. C. L. Chan and B. J. Sullivan, "Nonlinear model-based spatio-temporal filtering of image sequences," in *IEEE Proc. Int. Conf. Acoust., Speech, and Signal Proc.*, vol. 4, (Toronto, Canada), pp. 2989–2992, IEEE, May 1991.

26. J. M. Boyce, "Noise reduction of image sequences using adaptive motion compensated frame averaging," in *IEEE Proc. Int. Conf. Acoust., Speech, and Signal Proc.*, (San Fransisco, California, USA), pp. 461–464, 1992.

27. C. L. Chan, B. J. Sullivan, A. V. Sahakian, A. K. Katsaggelos, T. Frohlich, and E. Byrom, "Spatio-temporal filtering of angiographic image sequences corrupted by quantum mottle," in *Proc. SPIE Conf. Visual Commun. and Image Processing*, vol. 1450, (Boston, Massachussets, USA), pp. 208–217, Nov. 1991.

28. S. V. Fogel, "Estimation of velocity vector fields from time-varying image sequences," *CVGIP: Image Understanding*, vol. 53, pp. 253–287, May 1991.

29. M. I. Sezan, M. K. Özkan, and S. V. Fogel, "Temporally adaptive filtering of noisy image sequences using a robust motion estimation algorithm," in *Proc. ICASSP '91*, vol. 4, (Toronto, Canada), pp. 2429–2432, May 14–17, 1991.

30. M. K. Özkan, M. I. Sezan, and A. M. Tekalp, "Motion-adaptive weighted averaging for temporal filtering of noisy image sequences," in *Image Processing Algorithms and Techniques III*, vol. 1657, (San Jose, California, USA), pp. 201–212, SPIE, Feb. 10–13, 1992.

31. A. T. Erdem, M. I. Sezan, and M. K. Özkan, "Motion compensated multiframe Wiener restoration of blurred and noisy image sequences," in *IEEE Proc. Int. Conf. Acoust., Speech, and Signal Proc.*, vol. III, pp. 293–296, IEEE, 1992.

32. M. K. Özkan, M. I. Sezan, and A. M. Tekalp, "Adaptive motion-compensated filtering of noisy image sequences." To appear in IEEE Transactions on Circuits and Systems for Video Technology, Feb. 1993.

33. A. C. Kokaram and P. J. W. Rayner, "A system for the removal of impulsive noise in image sequences," in *Proc. SPIE Conf. Visual Commun. and Image Processing*, vol. 1818, (Boston, Massachusetts, USA), pp. 322–331, SPIE, Nov. 1992 1992.

34. J. W. Woods and J. Kim, "Motion compensated spatiotemporal Kalman filter," in *Motion analysis and image sequence processing* (R. L. Lagendijk and M. I. Sezan, eds.), ch. 12, Kluwer Academic Publishers, 1993.

35. R. P. Kleihorst, R. L. Lagendijk, and J. Biemond, "Motion estimation from noisy image sequences," in *1993 Picture coding symposium*, (Lausanne, Switzerland), March 17–19 1993.

36. R. P. Kleihorst, R. L. Lagendijk, and J. Biemond, "Noise reduction of severely corrupted image sequences," in *IEEE Proc. Int. Conf. Acoust., Speech, and Signal Proc.*, (Minneapolis, Minnesota, USA), April 27–30 1993.

37. S. Geman and D. E. McClure, "A nonlinear filter for film restoration and other problems in image processing," *CVGIP: Graphical models and image processing*, vol. 54, pp. 281–289, July 1992.

38. V. V. Digalakis, D. G. Manolakis, P. Lazaridis, and V. K. Ingle, "Enhancement

of digital angiographic images with misregistration correction and spatially adaptive matched filtering," in *Proc. SPIE Conf. Visual Commun. and Image Processing*, vol. 1818, (Boston, Massachussets, USA), pp. 1264–1270, SPIE, Nov. 1992.

39. C. L. Chan, A. K. Katsaggelos, and A. V. Sahakian, "Enhancement of low-dosage cine-angiographic image sequences using a modified expectation maximization algorithm," in *Proc. SPIE Conf. Visual Commun. and Image Processing*, vol. 1660, (Boston, Massachussets, USA), pp. 290–298, Nov. 1992.

40. R. P. Kleihorst, R. L. Lagendijk, and J. Biemond, "Order statistics supported filtering of low-dosage X-ray image sequences," in *Eight workshop on image and multidimensional signal processing*, (Cannes, France), September 8–10 1993.

41. J. N. Driessen, *Motion estimation for digital video*. Phd. thesis, Delft University of Technology, The Netherlands, Sept. 1992.

42. J. C. Brailean and A. K. Katsaggelos, "Simultaneous recursive displacement estimation and restoration of image sequences," in *Proc. IEEE Conf. on Sys., Man, & Cyber.*, (Chicago, Illinois, USA), pp. 1574–1579, IEEE, Nov. 1992.

43. J. C. Brailean and A. K. Katsaggelos, "Recursive displacement estimation and restoration of noisy-blurred image sequences," in *IEEE Proc. Int. Conf. Acoust., Speech, and Signal Proc.*, (Minneapolis, Minnesota, USA), IEEE, Apr. 1993.

44. M. K. Özkan, M. I. Sezan, and A. T. Erdem, "LMMSE restoration of blurred and noisy image sequences," in *Proc. SPIE Conf. Visual Commun. and Image Processing*, vol. 1606, pp. 743–754, SPIE, 1991.

45. M. K. Özkan, A. T. Erdem, M. I. Sezan, and A. M. Tekalp, "Efficient multiframe Wiener restoration of blurred and noisy image sequences," *IEEE Trans. Image Processing*, vol. 1, pp. 453–476, Oct. 1992.

46. J. C. Brailean, R. P. Kleihorst, S. N. Efstratiadis, A. K. Katsaggelos, and R. L. Lagendijk, "Image sequence noise filtering." To appear in IEEE Transactions on Circuits and Systems for Video Technology, 1993.

*Time-Varying Image Processing and
Moving Object Recognition, 3* – V. Cappellini (Ed.)
© 1994 Elsevier Science B.V. All rights reserved.

Even-median filters for noisy image restoration

L. Alparone, M. Barni, V. Cappellini

Dipartimento di Ingegneria Elettronica, University of Florence,
via di S. Marta, 3, 50139 Florence, ITALY.

Iterated medians of an even number (2x2) of image samples are proposed for unbiased and accurate edge-preserving removal of heavy impulsive noise. Both visual and MAE comparisons between noise-free and restored images prove superiority of the novel outline over the 3x3 median filter scheme.

1.   INTRODUCTION

Two-dimensional median filter is widely recognized as a powerful tool for impulsive noise removal in digital imagery. Its effectiveness depends on that abnormal values due to superimposed noise pulses do not affect the response, as for linear filters, but are simply discarded as extreme local values.

Median filter has been further specialized to spike noise removal, yet not affecting image details, like the weighted median introduced by Brownrigg [1], the threshold median proposed by Ip *et al.* [2], and the scheme outlined by Carlà *et al.* [3], in which only pixel values belonging to either of the local histogram tails are replaced with the median. This last algorithm has the advantage that neither weight nor threshold values need to be specified. Considerations that the presence of noise pulses in the local window can affect the spatial accuracy of the response in presence of signal edges, led to the adjusted version proposed by Davies [4]. Only noise pulses are processed, after being identified by comparing the difference between the current pixel value and the median, with a fraction of the level range inside the local window.

When dealing with median, or more generally with nonlinear filtering, the iterated application with small windows is usually preferable to the single application of the filter with a large window. In this way, since the convergence of median filter is faster for small local windows [5], comparable filtering capabilities can be achieved, with a better preservation of sharp edges and fine spatial details. This consideration suggests adopting an iterative scheme employing square local windows of size as small as possible, that is 2x2. However, odd-sized windows are usually adopted for median filtering, as well as for local filtering in general. Such a choice has a twofold explanation: the median of an even number of elements is generally not uniquely defined, but a *lower* median and an *upper* median exist; secondly, for even-sized windows the central point is sub-pixel displaced, thus making the output to be half-pixel shifted.

Two different filter versions can be achieved from the iterative application of alternately lower and upper even-median filters, depending on which one is taken first. Moreover, an even number of iterations of the 2x2 median can be arranged so that the half-pixel shifts produced by the even windows are accurately balanced. The resulting two images, or channels, are less blurred than the output of a 3x3 median filter, but are affected by correlated noise pulses, of different polarity in either channel. However, the chance of wrong values in corresponding positions, is equal to the probability of three noise spikes of the same color within the same 2x2 area, that is far lower then the impulse occurrence probability. Therefore, an accurately restored version is produced by combining the two channels on the basis of a reference image in which all noise pulses have been suppressed, despite of the detail preservation, like the output of a large-window median filter.

2. EVEN MEDIAN FILTERS

The median of a sorted set X with an *odd* number of elements $X = \{x_1, \dots, x_{2N+1}\}$ is the *central* element x_{N+1}. If the set contains an even number of elements, a central value does not exist; therefore the median of a sorted set X with an *even* number of elements $X = \{x_1, \dots, x_{2N}\}$, according to the classical definition [6] is taken as the average of x_N and x_{N+1}. However, some important features of the median filter are lost with this choice. In fact, since the output is no more constrained to assume a value already existing inside the window (closed operations), the edge preserving capability of the filter decreases.

A fundamental property of the median states that if x_m is the median of a set X with a number of elements M ($M=2N$ or $M=2N+1$), then the above definitions lead to

$$MIN[\sum_{j=1,M} \sum_{i=1}^{M} |x_i - x_j|] = \sum_{i=1}^{M} |x_i - x_m| = D_{MIN} \qquad (1)$$

Now let $x_{lm}=x_N$ be the *lower-median* and $x_{um}=x_{N+1}$ be the *upper-median* of the sorted *even* set X, then it can be proven that

$$\sum_{i=1}^{M} |x_i - x_{lm}| = \sum_{i=1}^{M} |x_i - x_{um}| = D_{MIN} \qquad (2)$$

Actually the two central values both minimize the average absolute difference from the other elements of the set, even if either of them is a biased estimate, the former in defect, the latter in excess, of the theoretical median value. In our approach, both the central values are defined as *even medians*.

The performance of even median filters will be discussed by considering the case of impulsive noise. The main drawback with the above definition of even-median is that dark and bright pulses are not identically treated. With reference to Figure 1 and 2, let us consider the case of, respectively, one or two opposite spikes within the local 2x2 window. Since the noisy pixels possess extreme values, either of the upper or the lower median can properly filter them out. On the contrary, when two impulses have the same polarity, the response depends on their color. The upper median removes only black spikes and the lower median only white spikes. Besides, the upper median enlarges correlated white pulses, as the lower median does on black spots.

The second reason for which even windows are usually rejected is that they have not a central point where the filter output can be placed. In the 2x2 case the output can be displaced at any of four positions, but in any case an anisotropic filter results. For example, if the output is placed in the upper-left position, the whole image will be shifted by half a pixel along the NW-SE direction. A solution consists in cascading an even number of filters so that an output image is produced that is not shifted with respect to the original.

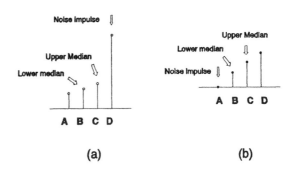

(a)   (b)

Figure 1 - Effects of a white (a) and black (b) noise spot on the even medians of a 2x2 sample set.

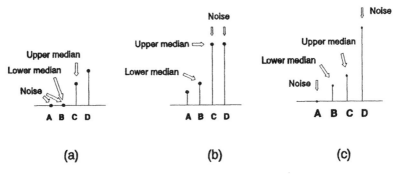

(a)   (b)   (c)

Figure 2 - Effects of two noise spots on the even medians of a 2x2 sample set: different colors (a); both white (b); both black (c).

For a two-step chain two different solutions are possible, depending on which one of the diagonal directions is chosen. In a four-iteration chain more sequences of displacements are feasible, as illustrated in Figure 3, all providing non-shifted outputs.

3.   DUAL-CHANNEL ITERATIVE SCHEME

The output images from the dual (*upper-first* and *lower-first*) chains can be combined together to achieve a more accurate filtering action on image structures. Let us focus the attention on black spots. Three cases are possible: both the chains remove them; none of the chains removes them; only the *upper-first* chain removes them. Analogously for white spots, the only difference being that the most effective chain is now the *lower-first* one. As a consequence, in most cases when a black spot is present in the *lower* of the dual images, the correct value can be found in the *upper* image. Conversely, the *lower* image usually contains the right values for pixels of the *upper* image corrupted by white spots. Clustered spots (3 or more pixels) are removed by neither of the chains.

The system we propose permits to exploit the above features to improve the filtering capability. First two channels are produced by applying upper-first and lower-first iterative 2x2 even-median chains. Then the outputs from the two channels are combined according to the output of a large window median filter (say 7x7), taken as reference image. Such a filter is capable of erasing practically all the spots, but blurs contours

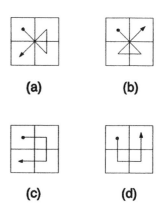

(a)   (b)

(c)   (d)

Figure 3 - Displacements inside a 2x2 window of a four times iterated filtering for overall shift balance

and destroys fine details. The combination consists in comparing the output of the first two channels and choosing, for each pixel, the value closer to the reference image. This strategy allows most of the noise spots which are still present at the output of the 2x2 median chains to be eliminated and accounts for the bias of each channel. In fact, when an white correlated spot is present in the upper image, the output of lower channel is likely to be noise-free and closer to the reference image. An analogous situation occurs in case of black correlated spots. The only noise impulses that are not filtered are those which neither of the channels is able to remove. To achieve a final image which is not shifted with respect to the original, the 2x2 medians must place their outputs in a proper sequence. The overall chain is outlined in Figure 4.

As twice iterating a 2x2 median filter better preserves edges than a single 3x3 filtering, four iterations of 2x2 medians are recommended, whenever two iterations of a 3x3 median are required, as for heavily noisy images. The four medians of the dual chains, drawn in Figure 5, must be alternately applied in complementary form, with outputs alternately displaced, for best shift balance.

4. EXPERIMENTAL RESULTS AND COMPARISONS

Tests have been performed to analyze the features of the dual-median scheme and to compare its results with those of the conventional (odd) median filter. Comparisons between a 3x3 median and the dual-channel 2x2 median filter have been carried out in terms of Mean Absolute Error (MAE) which is known to reflect image fidelity, as it appears to a human viewer, better than the usual Mean Square Error (MSE) [7].

A noisy version of the original 256x256, 8 bit "Girl" test image is produced by randomly replacing a fraction p of pixels with uniformly distributed 8-bit samples. Figure 6a

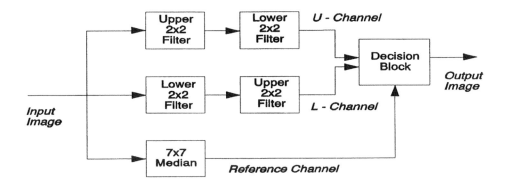

Figure 4 - Processing chain for 2-iterations dual channel even median filter

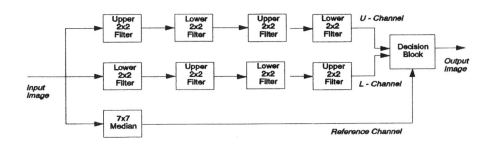

Figure 5 - Processing chain for 4-iterations dual channel even median filter

shows the original 256x256, 8 bit "Girl" image; in 6b a noisy version of 6a. Fig. 6c and 6d portray the outputs of U and L channels of Fig. 4, while the nonlinear combination of U and L is shown in 6e. The output of a 3x3 median applied to the same noisy image is displayed in 6f for comparison. The spike-removal capability of the odd and even schemes is similar, but the 2x2 filtered image is superior to the output of the 3x3 median in terms of detail preservation as well as of MAE, reported for both in the plots of Figure 7a versus the spike noise occurrence probability p. Similarly, the dual chain with four iterations of 2x2 median filter in each channel, as in Fig. 5, is compared with a twice iterated 3x3 median filter. The filtering capability is greater than for the scheme of Fig. 4 and comparable with odd median filter, but again the dual-channel filter output exhibits a lower MAE, as appears from the plots

Figure 6 - Visual results of even-median filters: (a)-(f) left-to-right and top-to-bottom. Original Girl image (a); noisy version of Girl with p=25% (b); output from lower (c) and upper (d) channels of Fig. 4; output of 2-iter. even median (e) and 3x3 median filters (f).

of Figure 7b. In Figure 8 the plots relative to the schemes of Fig. 4 and Fig. 5 are compared, again varying with p. For lower p the 2-iterations scheme is preferable, whereas for higher p the 4-iterations one is superior; the critical value of p is about 25%.

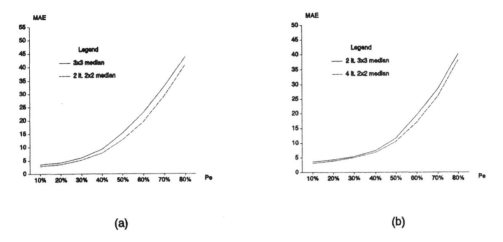

(a)   (b)

Figure 7 -   MAE vs spike occurrence probability p for (a) 2-iter. 2x2 median and 3x3 median; (b) 4-iter. 2x2 median and 2-iter. 3x3 median.

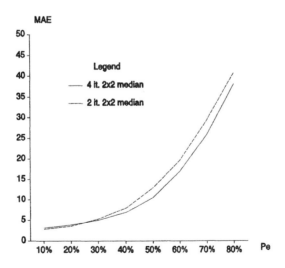

Figure 8 - MAE vs. spike occurrence probability p for 2 and 4 iteration even median filter

5. CONCLUDING REMARKS

An iterative scheme exploiting both the central values (lower and upper median) of pixels inside an even (2x2) window, instead of their average, has been proposed for accurate edge-preserving smoothing of digital images heavily degraded by impulsive noise. Two chains are produced, depending on which of the medians is applied first. A combination on the basis of a reference given by a large-window odd-median filter, yields a faithfully restored image. Both visual and MAE comparisons between noise-free and processed images, prove the validity of the scheme as a feasible alternative to conventional median filter.

REFERENCES

1   D.R.K. Brownrigg, The Weighted Median Filter, Communications of the ACM, Vol. 27, No. 8, (1984) 807-818.

2   H. Ip, D.J. Potter and D.S. Lebedev, Impulse noise cleaning by iterative threshold median filtering, Pattern Recognition Letters, Vol. 2, (1983) pp. 89-93.

3   R. Carlà, V.M. Sacco, and S. Baronti, Digital Techniques for Noise Reduction in APT NOAA Satellite Images, In Proceedings IGARSS '86, (1986) 995-1000.

4   E.R. Davies, Accurate filter for removing impulse noise from one- or two-dimensional signals, IEE PROCEEDINGS-E, Vol. 139, No. 2, (1992) 111-116.

5   J.P. Fitch, E.J. Coyle, and N.C. Gallagher, Root properties and convergence rates of median filters. IEEE Trans. Acoustic Speech and Signal Processing, Vol. ASSP-33, No .1, (1985) 230-239.

6   I. Pitas, and A.N. Venetsanopoulos, Nonlinear Digital Filters: Principles and Applications, Kluwer Academic, 1990

7   E.J. Coyle, J.H. Lin, and M. Gabbouj, Optimal stack filtering and the estimation and structural approaches to image processing, IEEE Trans. Acoustic Speech and Signal Processing, Vol. ASSP-37, No. 6, (1989) 2037-2066.

Time-Varying Image Processing and
Moving Object Recognition, 3 – V. Cappellini (Ed.)

Global Probabilistic Reinforcement of Straight Segments

C.S. Regazzoni

Department of Biophysical and Electronic Engineering, University of Genoa, Via all'Opera Pia 11A, I-16145, Genova, Italy

1. INTRODUCTION

In many image processing problems edge extraction is regarded as a binary decision on whether or not to identify a inter-pixel location as a local luminance discontinuity. In this case, the mapping of the input data space (i.e., differential local information in the presence of noise) into the solution space (i.e., the edge configuration space) may be non-unique.

A solution suggested in literature [1] lies in imposing local constraints on the shapes of the obtained contours. For example, edge detection is favoured near high-contrast edges by relaxing threshold conditions in the direction of already detected edge (continuity). This method is dependent on the presence of a sufficiently high local contrast between adjacent pixels.

Several well-known examples reported by psychologists seem to indicate that such a local contrast is not always necessary. Gestalt theory [2] provided evidence for the effectiveness of using other constraints to decide on the presence of an edge. From these examples, one can deduce that a decision depends on a global perceptual property (i.e., membership of a point in a group). In this paper, properties of interest are closureness, familiarity (i.e., the probability of the presence of an edge may increase, depending on the capability of the edge to complete a closed curve or a familiar shape), and symmetry (i.e., the probability of presence of a pair of elements, if such elements share similar relations with a third image point).

These additional constraints are characterized by a global nature, i.e., they cannot be satisfied by combining local image measures. Consequently, an edge-detection mechanism that takes into account complex global constraints should yield results more similar to those of human visual perception.

In this paper, a computational approach to embed global constraints in the edge-extraction phase is described. In particular, the problem of reinforcing of straight lines is addressed as a case discussion of specific interest for many applications.

A hierarchical three-level architecture is proposed to take into account both local and global constraints. The first level is associated with the 2D field of observed image variables. The second level is represented by a coupled field associated with the restored intensity field and its discontinuities. The third level is constituted by a multidimensional field whose elements represent straight segments, each characterized by its direction, distance from the origin, and the positions of its extrema.

From a computational point of view, edge detection is performed at the second level, and the third level includes an estimation phase to establish the presence of a global pattern (in the present case, a straight edge). These phases are repeated by activating, in an iterative way, the nodes at which they are performed. The results of pattern extraction are used to adaptively modulate the behaviour of the edge-detection algorithm: the probability of the presence of an edge may change depending on the membership of a point to a class of patterns to be reinforced. The pattern-extraction algorithm provides the network with the interpolation capabilities necessary to perform a feedback towards the edge-detection level. To this end, an image point may be considered as a part of a pattern of the considered class depending on the global content of the image, and not on its own properties only. A voting approach is used to detect a subset of patterns within the considered class which are characterized by a sufficiently high probability to be present in the image. Each element of this subset identifies a group of points in the image space: the points of such a group that did not vote for the pattern can be considered as interpolated points, and their reinforcement depends on the global image properties only.

In this paper, two methods to be used at levels 2 and 3, respectively are proposed. A modified version of the Deterministic Annealing (DA) algorithm [3] extended so as to take into account group-dependent expectations, is used as an edge detection method at level 2. The Direct Hough Transform (DHT) [4] is performed for detection and spatial interpolation at level 3.

It will be shown that the above approach can be described in terms of Bayesian Networks. The algorithms used at each level (node) of the network can be interpreted as special cases of local optimization inside BNs (i.e., Belief maximization [5]). A formal demonstration of this point is beyond the scope of this work, and has been included in a paper submitted for publication [6].

The present paper is organized into four sections. The proposed approach is described in the next section, pointing out the representation strategy used at each level of the network. The algorithms used at the three levels are detailed in section 3. In section 4, some results are reported to prove the feasibility of the approach. Finally, conclusions are drawn in section 5.

2. DESCRIPTION OF THE APPROACH

In this section, a probabilistic formulation of the edge-detection problem is provided, which is based on Bayesian networks [5] of fields of variables [8]. This formulation implies the definition of a three-level network, $N=\{n_r:r=1..3\}$, where n_r is a node associated with a 2D field of variables, say X_r.

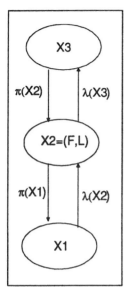

2.1 First level

At the first level, a 2D lattice $S_1=\{s_1{}^n=(x,y):n=1..N\}$, corresponding to the locations (pixels) in an image, is used to store a 2D field of variables $X_1=\{x_1{}^n:n=1..N\}$ representing intensity observations acquired with an imaging system.

These variables are supposed to be affected by noise and a blurring function, so they are related to the true intensity values by a relation of the type:

$$x_1^n = H(f,l) \circ \eta$$

where H is a non linear blurring operator which acts, in a different way, on the true intensity variables f and the related discontinuities l. The noise η can be of different types (e.g., Gaussian or χ–square distributed) and the operator $\circ$ should be invertible (e.g., addition or multiplication) .

2.2 Second level

The goal of the second level is to estimate the field $X_2=(F,L)$. To this end, the edge- detection problem is modeled by using the probabilistic representation provided by a coupled Markov Random Field [7,3], i.e., a coupled field of random variables whose properties can be described in terms of local probabilistic knowledge. $F=\{f_n=f(x,y):s_1{}^n=(x,y), n=1..N, f_n\in [0...M]\}$ is the true image intensity field defined over S_1. $L=(H,V)$ is a double field representing horizontal and vertical discontinuities detected in the image, as in [3]. L is the Line field and is defined over a pair of lattices made up of inter-pixel locations. To simplify the notation, in the rest of the paper we shall indicate the discontinuity field with $L=\{l_{nm}:C(s_1{}^n,s_1{}^m)=1,l_{nm}\in [0,1]\}$, where $C()$ is a clique function which assumes the value 1 only if two sites are neighbours of each other. The Markov property implies that the probability of a variable associated with a site depends only on a subset of field variables. In particular, one can write:

$P(f_n,l_{nm}/F,L)=P(f_n,l_{nm}/NF_n,NL_m)$

where $NF_n=\{f_j:s_j\in N_n,|s_n-s_j|<o\}$, $NL_n=\{l_{km}:|s_n-s_m|<o'\}$ where o,o' are the orders of the neighbourhood systems. It can be shown [7] that the probability $P(X_2)$ can be written as a Gibbs distribution, i.e., $P(X_2)=1/Z \exp(-U(X_2)/T)$, where T is the temperature of the field and Z is the partition function.

The problem to be solved at this level is to estimate the fields F and L, given the observation field X_1 and the expectation field composed of the set of straight lines estimated at level 3. In classical approaches, expectations are not taken into account, so the Maximum A-Posteriori (MAP) criterion can be applied [7].

In this case, one can use distributed local Belief maximization as a tool for obtaining the Most Probable Explanation estimate, which is a generalization of a MAP estimate, as shown in [8]. To this end, it is necessary to define, in a probabilistic way, messages exchanged by the nodes of the network as well as the local knowledge used at each node.

In particular, one can define three terms:

a) a local regularizing term $\gamma(X_2)=P(X_2)=1/Z_\gamma \exp(-U_\gamma(X_2))$, where $U_\gamma(X_2)$ is a non-linear energy function of the type:

$$U_\gamma(X_2) = \sum_n \sum_j \alpha(f_n - f_j)^2 (1 - l_{nj}) + \tau l_{nj}$$

which aims at favouring piecewise continuous solutions;

b) an evidence term $\lambda(X_2)=\max_{X1} [P(X_1/X_2)]=1/Z_\lambda \exp(-U_\lambda(X_1/F))$, where

$$U_\lambda(X_2) = \sum_n (x_1^n - H \cdot (F,L))^2$$

which relates observations (i.e., the input image) to the results of edge estimation by penalizing solutions not supported by data;

c) an expectation term $\pi(X_2)=\max_{X3} [P(X_3/L)]=1/Z_\pi \exp(-U_\pi(X_3/L))$, which aims at favouring solutions that are closer to the expectations (see next subsection).

Obtaining the configuration characterized by the maximum local Belief, i.e.,

$Bel(X_2)=\max_{X2} \gamma(X_2)\pi(X_2)\lambda(X_2)$

is equivalent to minimizing the energy function given by the expression:

$X_2^*=\min_{X2} [U_\gamma(X_1) +U_\lambda(G/X_1)+U_\pi(X_2/X_1)]=\min_{X2} [U_\gamma(X_1) +U_\lambda(G/X_1)+U_\pi(X_2/X_1)]/T$,

where the temperature $T=1/\beta$ has been inserted in order to use Mean Field Annealing [3] as an optimization method to solve the above expression.

2.3 Third level

At the third level, straight-line extraction is modeled as a node associated with a vectorial deterministic field of variables, X_3, defined over a hierarchical lattice. First, the lattice of lines without information about their extrema is expressed as $D=\{\eta=(\rho,\theta):\rho=1..R, \theta=1..\Theta\}$, where ρ denotes the distance of a line from the origin of the reference system, and θ its direction.

For each possible line η, the information about the extrema of a segment is represented in a space $E=\{t:t=1.. \max_k |s_1^k-P_{ref}(\eta)|,\}$, where $P_{ref}(\eta)$ is a reference point in the η direction, not necessarily belonging to the lattice S_1, and $s_1^k \in S_1$ is a generic point which verifies the system of generating equations $G()=0$ (for the η direction) used to define the global properties considered at level 3. In the case of straight lines, $G()$ is a system of three equations:

$g_1(D,L)=\rho-x\cos\theta-y\sin\theta$,

$g_2(D,L)=\phi-\theta+\pi/2$,

$g_3(E,L)=|s_1^n-P_{ref}(\eta)|-t$

where $(x,y)=s_1^n$, s_1^n is the site related to the line element l_{nm}, and ϕ is the measured gradient at the site s_1^n. The weighted version of the voting part of the Direct Hough Transform [4] used here to detect straight lines, can be expressed as:

$V(\rho,\theta,t)= \sum_{nm} l_{nm} \delta(G(s_1^n,\phi,\rho,\theta,t))$

where δ is the Kroenecker delta.

Each edge point votes according to a weight proportional to the edge belief computed at level 2 provided by the variable l_{nm}. Each point votes for a subset of solutions: this is possible by considering the subset of points $x'=x\pm\Delta x$, $y'=y\pm\Delta y$, $\phi'=\phi\pm\Delta\phi$ instead of the measured x,y,ϕ only. The three errors are related to one another, as $\Delta x=\Delta s \sin(\phi')$, and $\Delta y=\Delta s \cos(\phi')$, where Δs is an absolute distance shift, which is chosen as a DHT parameter. In this way, a point in the

E space will receive a number of contributions equivalent to the number of edge points in the range $s'=s\pm\Delta s$.

According to the DHT, the parameter space is then explored, in a hierarchical way, to detect straight segments by first searching for local maxima, say (ρ^*,θ^*), in the D space only, and then performing a local complete search within a set of parameter locations given by $\{\rho,\theta,t: \rho,\theta \in D'(\rho^*,\theta^*), t \in E\}$ containing spatial information concerning with most voted lines, where $D'(\rho^*,\theta^*)=\{(\omega,\psi):(\omega,\psi)\in D, |(\omega,\psi)-(\rho^*,\theta^*)| <o"\}$ is a local subset of lines with distance from the origin and direction similar to the one of the detected maximum.

The 3D field at level 3 is expressed by a binary field of variables $X_3=\{h(\rho,\theta,t):(\rho,\theta)\in D,t \in E, h(\rho,\theta,t)\in [0,1]\}$, and can be deduced from $V(\rho,\theta,t)$ by applying the following rules:

$h(\rho,\theta,t)=1$ if $(\rho^*,\theta^*)=(\rho,\theta)$, and $V(\rho,\theta,t)>L$.

The segments present along a line $\eta=(\rho,\theta)$ are identified by continuous subsets of $h(\rho,\theta,t)=1$ values along t: by inspecting E subspaces starting from the reference point $P_{ref}(\eta)$, the k-th transition from 0 to 1 occurring at location t is defined as the minimum extrema of the k-th segment, $t^k_{min}(\rho,\theta)$, while consequent transitions from 1 to 0 correspond to the maximum extrema, $t^k_{max}(\rho,\theta)$.

In [3], this rule is expressed in a probabilistic way, by demonstrating that the optimal configuration of X_3 can be found by minimizing an energy function as follows:

$X_3^*=\min_{X3} U_\lambda(X_3)+U_\gamma(X_3)$

In this way, the voting part of the DHT and the following binarization can be interpreted, in a probabilistic way, as a message $\lambda(X_3)$: the higher the number of votes, more probable the presence of a segment into a location. The term $\gamma(X_3)$ can be considered as a two-level constraint: first, a Winner-Takes-All (WTA) constraint that operates in the D space, favouring local maxima in D. Secondly, a continuity term in E, favouring the presence of a segment near locations where other segments have already been detected.

Once a configuration X_3^* has been fixed, it is possible to obtain a feedback towards the lower level by choosing a message $\pi(X_2)$ of the type:

$\pi(X_2)=1/Z_\pi \exp(\varepsilon\tau\Sigma_{\rho,\theta,t} \delta(G(s_1^n,\rho,\theta,t)) f(h(\rho,\theta,t)))$

where ε and τ are two parameters included in the range [0,1].

The function $f(h(\rho,\theta,t))$ is a weight assigned to a point (x,y) in the image plane which is defined as:

$$f(h(\rho,\vartheta,t)) = \begin{cases} 0 & \text{if } h(\rho,\vartheta,t) = 0 \\ \max\left(\sum_{t^k_{min}(\rho,\vartheta)}^{t^k_{max}(\rho,\vartheta)} \frac{V(\rho,\vartheta,t)}{\Delta s}, (t^k_{max}(\rho,\vartheta) - t^k_{min}(\rho,\vartheta)) \right) & \text{if } h(\rho,\vartheta,t) = 1, t \in \left[t^k_{min}(\rho,\vartheta), t^k_{max}(\rho,\vartheta)\right] \end{cases}$$

The weight corresponds to the maximum between the sum of contributes received by the segment which includes point (ρ,θ,t) and its length. In this way, the locations more voted at level 3 produce an higher expectation in the points included in the kernel of the generating equation G. Note that, if the segment direction θ is used, the argument of the exponential in the $\pi(X_2)$ message is the antitransformation of the DHT.

To understand how a feedback is feasible even at locations with low gradient magnitudes, it should be noted that, from the above definition, it is not necessary for a lattice point to be detected as an edge point to be characterized by a high value of $\pi(l_{nm})$. In particular, non-null expectations are assigned to the points (ρ^*,θ^*,t) anti-transformed from a local maximum

(ρ^*,θ^*) and characterized by t $\in$ [t_{min},t_{max}] for a certain segment detected along the straight line (ρ^*,θ^*), without considering their l_{nm} values.. Consequently, those points, which are not considered as edge points (i.e., $l_{nm} < 0.5$), will receive a positive feedback through a $\pi(X_2)$ message.

3. DISTRIBUTED PROBABILISTIC INFERENCE

The distributed inference mechanism can be summarized in the following steps
1- Set $\pi(X_r)=1$ (no expectation is available at the beginning of the process).
   Set X_1, considering the current image.
2- Compute the message $\lambda(X_1)=P(\eta)=P(X_1/F)$.
3- Compute the configuration $X_2^*(0)=(F^*(0),L^*(0))$ by using $\pi(X_2)=1$.
4- Compute DHT($L^*(k)$).
5- Compute $\pi(X_2)$
6- Compute $X_2^*(k)$ by using $\pi(X_2)$, as computed at step 5.
7- k=k+1; repeat steps 4-7, raising the process temperature of the process, represented by a parameter $\beta=1/T$ which multiplies the energy function.

It can be shown that this mechanism corresponds to extending the Mean Field Annealing technique proposed in [3] to solve the Most Probable Explanation (MPE) problem associated with multilevel Bayesian Networks of 2D fields [8], like the ones involved in the present problem. The mean field equations can be obtained by evaluating the expected value of each variable under the hypothesis that the neighbouring variables are frozen to their mean values.
   In particular, as in [3], the local estimates at steps 3 and 6, for a fixed β value, can be computed at different iterations p as:

$$l_{nm}^p = \frac{1}{1+\exp\left(\beta\left(\tau\times(1-\varepsilon)H\left(l_{nm}^{p-1}\right)\right)-\alpha\Delta f^2\right)}$$

$$f_m^{p+1} = f_m^p - \omega\left[\left(x_1^m - Hf_m^p\right)+2\times\alpha f_m^p\sum_{r\in N_m}\left(f_m^p - f_r^p\right)^2\left(1-l_{nr}\right)\right]$$

By repeating this process using increasing values of β, it can be shown that the mean field estimate converges to the MPE estimate, as occurs when the MAP criterion is used for single-level processes.

4. RESULTS

The proposed method exhibits a high degree of generality, in that the shape of the structure to be reinforced can be modified by appropriately changing the generating functions G and the parameter space of the DHT, so as to reinforce the shapes of objects and shapes of interest for an application domain. The algorithm was applied to extract edges from two types of images: visual images with additive Gaussian noise and Synthetic Aperture Radar (SAR) images affected by multiplicative speckle noise. A deterministic Mean Field approximation was used to compute the optimal configuration at level 2. Results were obtained on synthetic

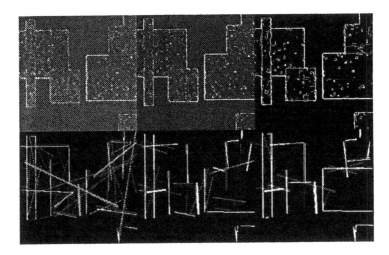

Figure 1. Synthetic Image: Evolution of the estimation process shown at the two level of the Bayesian Network for increasing β values. The line process L (top) and the corresponding anti-transformation of the straight-segments field (X_3) (bottom), i.e., $\pi(X_3)$ are shown. Black level corresponds to 0 value, white level to 1, and grey levels to intermediate probability levels.

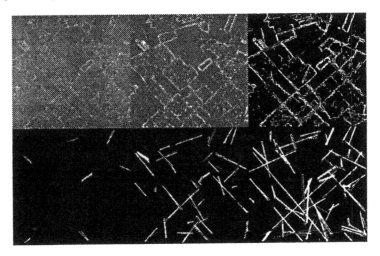

Figure 2. Real image. Evolution of the line process L at level 2 (top), and image-plane projection of detected segments at level 3 (bottom).

and real images which prove the advantages of the method over locally constrained approaches. Limitations of the method mainly lie in some errors introduced by the pattern-detection algorithm (i.e., the DHT): wrong positive feedbacks are sometimes provided to image locations not belonging to any actual group, due to spurious maxima detection in the

Hough space. Such problems require a further research step on the properties of the pattern-detection method, as well as automatic criteria for assessing weights of expectation and evidence term in the cost functional.

5. CONCLUSIONS

In this paper, a new reinforcement method was presented, useful to enhance, in an adaptive, domain-dependent way, straight edges, and, in general, patterns described by means of generating equations. The method is based on Bayesian Networks [5], providing a probabilistic distributed solution to the problem.

Two algorithms (i.e., DHT and DA) have been used to obtain the estimates necessary to solve the problem at the different levels of the proposed network. Results are satisfactory enough to extend the approach to multilevel structures of mutually reinforcing image-processing modules.

6. ACKNOWLEDGEMENTS

The author wishes to thank Drs. Monti, Pagliughi, and Dambra for helpful discussions and suggestions.

7. REFERENCES

1. J.F.Canny, Finding Edges and Lines in Images, PhD Thesis TR-720, AI Lab. MIT, Cambridge, Massachusetts, 1983.

2. G. Kanisza, Organization in Vision, New York, NY, Praeger, 1979.

3. D. Geiger and F.Girosi, Parallel and Deterministic Algorithms for MRF's: Surface Reconstruction, IEEE Transactions on Pattern Analysis and Machine Intelligence, vol.PAMI-13, No.5, pp.401-412, May 1991.

4. C.S.Regazzoni, G.L.Foresti, and V.Murino, A Distributed Hierarchical Regularization System for Recognition of Planar Surfaces, accepted for publication on Special Issue of Optical Engineering "From numerical to Symbolic Image Processing", 1993.

5. J.Pearl, Probabilistic Reasoning in Intelligent Systems, Morgan Kaufmann Publishers, San Mateo, California, 1988.

6. C.S.Regazzoni, A.N.Venetsanopoulos, and G.Vernazza, Group Membership Reinforcement of Straight Edges Based on Bayesian Networks, submitted to Transactions on Image Processing.

7 S.Geman and D.Geman, Stochastic Relaxation, Gibbs Distribution and the Bayesian Restoration of Images, IEEE Trans. on Pattern Analysis and Machine Intelligence, Vol.PAMI-6, pp.721-741.Nov.1984.

8. C.S.Regazzoni, F.Arduini, and G.Vernazza, A Multilevel GMRF-Based Approach to Image Segmentation and Restoration, Signal Processing, Vol.34, N.1, September 1993.

D
IMPLEMENTATION TECHNIQUES

Time-Varying Image Processing and
Moving Object Recognition, 3 – V. Cappellini (Ed.)
1994 Elsevier Science B.V.

105

Preliminary benchmark of low level image processing algorithms on the *ViP_1* parallel processor.

G. Gugliotta, A. Machì

IFCAI- Istituto di Fisica Cosmica ed Applicazioni dell' Informatica del CNR
Via M. Stabile 172 90139 Palermo, Italy.

The paper is a progress report on the implementation of the prototype of the parallel processor *ViP* (Versatile Image Processor).

The first section of the paper gives some technical details of the machine prototype ViP_1. The second one describes some techniques useful to optimize the parallel operation of the computational elements at the cluster and crate level; it also introduces a model for the evaluation of the machine performance in low-level image processing. The last section discusses some results of actual benchmark activity.

1. THE VERSATILE IMAGE PROCESSOR

ViP is a medium grain parallel processor, with scalable configuration between 4 and a few hundred nodes. ViP is not specialized, but well suitable for both image analysis and synthesis [1].

The machine is being developed at IFCAI, in Palermo in co-operation with the company TECNINT ltd., Verona, Italy, and is sponsored by Consiglio Nazionale delle Ricerche "Progetto Finalizzato Sistemi Informatici e Calcolo Parallelo".

The machine architecture is hierarchical; nodes are constututed by microprocessors and are organized in 3 levels: the cluster, the crate (multicluster) and the system levels. Different connection paths and various data transfer techniques are available at the three levels to match different data access and task distribution strategies.

At the bottom level a limited number of microprocessors are tightly coupled through a local bus on single board clusters; they execute independently tasks in MIMD or SPMD mode and concurrently access an on-board bank of shared RAM. Hardware implementation considerations (as board area, CPU clock cycle and RAM access time) restrict the cluster cardinality.

The intermediate level couples clusters and input-output units in crates (or multiclusters). *Fat* channels are provided to efficiently transfer blocks of data or to randomly access shared RAM distributed over the clusters.

Research sponsored by Italian National Research Council: *Progetto Finalizzato Sistemi Informatici e CalcoloParallelo* under Grant N° 212363/69/9107184

Multiclusters may be geographically spread distributed; *thin* channels connect them, at the highest level, in a whole system. Channels carry symbolic data and commands for high level services (as task control, file-service or program debugging). A communication microprocessor may implement the protocols required for message passing management.

The tree structure allows to scale the machine according to the application size and to partition it in different ways according to the task distribution model chosen.

The *ViP* logical hierarchical architecture is presently implemented using supercomputing RISC microprocessors as computational nodes and Video RAMs as cluster memories; programmable logic devices (MACHs, EPLDs) constitute the communication interfaces.

The system prototype (fig. 1) is composed by 4 clusters of 4 microprocessors each assembled in a crate with standard VME input-output units. The system is connected to a DOS PC computer through one AT-MXI and one MXI-VME interface from National Instruments Co.,USA.

The PC hosts the programming environment and performs task control and file-service.

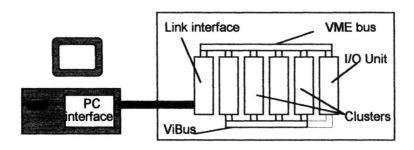

Fig 1: Simple block diagram of the ViP_1 prototype.

Each cluster (fig. 2) is composed by a single board including 4 Intel i860 super computing microprocessors, 4 Mby of dynamic Video RAM, two parallel bus interfaces and an arbiter. The microprocessors access the memory concurrently through a 64 bit wide local bus, using the VRAM parallel port.

The microprocessors operate at 32 MHz clock cycle and hold on chip both a 4 KBy instruction and an 8 KBy data cache. The data cache is large enough to contain one image line and the entire neighbourhood of its pixels.

Memory is arranged in two 64 bit banks; the banks are interleaved to allow fast access even using cheap 100 ns. VRAMs. The local bus sustains a throughput of 36 Mby/s during normal random access cycles; special cycles devoted to cache fill are going to be implemented; they will to increase the throughput to 80 Mby/s.

Two parallel busses connect the clusters: a 32 bit wide bus implemented according to the VME standard and a 64 bit wide custom Video bus (*ViBus*) presently under development. The first bus allows to integrate into the system third party units as a host embedded platform or a process controller, while the second allows very high speed data block transfers. The VME bus connects memories on clusters and input-output units from the parallel side only, while the *ViBus* connects them from both the parallel and serial ones.[1]

Every microprocessor on the cluster may be independently interrupted from the local, the VME or the ViBus. Hardware mailboxes are mapped into the cluster memory for this purpose.

The VME interface implements a pure slave protocol and allows a moderate sustained throughput of 15 Mby/s managed from external bus masters.

The ViBus presently offers concurrent random accesses to memory at a rate comparable with VME. After full implementation it will allow block transfer cycles between VRAM serial ports at a peak rate of 260 Mby/s.

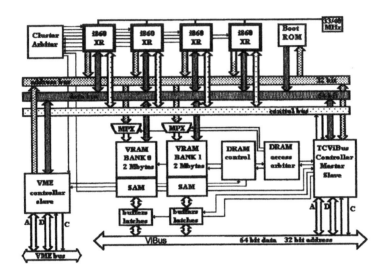

Fig. 2: Logic diagram of a V.I.P. Computational Unit.

2. PARALLEL PROGRAMMING, OPTIMIZATION AND PERFORMANCE EVALUATION

Punctual and local image processing algorithms (as threshold, histogram or homogeneous filters), apply the same sequence of operations to each pixel. They use as input data the values of the pixel itself and of its neighbour ones.

Simple parallel techniques (as *farmer-workers* task distribution and static data partitioning) are then applicable [2]: a controller (one microprocessor or the host computer) manages the assignment of different sections of the input image to each working microprocessor; and output data are collected into a common storage after the processing.

A simple model allows to estimate the time ET required to execute on the ViP prototype algorithms of this class.

The model takes in account three contributions to the execution time ET:
- the *transfer time* TT spent to move input data from the primary storage to the microprocessor's working memory (cluster local RAMs or microprocessor's caches) and to get output data back;
- the *load-store time* LT spent to read-write data from the working memories (or caches);
- the *compute time* CT effectively spent to operate on the data;

TT is the minimum absolute time required for the execution of any algorithm on a distributed or shared memory system. In fact, while LT and CT decrease when the data are distributed among the microprocessors, TT remains constant, because it involves the access to the global memory, a process essentially serial [3]. As an example, scattering a standard 512 by 512 pixel by 8 bit image to the processors through the VME bus and gathering it again takes about 35 ms. at 15 Mby/s. This time will be reduced on the ViP prototype to just 2 ms after ViBus full implementation.

The speed of cluster memory and the local bus arbitration affect LT. As for TT, a minimum may be evaluated. Assuming 4 clusters and a Read or Write cycle duration of 220 ns., one obtains for LT values between 7 and 56 ms for each read or write scan of the image; the variability is related to the format used to load or store the data, the minimum being obtained fully exploiting the 64 bits bus parallelism.

When data are accessed directly from the cluster memory, LT is equivalent to one scan time for histogram, 2 scan times (one read + one write) for threshold, 10 (9 read + 1 write) for a 3x3 convolution.

When i860 microprocessor internal cache is exploted as an intermediate data storage, the read contributions to LT are reduced in two ways: firstly because data are always loaded in cache using optimized cache fill cycles (4 by 64 bits readings); secondly because each pixel is read just once. The cache is in fact large enough (8 Kby) to contain several image rows and the value of a pixel is still there when searched for the processing of a neighbour one.

Cache enabling affects differently write contributions to LT, because the i860 microprocessor uses a write-back update policy; if a cache miss happens during a write cycle the datum is updated directly in memory. This is normally the case of not-in-place algorithms.

However, if the bus is not saturated, write cache-miss cycles overlap computation.

Packing of pixels before writes may help to reduce the bus load.

Finally, algorithm complexity, as well as microprocessor architecture and clock speed affect CT.

The i860 is a complex RISC microprocessor; it performs instruction pipelining and may execute in parallel (in the so-called dual-operation mode) one scalar instruction and two floating point ones [4]. Most scalar instructions are executed in one cycle, floating point ones require three cycles; instruction pipelining speeds-up processing because it allows to fetch and decode subsequent instructions while the completion of the two previous ones is still pending. Conditional branches, cache miss cycles, and data dependencies introduce wait states into the pipe and reduce efficiency.

Assuming optimal scalar performance (one scalar instruction for each clock cycle) and complete algorithm homogeneity over the image, CT is approximately equivalent to 8 ms multiplied by the number of machine instructions in the algorithm inner loop and divided by the number of microprocessors available.

Histogram, threshold or image difference require about 10 instructions per pixel, average filtering about 40, and a standard 3x3 convolution about 55 scalar and 35 floating point ones. Computation times between hundreds of milliseconds and a few seconds are then expected for the serial execution of algorithms of this class.

The contributions to the execution time sum differently, according to the level of bus saturation: if the local bus is not saturated, the execution time is equal to: $ET = TT + LT$ *(read)* $+ TC$, otherwise data access is dominant and $ET = TT + LT$ *(read + write)*.

The cluster bus is expected to saturate when the algorithm requires less than 15 instructions per pixel. Because of the ViBus higher throughput its occupancy is expected to be limited to less than 15% when feeding data to 4 clusters.

3. BENCHMARK DISCUSSION AND CONCLUSIONS

To verify the model, a threshold, a low-pass average filter and a general 3x3 convolution algorithm have been coded for parallel operation, optimized and run on the ViP processor, using a variable number of working microprocessors.

Table 1 shows the absolute execution times measured without the *TT* overhead.

Results are consistent with the model expectancies:

• once avoided the *TT* overhead, useful algorithms may be executed in less than *one frame time* (40 ms is the time required to acquire a standard image frame from a CCD camera).

• the basic convolution algorithm executes in one hundred ms. and may be still optimized using vectorization (i860 dual-instruction mode).

Benchmarks on other commonly used low level algorithms (transforms) are being performed to evaluate the possible applications of such a general purpose multiprocessor instead of dedicated hardware.

	1 proc.	*2 proc.*	*4 proc.*	*8 proc.*	*16 proc.*
threshold	95	49	38	21	13
average filter	336	170	118	60	31
3x3 convolution	1500	714	380	190	96

**Table 1: Algorithm execution times versus number of microprocessors
(ms. required to process an image 512x512 pixels by 8 bit each)**

REFERENCES

[1] Gugliotta G., Machì A.: "The Versatile Image Processor *ViP: (hardware Design)."* IAPR Int. Workshop on Machine Vision Applications; Tokyo Dec. 1992.

[2] A. Anzalone, G. Gugliotta and A. Machì' : "Study of workload partitioning on a PRAM-CREW multiprocessor system performing low-mid level image processing tasks." in: V. cantoni et als eds. Progress in Image Analysis ane Porcessing. World Scientific Singapore 1990.

[3] Anzalone A., Gerardi G., Lenzitti B., Machì A.: "Parallel implementation of low level image processing algorithms on a linear array of transputers and on a shared memory based multiprocessor" in: Supercomputing Tools for Science and Engineering; F. Angeli Milano 1989.

[4] N. Margulis: "i860 microprocessor architecture" Osborne-McGraw-Hill, Berkeley 1990.

Time-Varying Image Processing and
Moving Object Recognition, 3 – V. Cappellini (Ed.)
© 1994 Elsevier Science B.V. All rights reserved.

On the implementation of the polyspectra and cumulants via Volterra kernels

B.G. Mertzios[a] and A.N. Venetsanopoulos[b]

[a] Department of Electrical Engineering, Democritus University of Thrace
67 100 Xanthi, Hellas

[b] Department of Electrical and Computer Engineering, University of Toronto
Toronto, M5S 1A4, Canada

The relation of the polyspectra and cumulants with the nonlinear Volterra kernels and the M-D digital filters is explored. Thus, existing advantageous techniques and algorithms for the implementation of the nonlinear Volterra filters may be used for the fast implementation of polyspectra and cumulants. Also the z transform of cumulants may be implemented using matrix decomposition based structures.

1. INTRODUCTION

System modeling and identification have attracted considerable attention during the past twenty years, due to their great importance to a number of applications in diverse fields, such as communications, control systems, economics, ecology, seismology and astronomy [1]-[5]. Recently, polyspectra and cumulants (the counterparts of polyspectra in the time domain) were introduced in digital signal processing and provided the means for an effective identification of nonminimum phase systems [6]-[9]. The use of polyspectra exploits the observation that the spectrum of the output of order higher than two, contains information regarding both the phase and the magnitude of the Fourier transform of the system.

The weakness of these techniques lies in the implied computation cost, since high order statistics should be computed. Therefore, there is an intense interest to develop fast computation techniques for the easy and fast implementation of systems and filters. The Volterra kernels that are involved in the identification procedure may be seen as multidimensional operators [10]-[12]. Thus, after proper modifications, the existing fast and efficient techniques of multidimensional filters and systems may be applied for the systems' identification using polyspectra and cumulants.

Some work on the fast implementation of 1-D and 2-D quadratic and higher order Volterra digital filters, i.e. of their corresponding Volterra kernels has been already done [13]-[18]. This background is useful for the development of more complex structures for identification of high order polyspectra and cumulants. Also, various efficient implementation of 2-D and multidimensional (m-D) digital filters have been proposed during the last decade, which are based on appropriate matrix decompositions. The general form of these techniques is given in [11] and [12] for the 2-D and m-D digital filters respectively. Special matrix decomposition based implementations are the LU decomposition and the SVD decomposition that are extensively treated in the literature and are used for the fast implementation of 2-D quadratic Volterra digital filters [13],[14]. Since a *mth* order Volterra kernel is a m-D function, the matrix-based implementations which have been developed for the m-D filters may be used for the development of corresponding advantageous implementations for the Volterra kernels. Moreover, since polyspectra and cumulants are expressed in terms of Volterra kernels, the matrix-based implementation techniques may be extended for the implementation of polyspectra and cumulants.

2. DEFINITIONS

The *kth* order cumulant $c_{kx}(i_1,i_2,\ldots,i_{k-1})$ of a stationary random process $x(i)$ is defined as the coeefficient $(u_1,u_2,\ldots,u_{k-1})$ in the Taylor series expansion of the cumulant generating function [19]

$$K(u_1,u_2,\ldots,u_{k-1}) = \ln E\left\{ \exp\left[\sum_{j=1}^{k} u_j \, y(i + i_{j-1}) \right] \right\} \tag{1}$$

If $x(i)$ is Gaussian, then $c_{kx}(i_1,i_2,\ldots,i_{k-1}) \equiv 0$ for $k \geq 3$, hence the cumulants are insensitive to additive Gaussian noise $u(i)$ (white or colored) of unknown covariance function, i.e.

$$c_{kz}(i_1,i_2,\ldots,i_{k-1}) = c_{ky}(i_1,i_2,\ldots,i_{k-1}), \qquad k \geq 3 \tag{2}$$

for $z(i) = y(i) + u(i)$ and $y(i)$, $u(i)$ statistically independent. Therefore, higher order statistics are more robust to additive Gaussian noise than correlation, even if noise is colored.

The second order cumulant $c_{kx}(m)$ is the autocorrelation and provides a complete statistical information only for Gaussian processes. Higher order cumulants carry important information for non-Gaussian processes in linear systems or Gaussian processes in nonlinear systems and are used for system identification and detection of phase coupling.

The *kth* order cumulant $c_{kx}(i_1,i_2,\ldots,i_{k-1})$ of a stationary process $x(t)$ is a function of the k-1 lags $i_1,i_2,\ldots,i_{k-1}$ and assume the symmetric forms

$$c_{kx}(i_1,i_2,\ldots,i_{k-1}) = c_{kx}(i_2,i_1,\ldots,i_{k-1}) = \cdots$$

$$= c_{kx}(i_{k-1},i_{k-2},\ldots,i_1) = c_{kx}(-i_1,i_2-i_1,\ldots,i_{k-1}-i_1) = \cdots$$

$$= c_{kx}(-i_1, i_2-i_1, \ldots, i_{k-1}-i_1) \qquad (3)$$

However, these forms are not fully symmetric, since they carry phase information.

Assuming that $c_{kx}(i_1, i_2, \ldots, i_{k-1})$ are absolutely summable, the kth-order *spectrum* of $\{y(i)\}$ is defined as the $(k-1)th$ - dimensional discrete Fourier transform of the kth order cumulant, as follows:

$$S_{kx}(\omega_1, \omega_2, \ldots, \omega_{k-1}) = \sum_{\tau_1=-\infty}^{\infty} \cdots \sum_{\tau_k=-\infty}^{\infty} c_{kx}(i_1, i_2, \ldots, i_{k-1}) \exp\left[-j\sum_{m=1}^{k-1}\omega_m i_m\right]$$

$$= F\left[c_{kx}(i_1, i_2, \ldots, i_{k-1}) \right] \qquad (4)$$

3. CUMULANTS AND POLYSPECTRA IN LINEAR SYSTEMS

Consider the discrete time single-input, single output (SISO) linear and time invariant (LTI) model (Figure 1) with transfer function $H(z)$ and impulse response sequence $h(i)$, where $v(i)$ and $n(i)$ are noise signals. The system $H(z)$ is assumed causal and exponentially stable.

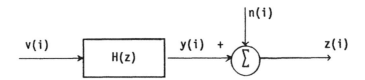

Figure 1. The model of a SISO LTI system with additive noise $n(i)$.

Let $v(i)$ and $n(i)$ be statistically independent white Gaussian noise signals with finite variances σ_v^2 and σ_n^2 respectively. Then the correlation and the spectrum of the output signal $z(i)$ are given by

$$r_z(i) = r_y(i) + r_n(i) = \sigma_v^2 \sum_{k=0}^{\infty} h(i)\, h(i+k) + \sigma_n^2\, \delta(i) \qquad (5)$$

$$S_z(\omega) = \sigma_v^2\, |H(\omega)|^2 + \sigma_n^2 \qquad (6)$$

It is seen from (5) and (6) that the phase information is not present in the correlation and spectrum functions. In the sequel assume that the input {v(i)} is an idependent, identically distributed (i.i.d.) non Gaussian sequence and the additive noise {n(i)} is Gaussian (colored or white) and independent of the output signal y(i). Then

$$c_{kv}(i_1, i_2, \ldots, i_{k-1}) = \gamma_{kv} \delta(i_1, i_2, \ldots, i_{k-1}) \qquad (7)$$

where γ_{kv} is the kth order cumulant of v(i) and $\delta(i_1, i_2, \ldots, i_{k-1})$ denotes the $(k-1)$th delta function. The form of c_{kv} and the above assumptions imply that the cumulants c_{ky} and c_{kz} are equal and may be used interchangeably. Specifically, the kth order cumulant of the output sequence is

$$c_{kz}(i_1, i_2, \ldots, i_{k-1}) = \gamma_{kv} \sum_{m=0}^{\infty} h(m) \, h(m+i_1) \ldots h(m+i_{k-1}) \qquad (8)$$

The kth spectrum of the output is

$$S_{kz}(\omega_1, \omega_2, \ldots, \omega_{k-1}) = \gamma_{kv} H(\omega_1) H(\omega_2) \ldots H(\omega_{k-1}) H\left[- \sum_{m=1}^{\infty} \omega_i\right] \qquad (9)$$

In the more general case where the additive noise {v(i)} is non-Gaussian, the kth order cumulant of z(i) is the multidimensional convolution of the kth order cumulant of v(i) with the deterministic correlation function of h(i), i.e.,

$$c_{kz}(i_1, i_2, \ldots, i_{k-1}) = \sum_{m_1} \sum_{m_2} \cdots \sum_{m_{k-1}} c_{kv}(i_1 - m_1, i_2 - m_2, \ldots, i_{k-1} - m_{k-1})$$
$$\cdot \, c_{kh}(m_1, m_2, \ldots, m_{k-1}) \qquad (10)$$

where

$$c_{kh}(m_1, m_2, \ldots, m_{k-1}) = \sum_{k_1 = 1} \cdot h(\tau) \, h(\tau + m_1) \ldots h(\tau + m_{k-1}) \qquad (11)$$

Equivalently, $c_{kz}(i_1, i_2, \ldots, i_{k-1})$ may be written in the form of a kth order Volterra kernel, as follows:

$$c_{kz}(i_1, i_2, \ldots, i_{k-1}) = \sum_{m_0} \sum_{m_1} \cdots \sum_{m_{k-1}} h(\tau - m_0) \, h(\tau - m_1 + i_1) \ldots h(\tau - m_{k-1} + i_{k-1})$$

$$\mathrm{cum} \, [v(m_0), v(m_1), \ldots, v(m_{k-1})] \qquad (12)$$

Moreover,

$$S_{kz}(\omega_1,\omega_2,\ldots,\omega_{k-1}) = S_{kv}(\omega_1,\omega_2,\ldots,\omega_{k-1})H(\omega_1)H(\omega_2)\ldots H(\omega_{k-1})\ H\left[-\sum_{m=1}^{}\omega_i\right] \tag{13}$$

Consider that the transfer function $H(z)$ is an exponentially stable rational function in z of the general form [9]

$$H(z) = \sum_{i=0}^{\infty} h(i)\ z^{-i} = \frac{B(z)}{A(z)} = \frac{\sum_{j=0}^{q} b(i)\ z^{-j}}{\sum_{i=0}^{p} a(j)\ z^{-j}} \tag{14}$$

The corresponding input-output difference equation is

$$\sum_{j=0}^{p} a(j)\ y(i-j) = \sum_{j=0}^{q} b(j)\ u(i-j) \tag{15}$$

while the system's output is corrupted by additive noise $v(i)$, i.e. $z(i)=y(i) + v(i)$.

The $(k-1)$-dimensional z transform of (8) gives

$$C_{kz}(z_1,z_2,\ldots,z_{k-1}) = \gamma_{kv}\ H(z_1)\ H(z_2)\ \ldots\ H(z_{k-1})\ H(z_1^{-1}\ z_2^{-1}\ldots z_{k-1}^{-1})$$

$$= \gamma_{kv}\ \frac{B(z_1,\ z_2,\ldots,z_{k-1})}{A(z_1,\ z_2,\ldots,z_{k-1})} \tag{16}$$

where

$$B(z_1,z_2,\ldots,z_{k-1}) = B(z_1)\ B(z_2)\ \ldots\ B(z_{k-1})\ B(z_1^{-1}\ z_2^{-1}\ldots z_{k-1}^{-1})$$

$$= \sum_{i_1,i_2,\ldots,i_{k-1}=-q}^{q} b_k(i_1,i_2,\ldots,i_{k-1})\ z_1^{-i_1} z_2^{-i_2}\ldots z_{k-1}^{-i_{k-1}} \tag{17a}$$

$$A(z_1,z_2,\ldots,z_{k-1}) = A(z_1)\ A(z_2)\ \ldots\ A(z_{k-1})\ A(z_1^{-1}\ z_2^{-1}\ldots z_{k-1}^{-1})$$

$$= \sum_{i_1,i_2,\ldots,i_{k-1}=-p}^{p} a_k(i_1,i_2,\ldots,i_{k-1})\ z_1^{-i_1} z_2^{-i_2}\ldots z_{k-1}^{-i_{k-1}} \tag{17b}$$

Moreover the coefficiets $a_k(i_1,i_2,\ldots,i_{k-1})$ and $b_k(i_1,i_2,\ldots,i_{k-1})$ are the k*th* order cumulants of the MA and AR parts of the model and are given by

$$b_k(i_1, i_2, \ldots, i_{k-1}) = \sum_{i=0}^{q} b(i) \; b(i+i_1) \; \ldots \; b(i+i_{k-1}) \tag{18a}$$

$$a_k(i_1, i_2, \ldots, i_{k-1}) = \sum_{i=0}^{p} a(i) \; a(i+i_1) \; \ldots \; a(i+i_{k-1}) \tag{18b}$$

4. RELATIONS OF POLYSPECTRA AND CUMULANTS WITH VOLTERRA KERNELS AND M-D DIGITAL FILTERS

The main result of the present paper is stated in the following two Lemmas.

Lemma 1

The kth order cumulant of the output process of a linear system, when additive non-Gaussian noise is present, may be written in the form of a kth order Volterra kernel, as it is seen by (12). In the special case of Gaussian (colored or white) additive noise, which is independent of the output signal $y(i)$, the kth order cumulant of the output process is given by a slice of a kth Volterra kernel with only one variable, as it is seen by (8).

∎

Various aspects on the relation of Volterra kernels, multidimensional systems and polyspectra have been recently studied [20]-[26]. In [15]-[17] systolic-type implementations of quadratic and higher order Volterra filters are presented. In [18] a general efficient model of the nonlinear finite extent Volterra digital filters, which exploits the symmetries of the coefficients in the Volterra kernels and constitutes the basis for fast implementations of the Volterra kernels via VLSI array processors (APs) is presented [27]. The use of systolic and wavefront APs, which take advantage of both parallelism and pipelining using the concept of the computational wavefront, may result in modular, regular implementations of Volterra systems featuring very high sampling rates, which are suitable for real-time image processing applications.

Lemma 2

The z-transform of a kth order cumulant of the output process of a linear system, when the additive noise is Gaussian (colored or white) and independent of the output signal, is a (k-1)-dimensional rational transfer function of the form (16).

∎

It results from Lemma 2 that all the existing implementation techniques for the M-D digital filters may be directly applied. In particular, the transfer function (16) may be implemented by applying known efficient matrix decomposition based techniques [11],[12].

It is known that the nonlinear Volterra kernels have a direct correspondence with the M-D digital filters; the main difference being that the Volterra kernels involve a number of inherent symmetries. The relations of polyspectra and cumulants with Volterra kernels and M-D digital filters is shown in Figure 2.

116

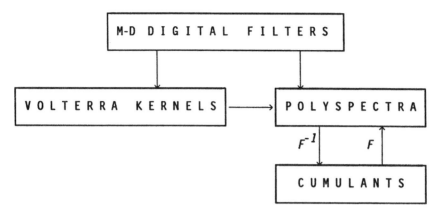

Figure 2. The relations of polyspectra, cumulants, Volterra kernels and
M-D digital filters, where ——→ means specialization.

REFERENCES

[1] H.W. Sorenson, Ed., *Kalman Filtering: Theory and Applications*, New York: IEEE Press, 1985.

[2] L. Ljung and T, Soderstrom, Theory and Practice of Recursive Identification, Cambridge, MA: MIT Press, 1983.

[3] A.H. Jazwinski, *Stochastic Process and Filtering Theory*, New York: Academic Press, 1970.

[4] G.C. Goodwin and R.L. Payne, *Dynamic System Identification: Experimental Design and Data Analysis*, New York: Academic Press, 1977.

[5] F.E. Daum, "Exact finite-dimensional nonlinear filters," *IEEE Trans. on Automat. Control*, vol. AC-31, No. 7, pp. 616- 622, July 1986.

[6] G.B. Giannakis, "Cumulants: A powerful tool in signal processing," Proc. IEEE, vol. 75, No. 9, pp. 1333-1334, Sept. 1987.

[7] J.M. Mendel, "Tutorial on higher order statistics (spectra) in signal processing and system theory," *Proc. IEEE*, vol. 79, No. 3, pp. 278- 305, March 1991.

[8] C.L. Nikias, "Higher-order spectral analysis," in Advances in Spectrum Analysis and Array Processing, S. Haykin (Ed.), Prentice-Hall, ch. 7, pp. 326-365, 1991.

[9] G. Giannakis, "On the identifiability of non-Gaussian ARMA models using cumulants, *IEEE Trans. Automat. Control*, vol. AC-35, No. 1, pp. 18-26, Jan. 1990.[10] M. Schetzen, *The Volterra and Wiener Theories of Nonlinear Systems*, New York, J. Wiley, 1980.

[11] A.N. Venetsanopoulos and B.G. Mertzios, "A decomposition theorem and its implications to the design and realization of two-dimensional filters," *IEEE Trans. on Acoustics, Speech and Signal Processing*, vol. ASSP-33, pp. 1562-1575, December 1985.

[12] B.G. Mertzios and A.N. Venetsanopoulos, "Modular realization of M-dimensional filters," *Signal Processing*, vol. 7, pp. 351-369, December 1984.

[13] B.G. Mertzios, G.L. Sicuranza and A.N. Venetsanopoulos, "Efficient

structures for two-dimensional quadratic filters," *Photogrammetria*, vol. 43, No. 3/4, pp. 157-166, March 1989.

[14] B.G. Mertzios, G.L. Sicuranza and A.N. Venetsanopoulos, "Efficient structures for 2-D quadratic digital filters," *IEEE Trans. on Acoustics, Speech and Signal Processing*, vol. ASSP-37, No. 5, pp. 765-768, May 1989.

[15] B.G. Mertzios and A.N. Venetsanopoulos, "Implementation of quadratic digital filters via VLSI array processors," *Archiv fur Elektronik und Ubertragungstechnik (AEU)*, vol. 43, No. 3, pp. 153-157, May/June 1989.

[16] B.G. Mertzios and A.N. Venetsanopoulos, "Fast implementation of nonlinear digital filters via systolic arrays," *Proc. of the 1990 Bilkent International Conference on New Trends in Communication, Control and Signal Processing*, pp. 1135-1141, Ankara, Turkey, July 2-5, 1990.

[17] B.G. Mertzios and A.N. Venetsanopoulos, "Systolic implementation of nonlinear volterra digital filters using pyramids," *Proc. of the Canadian Conference on Electrical and Computer Engineering*, pp. WM4.28.1-4, Toronto, Canada, September 13-16, 1992.

[18] B.G. Mertzios, "Parallel modeling and structure of nonlinear volterra discrete systems," *IEEE Trans. on Circuits and Systems*. To appear.

[19] M. Rosenblatt, "Cumulants and cumulant spectra," in *Time Series Analysis in the Frequency Domain*, D. Brillinger and P. Krishnaiah (Eds.), North Holland Press, 1985.

[20] S.A. Alshebeili, A.E. Cetin and A.N. Venetsanopoulos, "An adaptive system identification method based on bispectrum," *IEEE Trans. Circuits Systems*, vol. CAS-38, No. 8, pp. 967- 969, Aug. 1991.

[21] S.A. Alshebeili, A.N. Venetsanopoulos and A.E. Cetin, "Reconstruction of FIR systems using slices of high order spectra," *Proc. Intern. Conf. on Digital Signal Processing*, pp. 126-132, Florence, Italy, Sept, 1991.

[22] S.A. Alshebeili and A.N. Venetsanopoulos, "Parameter estimation and spectral analysis of the discrete nonlinear second-order Wiener filter," *Circuits, Systems and Signal Processing*," vol. 10, No. 1, pp. 31-51, 1991.

[23] S.A. Alshebeili, *Volterra Type Systems and Polyspectra*, Ph.D. Thesis, Department of Electrical Engineering, University of Toronto, 1991.

[24] Y. Zhang, D. Hatzinakos and A.N. Venetsanopoulos, "Bootstrapping techniques in the estimation of higher order cumulants from short data records," *Proc. of the Int. Conf. on Acoustics, Speech and Signal Processing*,", Minneapolis, Minnesota, April 1993.

[25] Y. Zhang, D. Hatzinakos and A.N. Venetsanopoulos, "Identification of multichannel and multidimensional systems using cumulants: Application to colour images," *Proc. of the Int. Conf. on Image Processing: Theory and Applications*, San Remo, Italy, June 14-16, 1993.

[26] S.A. ALshebeili, B.G. Mertzios and A.N. Venetsanopoulos, "Efficient implementation of nonlinear volterra digital systems via laguerre orthogonal functions," *Proceedings of the Sixth European Signal Processing Conference EUSIPCO-92*, pp. 685-688, Brussels, Belgium, August 25-28, 1992.

[27] S.Y. Kung, *VLSI Array Processors*, Englewood Cliffs, NJ: Prentice-Hall, 1987.

Time-Varying Image Processing and
Moving Object Recognition, 3 – V. Cappellini (Ed.)

A multitracking system for trajectory analysis of people in a restricted area

J. Aranda, C. León, M. Frigola.

Dept. of Automatic Control & Computer Engineering
Universitat Politècnica de Catalunya
Pau Gargallo, 5 08028-Barcelona (Spain)

In this paper we present a low cost, real time image processing system which permits to detect and to track, in an individualized way, a set of people that comes into and moves through a supervised restricted area. The developed multitracking system can track up to 40 individualized people trajectories with a sampling time of 20 ms. With this data we can obtain statistical information about acceptance of different parts and people's behavior in some public exhibitions or museum rooms.

1. INTRODUCTION

This paper describes how computer vision can be used to know the acceptance of different parts in an exposition or a museum room, and also to get some information about visitors' behavior. Up to now it was possible to know automatically some information such as the input and output flow, and then the occupation level of the room along the day.

This system makes possible to detect and to track in real time and in an individualized way every individual that comes in the watched room until they leave it. The system uses the image taken from a zenithal camera in the center of the room ceiling, which allows to observe whole the interesting area without people overlapping.

The system can get the same information as the present automatic detection systems and furthermore it offers some extra information such as the number of people and permanence time in every part of the exposition, trajectories followed by people, etc

2. SYSTEM DESCRIPTION

With the aim to track people within the room we have developed a visual tracking system consisting of a PC fitted with a low cost specific image preprocessor. The image preprocessor gives to the PC only the contour pixels coordinates of the moving objects. This information is extracted at pixel rate from the input image in two steps, first, contour extraction and second, elimination of contour pixels that belong to the scene background (Figure 1).

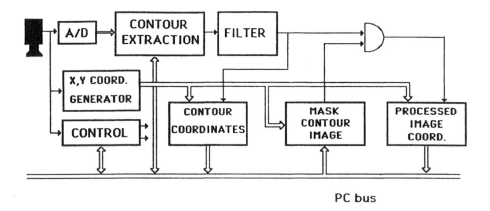

Fig 1. Developed image preprocessor.

The video signal is sampled and digitized in order to get an 8 bits 256x256 pixels image (Figure 2a). In order to reduce this amount of information keeping the relevant data that the tracking algorithm needs, the system performs a Sobel contour extraction with two real time 3x3 convolvers plus a LUT that generates a binary contour image. This image is filtered to eliminate sparse noisy pixels (Figure 2b). When a contour pixel is detected, its coordinates are written in a FIFO memory shared with the PC. In this way the necessary reading time spent by the computer is minimized [1].

In other tracking systems usually the target is detected from the background by simple image difference [2]. In the system we present the elimination of the contour pixels corresponding to static objects and to the image background, is attained at pixel rate by eliminating from the contour image those pixels corresponding to a binary mask (Figure 2c). This mask is generated by means of recursive accumulation, with a high time constant, of the successive contour images acquired every 20 ms. In this way the mask is continuously updated and adapts itself to the image evolution produced by the day-night cycle in outdoor rooms. This mask is calculated by the PC, which counts/discounts the number of appearances of every pixel in the image. In this way a background image is continuously generated, including all the static points that afterwards will be eliminated. The PC updates slowly but continuously this mask, which is adapted to light fluctuations or shadows that can occur along the day.

The pixels coordinates that are not eliminated from this mask are written in the second FIFO memory and constitute the optimum and necessary data to simplify and speed up the tracking algorithm (Figure 2d).

120

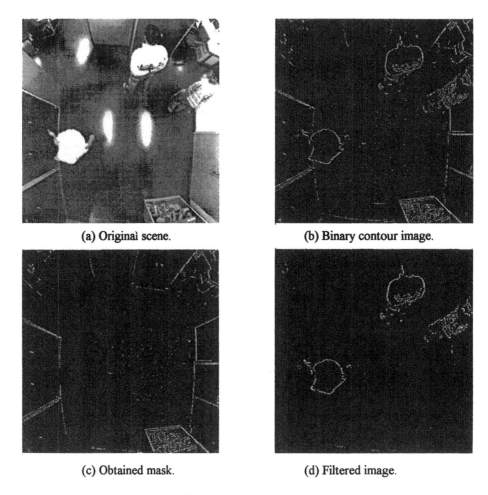

(a) Original scene.

(b) Binary contour image.

(c) Obtained mask.

(d) Filtered image.

Fig. 2. Difference steps of the image process

3. TRACKING ALGORITHM

Once the image has been acquired and processed by the specific image preprocessor, the PC receives a filtered image that contains only the information required by the tracking algorithm. This data consists of the coordinates (x,y) of the image contour pixels corresponding to the new objects that appear in the scene.

From this information, the algorithm looks for new objects coming into the scene through those regions of the image selected as entrance points. These regions or windows define the beginning of the trajectories that must be registered and which are selected by the system user in a previous phase. The end of the analyzed trajectories has not to be defined by the system user since the system automatically stops to follow them when people leave the room.

Every time a change in contour information is detected in any entrance point, it is interpreted as a new person coming into the scene if it verifies some simple coherence conditions as the number of contour pixels and their distribution. It is not necessary to use sophisticated recognition algorithms [3] because the designed image preprocessor gives only the relevant data of the image. When a visitor has been detected, the tracking system begins to track it using a variable size window that contains the visitor. This means that the tracking algorithm will have as many open windows as visitors tracked simultaneously.

The people's position is calculated from the centroid of the contour pixels coordinates within the tracking window (Figure 3). In order to expedite the speed of calculations, an efficient algorithm using the Hadamard transform method is used here [4][5]. This position is taken as the new center for the tracking window in the next sampling time and the centroid is calculated again. The tracking window size is determined both from the people size plus some extra space calculated from people speed and the sampling time. This ensures that the tracking window always encloses the person to be tracked. Due to the short sampling time, 20 ms, and the relative slow visitors' speed the tracking window size is only a little bigger than the area occupied by a visitor and it is not necessary to use trajectory prediction algorithms in the system. Some overlapping can appear between two tracking windows that can produce some distortion in the measure of the positions of two visitors. This is partially solved by the assignment of the contour pixels in the overlapping area to only one of the tracking windows.

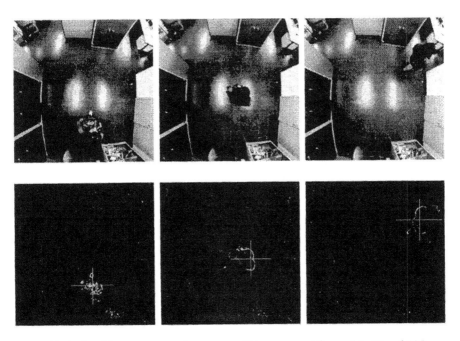

Fig 3. Tracking sequence of a person within a room at frames 18, 45 and 105.

In order to permit the statistical analysis of trajectories the system would have to save the individualized position of every visitor at every new sample. This would generate a high volume of data and also would produce a waste of time. On the other hand, the position resolution given by the system (256x256) is higher than the needed for this application. So the visitors' position inside the supervised area are discretized in 16x16 regions and their movements are coded as incremental displacements to neighboring regions. This discretization permits an important reduction of the information so that it is only necessary to save the permanence time in the region and the direction of change when people move from one region to another. In this way the PC saves only the structure < person_label, direction, permanence_time > in a file when one person moves of region. This file includes all the trajectories information to perform an exhaustive analysis about the visitors' behavior.

We also have to take into account some usual queries that will be requested by users in order to answer them faster. For example, if the user inquires the number of people that have passed through a given region of the area, the trajectory analyzer would have to read all the trajectories file generated. Therefore, in order to reduce the response time to this kind of questions, the system besides saving the data provided by the tracking algorithm, computes also the statistical accumulative values defined by the user. In this way, the system updates two files, the one which contains the global trajectories information and the second that contains the accumulative values that can be provided by a simple memory access.

4. RESULTS

The system has been tested in two different scenarios, first in an indoors environment, our lab, with artificial light and second in the faculty garden, a free space area with natural light.

In the first experiment the camera was installed in the middle of the ceiling, 3 meters high, with a fish-eye objective that permitted to watch a 4x4 meters area. One entrance point was defined which indicates the beginning of all the controlled trajectories. Contour extraction and mask filtering worked well even when light from windows changed the illumination conditions. Direct reflections of light on the floor were not a problem to detect and track people that came into this restricted area (Figure 3). Figure 4 and 5 show some statistical tracking results from this area along one day.

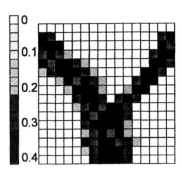 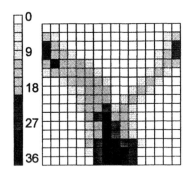

Fig 4a. Mean permanence time   Fig 4b. Number of people.

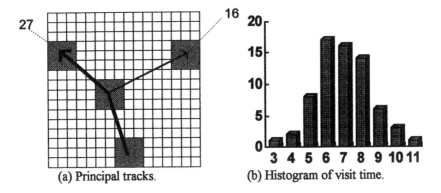

(a) Principal tracks. (b) Histogram of visit time.

Fig 5. Some statistical results of the tracking system.

In the second experiment in a free area, the illumination problem was resolved in the same way by adapting the binaryzation threshold of the contour extraction and the time constant of the recursive mask to the new situation. A new problem appeared with contour position changes due to little movements of the background objects and the own camera produced mainly by the wind. Once again, regulation of the time constant of the recursive mask solved this problem but produced strong contour masks which disable a bigger quantity of relevant contour coordinates (Figure 6).

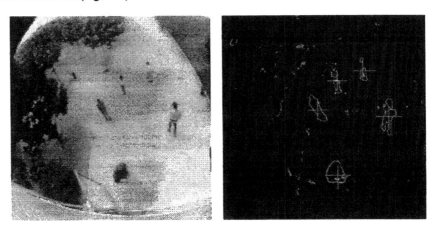

Fig 6. A multitracking scene in the faculty garden.

In this experiment the camera was situated over the entrance point and oriented towards the area to be analyzed. Perspective distortion produced some partial occlusions between people. These partial occlusions caused the temporally loose of some people, if the number of contour coordinates remaining inside the tracking window was very little. Nevertheless this loose of information is not significant for the statistical results about the people's behavior.

REFERENCES

1. A. B. Martínez, J. M. Asensio and J. Aranda, A real time vision system as an aid in learning tasks in Robotics, IFAC Symposium on Intelligent Components and Instruments for Control Applications SICICA'92, Málaga, Spain, May (1992).

2. L. Chen and S. Chang, A video tracking system with adaptive predictors, Pattern Recognition, Vol. 25, 1171-1180 (1992).

3. D. W. Murray, D. A. Castelow, B. F. Buxton, From image sequences to recognized moving polyhedral objects, International Journal of Computer Vision 3, 181-208 (1989).

4. S. H. Lai. and S. Chang, Estimation of 3-D translational motion parameters via Hadamard transform, Pattern Recognition Lett. 8, 341-345 (1988).

5. C. W. Fu and S. Chang, A motion estimation algorithm under time-varing illumination, Pattern Recognition Lett. 10, 195-199 (1989).

6. J. Amat, J. Aranda and A. Casals, A tracking system for robot servopositioning in moving parts operations, 23th International Symposium on Industrial Robots ISIR'92, Barcelona, Spain, October (1992).

7. A. L. Gilbert, M. K. Giles, G. M. Flachs, R. B. Rogers and H.U. Yee, A real time video tracking system, IEEE Trans. Pattern Analysis Mach. Intell. 2, 47-56 (1980).

8. C. Harris and C. Stennet, RAPID- A video rate object tracker, in Proc. British Machine Vision Conference, Oxford, UK, Sept., 24-27 (1990).

E
COMPUTER VISION

Time-Varying Image Processing and
Moving Object Recognition, 3 – V. Cappellini (Ed.)
127

Stereoscopic Correcting
Pose of 3D Model of Object *

Masanobu Yamamoto and Kenzou Ikeda

Department of Information Engineering, Niigata University

8050 Ikarashi 2-nocho, Niigata-city, 950-21, JAPAN

Abstract

When a 3D model is slightly displaced from the object, this paper describes a method for fitting 3D model to the object. The fitting procedure is established by minimizing difference between the depth measured by stereo camera and the depth given from the 3D model. The advantages of the method are a high accuracy of fitting in the depth direction, and stereo correspondence guided by the 3D model.

1 Introduction

Measuring a pose of the objects in the 3-D environment is one of the most important tasks in robotics. The large number of methods using image sensors have been proposed to get 3D information of the objects [6][4][5]. However, after getting the 3D information, if the objects accidentally move or the image sensor moves, the relative pose between the sensor and objects changes. The pose of the objects has to be measured again. If the movement is small, it is a problem to correct the current pose of the object. The problem is equal to one fitting the 3D model to the object. There are several approaches for the problem. A typical one is a method fitting 3D model to 2D image features[1]. The other one is a method of fitting 3D model to 3D coordinates on the object from a range finder[2][7]. The former results into an inaccurate fitting in the depth direction. The later needs a large scale system, in the case of moving object, very expensive system[7]. This paper presents a method using stereo camera system. The method has two advantages. First, an accuracy of fitting in the depth direction can be guaranteed by depth information from the stereo camera system. Second, the stereo camera system has much portability. We describe fitting procedure using Newton's method in Section 2. Section 3 gives stereo image corresponding guided by 3D model to get depth information. Section 4 describes experimental results.

---

*This work was supported in part by the SANKYO SEIKI MFG. CO., LTD..

2   Correcting Pose of 3D Model by Depth

Let a coordinate system of the scene be a Cartesian (x, y, z) one. A camera can observe objects of which projection lies on a image plane, $z = 1$, with the origin of the coordinate as the projection center. An object point $\mathbf{p} = (x, y, z)$ is projected on the image point (X, Y).

$$\begin{cases} X = x/z \\ Y = y/z \end{cases} \tag{1}$$

We suppose that a stereo camera system can measure z-coordinate, i.e. depth, of the object point corresponding to the image point (X, Y), and 3D model of the object also gives a depth information $H(X, Y)$. If the 3D model does not fit the object, $H(X, Y)$ is not always equal to $z = z(X, Y)$.

We describe a method to correct a pose of the 3D model so that the 3D model may fit the object. The pose displacement of the 3D model is represented by the translation and rotation around the axis passing through , $\mathbf{p}_c = (x_c, y_c, z_c)$, the center of the object. Supposing the displacement to be very small, the pose corrections can be represented by the translation vector $\mathbf{S} = (S_x, S_y, S_z)$ and rotation one $\mathbf{Q} = (Q_x, Q_y, Q_z)$. Given the corrections, we express the depth map from the corrected 3D model as

$$z = G(X, Y, \mathbf{S}, \mathbf{Q}). \tag{2}$$

No correction means

$$G(X, Y, \mathbf{0}, \mathbf{0}) = H(X, Y). \tag{3}$$

The depth from corrected 3D model is equal to the depth from stereo camera system.

$$G(X, Y, \mathbf{S}, \mathbf{Q}) - z(X, Y) = 0 \tag{4}$$

Conversely, to obtain the pose corrections, we have to solve a system of the above nonlinear equations. However, it is difficult to obtain the closed-form solution to satisfy the system of the equations because of nonlinearity. Minimizing the L_2 norm measure of the left hand side of eq.(4),

$$\sum_{X,Y} (G(X, Y, \mathbf{S}, \mathbf{Q}) - z(X, Y))^2 \tag{5}$$

we can get a least-square solution. Starting from an appropriate initial guess, the Newton method can give an optimal solution.

The initial guess is set as $\mathbf{S}^{(0)}, \mathbf{Q}^{(0)}$. Assuming $\mathbf{S}, \mathbf{Q}$ are very small, $\mathbf{S}^{(0)} = \mathbf{0}, \mathbf{Q}^{(0)} = \mathbf{0}$. The optimal solution is given by the iterative form

$$\begin{aligned} \mathbf{S}^{(n+1)} &= \mathbf{S}^{(n)} + \Delta\mathbf{S} \\ \mathbf{Q}^{(n+1)} &= \mathbf{Q}^{(n)} + \Delta\mathbf{Q}. \end{aligned}$$

The small corrections to $\mathbf{Q}$ and $\mathbf{S}$,

$$\begin{aligned} \Delta\mathbf{S} &= (\Delta S_x, \Delta S_y, \Delta S_z)^T \\ \Delta\mathbf{Q} &= (\Delta Q_x, \Delta Q_y, \Delta Q_z)^T, \end{aligned}$$

are obtained from the following system of linear equations.

$$J(\mathbf{S}^{(n)}, \mathbf{Q}^{(n)}) \begin{pmatrix} \Delta\mathbf{S} \\ \Delta\mathbf{Q} \end{pmatrix} = \mathbf{e}^{(n)} \tag{6}$$

where $J(\mathbf{S}, \mathbf{Q})$ is Jacobian matrix,

$$J(\mathbf{S}, \mathbf{Q}) = \begin{pmatrix} \vdots \\ \mathbf{J}_{X,Y}(\mathbf{S}, \mathbf{Q}) \\ \vdots \end{pmatrix}$$

$$\mathbf{J}_{X,Y}(\mathbf{S}, \mathbf{Q}) = \frac{\partial G(X, Y, \mathbf{S}, \mathbf{Q})}{\partial(\mathbf{S}, \mathbf{Q})}$$

$$\mathbf{e}^{(n)} = \begin{pmatrix} \vdots \\ e^{(n)}_{X,Y} \\ \vdots \end{pmatrix}$$

$$e^{(n)}_{X,Y} = z(X, Y) - G(X, Y, \mathbf{S}^{(n)}, \mathbf{Q}^{(n)})$$

To evaluate an component of the Jacobian matrix $\mathbf{J}_{X,Y}(\mathbf{S}, \mathbf{Q})$, we have to obtain a partial derivatives of $G(X, Y, \mathbf{S}, \mathbf{Q})$. Unfortunately, we cannot directly calculate the partial derivatives because $G(X, Y, \mathbf{S}, \mathbf{Q})$ in eq.(4) is not yet specified.

We describe how to get the partial derivatives. The pose corrections $\mathbf{S}, \mathbf{Q}$ is represented by Cartesian coordinates (x, y, z), while the camera observes the scene by the camera coordinates (X, Y, z). Substituting X and Y of eq.(4) into the depth map $z = H(X, Y)$, and solving result with respect to z gives a depth map by Cartesian coordinates.

$$z = h(x, y) \tag{7}$$

Correcting the pose displacements of 3D model, a point $\mathbf{p} = (x, y, z)$ on the incorrect 3D model moves to $\mathbf{p}' = (x', y', z')$.

$$\mathbf{p}' = \mathbf{p} + \mathbf{Q} \times (\mathbf{p} - \mathbf{p}_c) + \mathbf{S} \tag{8}$$

Solving eq.(8) with respect to x, y, z, and neglecting the products of the components of $\mathbf{Q}, \mathbf{S}$ because these products can be regarded as very small second and more order terms, the $\mathbf{p}$ is approximated by

$$\mathbf{p} = \mathbf{p}' - \mathbf{Q} \times (\mathbf{p}' - \mathbf{p}_c) - \mathbf{S}. \tag{9}$$

Substituting eq.(9) into eq.(7), and rewriting terms x', y', z' with x, y, z, respectively, we obtain

$$z + Q_y(x - x_c) - Q_x(y - y_c) - S_z = \\ h(x + Q_z(y - y_c) - Q_y(z - z_c) - S_x, y - Q_z(x - x_c) + Q_x(z - z_c) - S_y). \tag{10}$$

Solving eq.(10) with respect to z, the correct depth map

$$z = g(x, y, \mathbf{S}, \mathbf{Q}) \tag{11}$$

is obtained. In the case of no correcting,

$$g(x, y, 0, 0) = h(x, y).$$

Furthermore, substituting $x = Xz, y = Yz$ into eq.(11) gives

$$z + Q_y(Xz - x_c) - Q_x(Yz - y_c) - S_z =$$
$$h(Xz + Q_z(Yz - y_c) - Q_y(z - z_c) - S_x, Yz - Q_z(Xz - x_c) + Q_x(z - z_c) - S_y) \quad (12)$$

Solving eq.(12) with respect to z gives

$$z = G(X, Y, \mathbf{S}, \mathbf{Q}). \quad (13)$$

In the case of no correcting,

$$G(X, Y, 0, 0) = H(X, Y).$$

Partially differentiating the eq.(12) with respect to $\mathbf{S}, \mathbf{Q}$ gives components of the Jacobian matrix,

$$\frac{\partial G}{\partial S_x} = \frac{h_x}{k} \quad (14)$$

$$\frac{\partial G}{\partial S_y} = \frac{h_y}{k} \quad (15)$$

$$\frac{\partial G}{\partial S_z} = \frac{-1}{k} \quad (16)$$

$$\frac{\partial G}{\partial Q_x} = -\frac{y - y_c + (z - z_c)h_y}{k} \quad (17)$$

$$\frac{\partial G}{\partial Q_y} = \frac{x - x_c + (z - z_c)h_x}{k} \quad (18)$$

$$\frac{\partial G}{\partial Q_z} = \frac{(x - x_c)h_y - (y - y_c)h_x}{k}, \quad (19)$$

where

$$k = Xh_x + Yh_y - 1 + (Y + h_y)Q_x - (X + h_x)Q_y + (Yh_x - Xh_y)Q_z.$$

The surface gradient (h_x, h_y) on the 3D model is calculated at $(x, y) = (Xz + (Yz - y_c)Q_z - (z - z_c)Q_y - S_x, Yz - (Xz - x_c)Q_z + (z - z_c)Q_x - S_y)$.

Correcting the pose of 3D model by $\Delta \mathbf{S}, \Delta \mathbf{Q}$ at every iteration, the components of the Jacobian matrix may be simplified. The $n + 1$th iteration is calculated based on the revised 3D model after the nth iteration. The Jacobian matrix can be updated by setting $\mathbf{S}^{(n)} = \mathbf{0}, \mathbf{Q}^{(n)} = \mathbf{0}$ since the 3D model has been revised. Therefore, the linear equation of the system (6) is described in detail

$$\frac{1}{k}(\mathbf{c} \cdot \Delta \mathbf{S} + (\mathbf{d} - \mathbf{d}_c) \cdot \Delta \mathbf{Q}) = e_{X,Y}^{(n)}, \quad (20)$$

where

$$c = \begin{pmatrix} h_x \\ h_y \\ -1 \end{pmatrix}$$

$$d = \begin{pmatrix} -y - zh_y \\ x + zh_x \\ xh_y - yh_x \end{pmatrix}$$

$$d_c = \begin{pmatrix} -y_c - z_ch_y \\ x_c + z_ch_x \\ x_ch_y - y_ch_x \end{pmatrix}$$

$$k = Xh_x + Yh_y - 1$$

To evaluate coefficients of the equation, we need

(1) Cartesian coordinates (x, y, z) of a point on 3D model corresponding to a image point (X, Y), and

(2) the gradient (h_x, h_y) on the 3D model.

These quantities are given by a geometrical modeler, namely SOLVER[3].

3 Matching Stereo Images

This Section describes how to get depth information of the object using a stereo camera system.

The main issue in stereo camera system is corresponding between left and right camera images. This paper uses a standard stereo camera system of which optical axes are in parallel. This type of stereo camera system has advantages that it is easy to find corresponding point since an epipolar line is parallel with horizontal scanning line in images. Let the image coordinates of a feature point P_l on the left image be (x_l, y_l), and the image coordinates of the corresponding point, P_r, on the right image (x_r, y_r). The both coordinates satisfy

$$x_r < x_l$$
$$y_r = y_l.$$

The relation is used to find corresponding points in stereo image as an epipolar constraint.

The horizontal coordinate x of the feature point is measured as a distance from the left side of the image plane. To measure a depth information more accurately, we have to make a baseline length of stereo camera be long. Then, the number of potential matches increases since x_l extremely differs from x_r. There may be a possibility that one camera cannot see the feature point to be matched due to the occlusion. These make stereo correspondence problem be difficult.

To cope with these problem, we can fortunately use 3D model which is located near the object. Figure 1 shows left and right images on which projections of 3D model of the

Figure 1: The stereo pair of images: Left image(upper), Right image(down).

object are superimposed, where the object is a cube painted with a checker pattern. If the feature point is on the projection of 3D model, we can estimate a depth information corresponding to the feature point in the image from the 3D model. Exactly speaking, the depth is not always correct because the 3D model does not fit the object. However, assuming that displacement between the model and object is very small, the depth is nearly equal to the correct one. Therefore, we can expect a projection of the feature point on the other image. The correct corresponding image point exists in the neighborhood of the projection. We can also see which facet of the 3D model the feature point is on. If the facet cannot be seen from the other camera, the corresponding feature point is occluded.

Here is a procedure to search for corresponding point on a pair of stereo images

1. Find a facet on which a feature point is in the left image. Let a distance of the feature point measured from the left boarder of the facet image be the horizontal coordinate x_l'.

2. If the facet cannot be seen from the right camera, the object point corresponding to the feature point is occluded. In this case, there is no corresponding feature point in the right image. Otherwise, the facet is projected in the right image. Locate a point x_l' away from the left boarder of the facet projection on the epipolar line in the right image. The point is used as a reference point to search corresponding point.

3. Find potential matches by searching in the neighborhood of the reference point in the right image.

4. Only one match, the point should be the corresponding point. In the case of a large number of potential matches, compare an intensity distribution near by each potential match with one near by the feature point in the left image. If the intensity-based pattern matching results only one feature point, the point should be the corresponding point. Otherwise, among the remaining potential matches the nearest one to the reference point should be corresponding point. The intensity pattern used here is reduced into two categories. One is that the left side intensity of the feature point is darker than the right side one. Another is reverse.

The algorithm may result into false correspondence. However, note that there is few number of the false correspondences as a result of the corresponding experiments. A large number of correct correspondences makes 3D model converge to the object exactly.

4 Experimental Results

Using depth information from stereo camera system, it shows that our algorithm fits 3D model with the objects.

The standard stereo camera system has 20 *cm* baseline. Figure 1 depicts left and right camera images. The size and gray level of each image are 638×480 and 256, respectively. We chose a cube as a target object. A model of the object, i.e. cube, is made in advance using a geometrical modeling system, namely, SOLVER[3]. The Sobel edge detector and thinning is used to extract edges in a pair of stereo images. The feature points to be corresponded are intersections of edges and epipolar lines. Figure 2 shows the first 7 iterations of fitting 3D model to the object by our method presented here.

5 Concluding Remarks

We presented a method for fitting 3D model to object by stereo camera system. A small error of fitting remains after many iterations for convergence. The main causes of the error are that optic axes of the stereo camera system are not always in parallel, and the accuracy of depth is not satisfactory. To cope with these error, it needs an accurate calibration of stereo camera system, and determining edge coordinates in sub-pixel accuracy.

134

Figure 2: After the seven iteration of convergence, the 3D model is superimposed on the left image.

References

[1] D.G.Lowe: Solving for the parameters of object models from image description, Proc. ARPA Image Understanding Workshop, pp.121-127, (1980)

[2] T.Hasegawa and S.Kameyama: Geometric modeling of manipulation environment with interactive teaching and automated accuracy improvement, Proc. 20th ISIR, pp.419-426, (1989)

[3] K.Koshikawa and Y.Shirai: A 3D modeler for vision research, Proc.ICAR, pp.185-190, (1985)

[4] M.Oshima and Y.Shirai: Object recognition using three-dimensional information, IEEE vol.PAMI-5, pp.353-361, (1983)

[5] Y.Ohta and T.Kanade: Stereo by intra- and inter scanline search using dynamic programming, IEEE vol.PAMI-7, pp.139-154, (1985)

[6] K.Ikeuchi and B.K.P.Horn: Numerical shape from shading and occluding boundaries, Artificial Intelligence, vol.17, no.1-3, pp.141-185, (1981)

[7] M.Yamamoto, P.Boulanger, J.-A.Beraldin, M.Rioux and J.Domey: Direct estimation of deformable motion parameters from range image sequence, Proc. 3rd ICCV, Osaka, pp.460-464, (1990)

Time-Varying Image Processing and
Moving Object Recognition, 3 – V. Cappellini (Ed.)
1994 Elsevier Science B.V.

Camera Calibration and Error Analysis. An application to Binocular and Trinocular Stereoscopic System.

F. Pedersini, S. Tubaro, F. Rocca

Dipartimento di Elettronica e Informazione, Politecnico di Milano
P.zza L. da Vinci, 32 - I-20133 Milano, Italy
Phone: +39 +2 2399 3577/3603 Fax: +39 +2 2399 3413
e-mail: pedersin@ipmel2.elet.polimi.it

ABSTRACT

The calibration of a 3D vision system is the process of determining the internal geometric and optical characteristics (intrinsic parameters) of the cameras and their position and orientation with respect to an external co-ordinate system (extrinsic parameters). An accurate calibration is essential to obtain the best performances of the vision systems. In the paper a technique to characterize and use a low-cost binocular and trinocular stereoscopic system is presented. First a good camera model that takes into account also lens distortion has been introduced; then the relation between calibration accuracy and calibration condition is found. Furthermore, good calibration conditions have been detected and the variance of the estimated parameters is determined. The calibrated vision system, both in binocular and trinocular configurations, has been used for accurate measurements of rigid objects obtaining good results and confirming the results of the error modelling procedure.

1. THE CAMERA MODEL

In many applications, a TV camera is modelled as ideal pin-hole image acquisition systems [1], and therefore only its extrinsic parameters (position and orientation) must be estimated. For a more precise modelling of low-cost cameras we must take into account of the geometrical distortions concerning the position of the image points on the image plane (fig. 1). These distortions are caused by imperfect lens shapes and by improper lens and camera assembly and are modelled in a very complex way [2]. However normally some terms are much more important than others: the radial distortion of the lens (see fig. 2) and the displacement between the intersection of the optical axis with the image plane and the center of the photosensitive surface; moreover, the parameters relative to the sampling of the analog video signal available at the camera output must be taken into account.

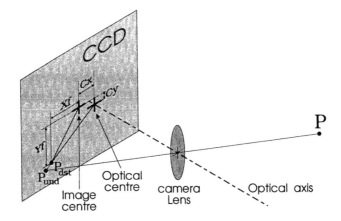

Figure 1. The camera model. P_{und} is the ideal (undistorted) projection of the point P, while P_{dst} takes into account lens distortion.

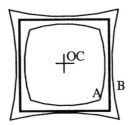

Figure 2: Effects of optical radial distortion. OC is the projection of the optical axis on the image plane. The solid lines represent the boundaries of the undistorted projection of a square parallel to the image plane, while the dashed ones show the effects of a) positive and b) negative radial distortion.

2. SYSTEM CALIBRATION

The calibration of an image-acquisition system is a set of operations that allows to estimate all the parameters describing the acquisition process. Some acquisitions of a sample object (the calibration target) are taken at first. The positions of some "fiducial points" on the images are detected and then used to estimate the parameters of the model. The values of the calibration parameters are the solutions of an overdetermined system of non-linear equations describing the relationship between 3-D fiducial marks coordinates and correspondent image coordinates. An efficient and well-tested calibration procedure is the one proposed by Tsai [3]. Its efficiency is related to having subdivided the global resolutive non-linear system in two independent subsystems, of which only one is non-linear. This method is theoretically correct only under the following assumptions:

- lens distortion has only a radial component;
- the Optical Center (intersection between the image plane and the optical axis of the lens) lies in the image center or, at least, its position is known.

The optical distortion of most part of the lenses can be considered radial, while the other assumption is generally contradicted in practice. Tsai's method gives in any case satisfactory performances, since an error in the position of the Optical Center (O.C.) changes calibration results very weakly. Nevertheless, knowledge of O.C. position is necessary to obtain, after calibration, very good performances in the back-projection of correspondent points.

For this reason we have developed a new calibration method, able to determine also the position of the Optical Center. Estimation of its position is a critical operation because the sensitivity of the available data (the target images) versus optical center shifts is very low. A very precise target and accurate detection of the fiducial points on the images are required since little inaccuracies on data modify in a significant way the calibration results. The calibration set up is presented in fig.3. The parameters to be calibrated are: the position and orientation of the camera with respect the target 3-D co-ordinate system; the lens focal length; the radial distortion coefficient; the position of the O.C. (image co-ordinates), and the equivalent horizontal size of the pixel acquired by the frame grabber (called d_{ex}). The aspect ratio of the CCD pixel cell is assumed known. If d_{ex} is a-priori known, only a coplanar set of fiducial points is needed. This is possible with digital output cameras or when the frame grabber is synchronized by the camera's pixel clock.

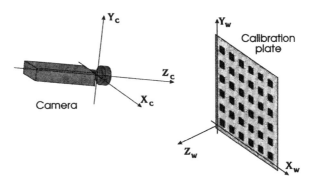

Figure 3. General system calibration setup.

When even the position of the optical center (OC) is an unknown, a global non-linear optimization algorithm (like simplex or conjugate gradient) must be used. In this case, Tsai's technique (with OC considered in the image center) is exploited for the estimation of the first guess.

In our experiments, the calibration plate is an array of black squares on a white background. The test points are vertices of the squares. Their 3-D position is known with an accuracy of about 10 μm.

The correspondent image points are localized with sub-pxl. accuracy as the intersection of adjacent side lines, determined by a least-square linear fit of edge points. Each edge point location is determined by a third order polynomial approximation of the intensity profile across a row or a column.

2.1. Calibration Performances

The analysis of the calibration problem has indicated that the parameter Estimation is an ill-conditioned problem. Each parameter is estimated with a different degree of precision, depending on its sensitivity with respect to the Data. The intrinsic parameters are much more critical than the extrinsic ones.

From a series of calibrations of a TV camera (target with about 400 fiducial points) we have obtained, for the intrinsic parameters, the results presented in Table 1. They confirm the non uniform accuracy of the parameter estimation.

$f = 16.553\ mm$	$K_d = 1.248\ 10^{-3}\ mm^{-2}$	$Cx = 5.99\ pxl$	$Cy = 13.51\ pxl$
$\sigma_f = 4.61\ 10^{-3}\ mm$	$\sigma_{Kd} = 1.03\ 10^{-6}\ mm^{-2}$	$\sigma_{Cx} = 0.196\ pxl$	$\sigma_{Cy} = 0.272\ pxl$
$\sigma_f/f = 2.78\ 10^{-4}$	$\sigma_{Kd}/K_d = 8.25\ 10^{-4}$	$\sigma_{Cx}/C_x = 3.27\ 10^{-2}$	$\sigma_{Cy}/C_y = 2.01\ 10^{-2}$

Table 1. Results of a multiple set of calibration experiments.

The mean error between the detected image positions of the fiducial points and those obtained by the model is less than 0.1 pxl.

2.2. Camera calibration as an inverse problem.

Camera calibration is a typical non-linear inverse problem. The corresponding forward process is represented in fig. 4.

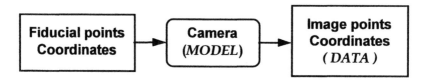

Figure 4. Camera acquisition scheme.

By performing a linearization of the inverse problem, the Model Parameter Variances can be evaluated on the basis of those relative to the errors in the image point locations.

Assuming a Gaussian distribution of these errors we have:

$$\sigma_{Model} = Q(\sigma_{Data})$$

where Q(.) is a linear function [4].

We have taken into account errors due to non-ideal 3-D positions of the fiducial marks and errors in detection of their correspondent image co-ordinates. The errors on image coordinates are both due to the acquisition noise and to the effects of the limited extension of each photosensor, with respect to the entire CCD cell. Let's consider a straight black-to-white light transition on the CCD surface (fig. 5); wherever the transition doesn't fall into a photosensitive area, the local estimation of its position is surely affected by uncertainty.

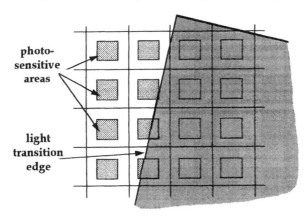

photo-
sensitive
areas

light
transition
edge

Figure 5. Representation of the problem related to the correct localization of luminance transition with a CCD sensor.

The comparison between the error analysis and the experimental calibration results have shown that the detected image coordinates of the fiducial marks are affected by an rms error of about 0.05 pixel.

Moreover the absence of systematic drifts in the differences between the real image coordinates of the calibration points and those predicted by the evaluated camera model is a proof of a correct camera modellization and an accurate parameter estimation. The error analysis has also shown the absolute necessity of the use of very accurate target to obtain effective calibration, because uncertainty on the 3-D position of the fiducial points greatly affects the performances of the entire process.

Regarding the stereoscopic system, calibration is performed independently for each camera, to estimate intrinsic parameters. In the second phase, calibration is run with the same target for both cameras, to calculate their relative position and orientation, and then the operator adjusts the system to obtain a suitable geometry. This operation is repeated iteratively until the calibration results show that the desired geometry is reached (for example near-parallel optical axes).

3. EXPERIMENTS WITH CALIBRATED SYSTEMS

Some practical tests have been performed with two- and three-camera system. In both cases we have used edges as features to be matched. The binocular system has been calibrated with roughly parallel optical axes. After calibration, the system has been employed to measure a mechanical part: a vehicle suspension coil spring, with variable radius and pitch.

With our system the typical relative positioning error is 0.2% in the plane orthogonal to the optical axis and 1% along the axis. This anisotropy is due to the small parallax angle formed by the cameras.

Since the stereo correspondent point is found as intersection between the epipolar line and an edge, generally such binocular systems can give good results only when the intersection angle is significant (i.e. greater than 45 degrees). This problem is completely overcome by using three cameras. In this case, if cameras are properly placed, choosing a feature on one image there is, at least on one other image, a significant intersection angle between the epipolar line and the local direction of the correspondent edge. This property is exploited by the matching strategy. Starting from one image, for any considered edge point, the local edge direction and the two epipolar lines, relative to the other two images, are evaluated. The first stereo matched point is searched in the image relative to the epipolar line forming the greatest intersection angle with the starting edge (in fig. 6, starting from the left, the right is chosen). The intersection on the third image is then redundant, so it's used to check the matching correctness and to improve the accuracy in back-projection. In the determination of the epipolar lines and for 3-D backprojections the position of the camera O.C.s and the optical radial distortions have been taken into account.

Experiments have been carried out with a trinocular system on several kinds of close-range scenes, obtaining very promising results. Some reconstruction results are shown in figures 7,8,9. In measurements of sample objects with known dimensions, we have obtained a relative accuracy of about 2 parts over 10,000. Moreover, considering separately the 3 back-projections obtained by the 3 stereo couples, the world co-ordinates of these points differ about 1 part over 10,000. It shows the accuracy of the trinocular system modellization and calibration.

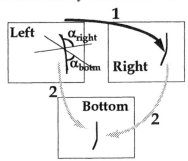

Figure 6. Trinocular matching strategy. Since α_{right} is greater than α_{botm}, the first stereo point is taken from the right image.

4. CONCLUSION AND FURTHER RESEARCHES.

The results of our experiments have shown the great importance of an accurate calibration to obtain good 3-D measures with a low-cost stereoscopic system. Precision in measurements depends much more on the accuracy in determining the model that describes the system than on the quality of the instruments that build up the system itself. The accuracy of the 3-D measurements carried out with both binocular and trinocular systems show the correctness of the proposed camera model, in conjunction with the developed calibration procedure.

Further research will be oriented to improve the calibration performances taking into account further knowledge on the CCD geometry and a more precise model of the image acquisition process. Research will be also oriented on the use of a calibrated trinocular system for 3-D scene description using as matching features not only edges, but also luminance profile of small image regions.

REFERENCES
[1]   N. Ayache - *Artificial vision for Mobile Robots* - MIT Press, 1990.
[2]   J. Weng, P. Cohen, M. Herniou - *Camera Calibration with Distortion Models and Accuracy Evaluation* - IEEE Trans. on Pattern Anal. and Machine Intelligence, Vol. PAMI-14, No. 10, Oct 1992, pp. 965-980.
[3]   R. Y. Tsai - *A versatile camera calibration technique for high-accuracy 3D machine vision metrology using off-the-shelf TV cameras and lenses* - IEEE Journal on Robotics and Automation, Vol. RA-3, No. 4, Aug 1987, pp. 323-344.
[4]   A. Tarantola - *Inverse Problem Theory* - Elsiever, 1987-88, ISBN 0-444-42765-1.

Figure 7. A close range scene. Original image (left).

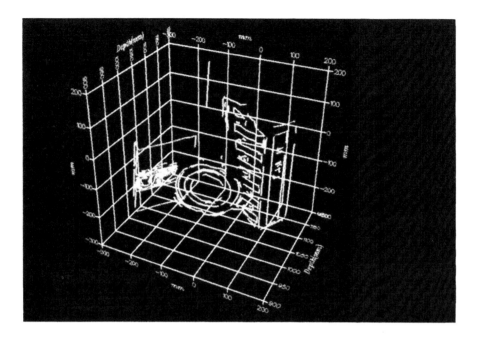

Figure 8. 3-D reconstruction obtained by back-projection of the matched edges.

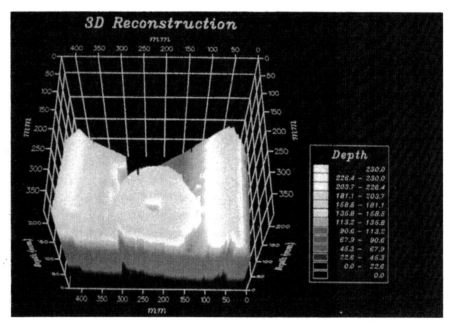

Figure 9. 3-D volumetric reconstruction of the considered image. Color (gray) information is the distance (in mm.) from the camera system.

Time-Varying Image Processing and
Moving Object Recognition, 3 – V. Cappellini (Ed.)
143

Stereo Vision Motion Compensation

Dr D R Broome
University College London

1. INTRODUCTION

A vision based motion compensation system is currently being evaluated as part of the UK ARM (Advanced Robotic Manipulator) project. This project is being conducted by Slingsby Engineering Ltd, Camera Alive Ltd, Technical Software Consultants Ltd and University College London with sponsorship from Mobil North Sea Ltd and the Offshore Supplies Office. The aim is to produce a new eight degree of freedom manipulator for mounting on an ROV (Remotely Operated Vehicle) to be used for the underwater inspection of offshore structures. The main aim of the ARM vision project is to produce a laboratory demonstrator system which will use the Camera Alive stereo vision system to measure relative motion between the manipulator "shoulder" and the workpiece and to use this information in the manipulator controller to compensate for this motion allowing the end effector to accurately follow a predefined path. Initial trials are being conducted with PUMA 560 and the Camera Alive system following a target moving in two dimensions. This will permit evaluation of the motion compensation algorithms.

The manipulator is mounted on Remote Operated Vehicle (ROV) which is 'flown' into the oil rig structure and fixes on to the tubular to be inspected using 'sticky' feet. The tidal flow and at shallow depths, the wave induced water currents, will disturb the vehicle against these compliant attachments and hence the shoulder end of the manipulator also moves. This directly affects the accuracy which it is possible to achieve in tracking the arm around nodal welds for cleaning or inspection - even if arm absolute accuracy is good, and the workpiece location is accurately known. The project is using a laboratory test set up to establish the accuracies possible for using stereo vision feedback. The target is mounted on a vertical x-y table which can be moved at various motion amplitudes and frequencies representative of sea induced motions. Initial tests on the PUMA 560 have been conducted without the stereo vision system to calibrate base performance of the arm control system. Here the arm has been required to track various forms of target namely

(i)   remain at a point (under motion)
(ii)   follow a line in a vertical plane
(iii)   track round a closed contour

The errors are measured using a 'slender finger' attached to the robot end plate, free to move in aximuth and elevation (with potentiometers reading each) - with the free end resting on the moving target hence any errors in the arm tracking the target are represented by the potentiometer readings - and an error value can be expressed as the integrated sum of the squares of these two error signals.

Initial tests currently underway, use the x and y demand signals to both drive the x-y table *AND* to be used by the arm controller as absolute measures of target position. These tests have given a basic calibration of the arm system. Next the stereo vision system will be introduced, set up to 'see' targets on the x-y table and a similar set of tests will be conducted to quantify the ability of the vision system to perform tracking under various forms of input waveform, amplitude and frequency.

2.   COORDINATE SYSTEMS AND TRANSFORMATION

2.1   Co-ordinate Systems

Figure 1 shows the basic co-ordinate systems involved. These are:

i.   ROV co-ordinate system centred at the CG of the ROV.
ii.   Arm shoulder co-ordinate system centred at the shoulder of the arm.
iii.   Camera co-ordinate system centred at the centre of the left-hand camera lens.
iv.   Workpiece co-ordinate system.

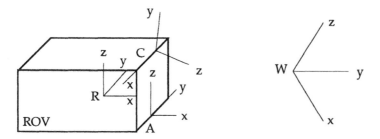

Figure 1 - Basic Co-ordinate Systems

The ROV co-ordinate system is used primarily as a reference for the other co-ordinate systems. In almost all cases it is possible to eliminate values represented in this co-ordinate system by expressing them in one of the other systems. The arm shoulder co-ordinate system is used by the inverse kinematics algorithms to calculate the arm joint angles necessary to position the tip at a required location and orientation. All information from the stereo camera system is presented in the camera co-ordinate system. The workpiece co-ordinate system is attached in the workpiece.

The workpiece co-ordinate system is fixed in space and the other three system will move as the ROV moves under the action of current and waves. It should be noted that the arm shoulder and camera co-ordinate systems may not be located at fixed positions relative to the ROV system as the camera may be mounted on a pan/tilt mechanism and the arm may be mounted on a primary deployment system. However this possibility is ignored for this phase of the project. The workpiece, arm and ROV co-ordinate systems use right-handed co-ordinate systems but the camera uses a let-handed system as shown in Figure 1.

2.2   Homogeneous Transformation Matrices

Homogeneous transformation matrices are used to specify transformations. These are 4x4 matrices defined as:

$$T = [\begin{matrix} R & p \\ 0 & 1 \end{matrix}]$$

where R is a 3x3 matrix representing rotations of the system from the base co-ordinate system and p is a 3x1 vector representing the translation of the system from the base co-ordinate system.

Individual points are represented by 4x1 extended vectors with the 4th element set to unity.

2.3   Transformations Between Co-ordinate Systems

Transformations between the basic co-ordinate systems are defined by the following homogeneous transformation matrices:

T_{cw} Transformation between camera and workpiece co-ordinate systems.
T_{rc} Transformation between ROV and camera co-ordinate systems.
T_{ra} Transformation between ROV and arm shoulder co-ordinate systems.
T_{aw} Transformation between arm shoulder and workpiece co-ordinate systems.

These transformation matrices may be combined to form more complex transformations.

Considering for example, T_{aw}, this may be written as:

$$T_{aw} = (\begin{matrix} x_{aw} & y_{aw} & z_{aw} & p_{aw} \\ 0 & 0 & 0 & 1 \end{matrix})$$

where x_{aw}, y_{aw} and z_{aw} are the unit direction vectors representing the x, y and z axes of the workpiece co-ordinate system expressed in the arm co-ordinate system, and p_{aw} is the position vector representing the origin of the workpiece co-ordinate system in arm co-ordinates.

As an example of the use of these transformation matrices, consider a point with co-ordinates p_w expressed in the workpiece co-ordinate system. The co-ordinates of this point in the ROV co-ordinate system may be calculated as:

$$p_r = T_{ra} T_{aw} p_w$$

or

$$p_r = T_{rc}\, T_{cw}\, p_w$$

3.   WORKPIECE MODELLING

It will be assumed, that geometrical information on the workpiece is know, eg. from CAD drawings, etc, which have been entered into the workpiece database. This information fully defines the workpiece co-ordinate system. A number of reflective targets will be fixed to the workpiece which will be used for ROV motion compensation. However, the precise location of these targets will be unknown.

For ROV motion compensation, the main aim of the workpiece modelling stage is to locate the 3-D co-ordinates of the target positions relative to the workpiece co-ordinate system, as will be shown in Section 4. This first requires estimation of the camera/workpiece transformation, ie. finding the location and orientation of the workpiece co-ordinate system in camera co-ordinates. This will be done by extracting easily detectable features such as workpiece edges from the stereo images and fitting these to the known workpiece, eventually allowing workpiece definition features such as cylinder axes to be located in the camera co-ordinate system. 3-D camera system co-ordinates for the targets must also be calculated as this time, and so the targets must be in position before starting the workpiece modelling phase. From this information it is possible to calculate the co-ordinates of the targets in the workpiece co-ordinate system. Note that this can be an off-line operation provided that all the necessary information can be obtained from a single stereo image pair.

3.1   Determination of Workpiece Co-ordinate System

The Camera Alive system and software will calculate basic position and direction information for the workpiece entities in the camera co-ordinate system. From this information we must calculate the workpiece location and orientation relative to the arm and the co-ordinates of the targets in the workpiece co-ordinate system. This requires the calculation and application of a number of transformation matrices to the basic information.

3.1.1   Calculation of Workpiece Co-ordinate System

The first requirement of the Workpiece Modelling stage is to calculate the arm to workpiece transformation matrix, T_{aw}, from the stereo image data. This will allow the ARM software to produce a graphical representation of the ROV, arm ad workpiece in the relative positions at the instant the stereo image was acquired. The Camera Alive software will extract sufficient information to allow the position and orientation of the workpiece to be calculated in the camera co-ordinate system, ie. defining the transformation matrix, T_{cw}. Note that this matrix is composed of the three unit vectors representing the axes of the workpiece system and the position vector of the origin of the workpiece system, ie.

$$T_{cw} = \begin{pmatrix} x & y & z & p \\ 0 & 0 & 0 & 1 \end{pmatrix}$$

3.1.2 Transformation to Arm Shoulder Co-ordinate System

Once the Workpiece co-ordinate system has been established in camera co-ordinates, T_{cw}, it must be transformed to the arm shoulder co-ordinate system before it can be used by the software for graphical display. This is done using a chain of transformations using the ROV co-ordinate system as a reference frame. In the ROV frame the workpiece system can be defined by the following transformation:

$$T_{rw} = T_{rc} \, T_{cw}$$

$$T_{rw} = T_{ra} \, T_{aw}$$

Thus:

$$T_{rc} \, T_{cw} = T_{ra} \, T_{aw}$$

and

$$T_{aw} = (T_{ra}^{-1} \, T_{rc}) \, T_{cw}$$

Note that the transformation matrix $T_{ra}^{-1} \, T_{rc}$ is determined by the geometrical layout of the ROV, camera and arm, but may include time dependent terms due to camera tilt/pan mechanisms or the arm primary deployment system. In theory it should be possible to perform measurements to calculate this transformation. However, in practice direct measurements are unlikely to be sufficiently accurate and may change over time due to damage, etc. It is proposed that a self-calibrating system is used to calculate this matrix at the start of any mission and this procedure is explained in more detail in a later section.

Once the transformation matrix T_{aw} has been found, the software can display the workpiece graphically. Note that for the moment the display will assume that the ROV position is fixed relative to the workpiece.

3.2 Calculation of Target Positions

For ROV Motion Compensation, the Workpiece Modelling stage is required to determine the positions of reflective targets in the Workpiece co-ordinate system. The Camera Alive system will calculate the positions of these targets in the camera co-ordinate system and a transformation is required to convert these to workpiece co-ordinates.

Consider a target with workpiece co-ordinates p_w. This point can be expressed in camera co-ordinates as:

$$p_c = T_{cw} \, p_w$$

148

Thus the workpiece co-ordinates can be calculated directly from the camera co-ordinates as:

$$p_w = T_{cw}^{-1} p_c$$

Note that Tcw is known from the workpiece co-ordinate system determination described in section 3.1.1.

4. ROV MOTION COMPENSATION

Algorithms for ROV motion compensation are considered in this section. It is assumed that the workpiece co-ordinates of a number of targets are known and that the Camera Alive system will return the camera co-ordinates of these targets in real-time. It is also assumed that the transformation matrix between the camera and arm shoulder co-ordinate systems is also known, either by direct measurement or through a calibration procedure.

4.1 Compensation for Workpiece Demands

The ultimate aim of motion compensation is to convert a tip position demand, expressed in the workpiece co-ordinate system, to arm shoulder co-ordinates so that the inverse kinematics algorithms may be used to calculate the joint angles required to achieve this demand. Movements of the ROV will modify the demanded position in arm shoulder co-ordinates even if the required absolute position (in workpiece co-ordinates) is unchanged. ROV movements also alter camera view of the workpiece.

Consider a demanded tip position and orientation (in workpiece co-ordinates), T_w, defined as:

$$T_w = \begin{pmatrix} x_w & y_x & z_w & p_w \\ 0 & 0 & 0 & 1 \end{pmatrix}$$

This demand may be transformed to the arm shoulder co-ordinate system via a transformation, T_{aw}, ie.

$$T_a = T_{aw} T_w$$

This transformation will vary as the ROV moves and must be calculated from the effective motion of the targets measured by the Camera Alive stereo camera system. It can be calculated by considering a chain of transformations.

Considering a target with workpiece co-ordinates, x_w.

At any instant, the camera system will measure the co-ordinates of this point as x_c through a transformation, T_{cw}.

This transformation matrix can be calculated provided at least 4 target positions are known. This is done by forming target position matrices X_c and X_w where:

$$X_c = (x_{c1} \ x_{c2} \ \cdots \ x_{cn})$$

and similarly for X_w. The transformation matrix can then be calculated as:

$$T_{cw} = X_c \ X_w^{-1}$$

Note that is more than 4 target points are known, these can be included in X_c and X_w to form an over-determined set of equations which can be solved in a least-squares sense to give the best fitting 4x4 transformation matrix.

The relationship between the ROV and camera co-ordinate systems can also be described by a transformation matrix T_{rc} which is determined by the position of the camera in the ROV co-ordinate system. This can be combined with the workpiece transformation matrix to form a transformation relating the ROV co-ordinate system to the workpiece co-ordinate system, ie.

$$T_{rw} = T_{rc} \ T_{cw}$$

A similar transformation matrix can be found relating the arm shoulder co-ordinate system to the ROV co-ordinate system, T_{ra}.

These transformation matrices can be used to calculate motion compensated demands as follows. The demanded position and orientation has ROV co-ordinates:

$$T_r = T_{rc} \ T_{cw} \ T_w$$

This can also be calculated from the arm co-ordinate system, ie.

$$T_r = T_{ra} \ T_a$$

Consequently the full transformation between the demands in workpiece co-ordinates and in the arm co-ordinate system is given by:

$$T_a = (T_{ra}^{-1} \ T_{rc}) \ T_{cw} \ T_w$$

This position and orientation is the modified end effector demand which will be used directly to calculated arm joint angles via inverse kinematics. Note that T_{cw} will vary in real-time but T_{ra} and T_{rc} are fixed unless the camera is mounted on a pan/tilt unit or the arm is mounted on a primary deployment system.

The transformation $T_{ra}^{-1} \ T_{rc}$ can be measured directly or may be obtained through a calibration procedure detailed in section 5.

4.2   Compensation for Arm Demands

In many cases, the arm will be controlled in World co-ordinates, ie. relative to the base of the arm, rather than directly in workpiece co-ordinates. However the intention is still the same, ie. the tip should be controlled relative to the workpiece. The most obvious case is for direct manual control, or simply when the arm is 'frozen'. In these cases it is necessary to use the initial camera/workpiece transformation, T_{cw0}. to interpret arm commands and to apply motion compensation.

Assume that the arm is commanded to hold a position and orientation, T'_a. By this we mean that T'_a should be regarded as the arm position and orientation measured in the arm co-ordinate system at the instant that the workpiece modelling stereo image was acquired. We may then calculate the actual workpiece co-ordinates of this point from:

$$T_r = T_{ra} T_a'$$

$$T_r = T_{rc} T_{cw0} T_w$$

ie

$$T_{rc} T_{cw0} T_w = T_{ra} T'_a$$

and so

$$T_w = T_{cw0}^{-1} T_{rc}^{-1} T_{ra} T_a'$$

The motion compensated arm co-ordinates are then calculated as:

$$T_a = (T_{ra}^{-1} T_{rc}) T_{cw} (T_{cw0}^{-1}) T_{rc})^{-1} T'_a$$

which may also be written as

$$T_a = (T_{ra}^{-1} T_{rc}) T_{cw} T_{cw0}^{-1} (T_{ra}^{-1} T_{rc})^{-1} T'_a$$

Note that T_{cw} is the only time dependent term in the above equation.

5.   SELF-CALIBRATION

For motion compensation, the compound transformation, $T_{ra}^{-1} T_{rc}$, relating the camera to the arm co-ordinate systems is used extensively. In theory the two component transformations can be measured directly but these measurements are unlikely to be sufficiently accurate. For this reason, it is advantageous that a self-calibrating method to determine this transformation is used. This procedure would be used to set up the system before it is launched, and may be repeated at any time on request, eg. if the ROV has collided with the structure the camera or arm system may move slightly.

The self-calibration procedure requires a number of targets at known positions fixed to the arm at positions x_{ai}, $i=1,2,...,n$. The vision system is then used to measure these positions in camera co-ordinates, x_{ci}, $i=1,2,...,n$. These co-ordinates are related by:

$$x_{ai} = T_{ra}^{-1} \, T_{rc} \, x_{ci}$$

Forming matrices X_a and X_c, where

$$X_a = (x_{a1} \ x_{a2} \ ... \ x_{an})$$

and similarly for X_c, we can form the overdetermined set of equations:

$$X_a = (T_{ra}^{-1} \, T_{rc}) \, X_c$$

which may be solved in a least squares sense for $T_{ra}^{-1} \, T_{rc}$. Note that this is the constant matrix used by the motion compensation algorithms given earlier.

6.   IMPLEMENTATION

The algorithms described have been programmed and initial testing performed using simulated camera data. A camera simulation programme produces xyz data of 5 target positions at a 10Hz update rate and this is passed via a serial link to the windows based arm control software which in turn sends arm control demands to a 68000 based controller connected to a PUMA560 robot. Arm motion responses correspond well with the simulated target position variations at the application frequency and amplitude.

At the same time the PUMA based tracking performance has been extensively investigated using a vertically orientated x-y plotter executing typical ROV motions. The PUMA end effector carries a gimboled slender pointer with its free end connected to the x-y plotter so that any tracking errors can be measured from pan and tilt potentiometers. The rms error of approximately 1mm is mainly due to position sensor noise and backlash in the physical connection between the pointer and the plotter system. A carbon fibre sheet with back reflecting targets has been mounted on the plotter, and the next phase of the project will be to conduct the same set of tests using stereo vision feedback. These tests will give an accurate measure of the speed/accuracy limitations of machine vision use for this purpose - and will be reported at a subsequent meeting.

7.   CONCLUSIONS

The work described in this paper is the theoretical preparation for assessing the suitability of using stereo image processing for the motion compensation of ROVs. The software developed from this theory has been tested in real time using simulated vision data. The initial calibration of a PUMA560 test robot has been performed on a realistic range of motion frequencies and amplitudes. Quantitative results of current tests with the stereo vision will be reported soon.

Time-Varying Image Processing and
Moving Object Recognition, 3 – V. Cappellini (Ed.)
© 1994 Elsevier Science B.V. All rights reserved.

An Autonomous Mobile Robot Prototype for Navigation in Indoor Environments

G. Garibotto, M. Ilic and S. Masciangelo

Elsag Bailey
via Puccini 2, 16154 Genova, Italy
and
TELEROBOT
via Hermada 6, 16154 Genova, Italy

The paper describes a mobile robot prototype able to navigate autonomously along predetermined routes avoiding unexpected obstacles. An overview of the technical hardware and software components of the system is given. Ultrasonic sensors are used to implement a reflexive obstacle avoidance strategy, whilst Computer Vision provides methods to recover the global robot's position and attitudes. Experiments carried out in an office-like indoor environment are also reported.

1 INTRODUCTION

The interest in indoor mobile robots is increasing both in the research community, since it is a common test bed for a number of disciplines, ranging from Artificial Intelligence to Computer Vision, from Controls to multi-sensory Data Fusion, and in the industrial application sector. In fact the first products which can be really considered autonomous mobile robots are coming up to the market in a variety of applications.

Telerobot is a business-oriented unit formed by Elsag Bailey and Ansaldo, whose aim is to develop products in the field of Advanced Robotics, including the exploitation of autonomous mobile robots in the Service applicative sector. Telerobot, therefore, started off a plan for the integration on a mobile platform of previous research results obtained by an Elsag Bailey research team mainly in collaboration with the University of Genoa under the framework of the ESPRIT II P2502 VOILA project. The VOILA project was mainly oriented to investigate Computer Vision as a support to navigation [1], and a number of vision modules and demonstrations have been carried out.

An analysis of the potential applications of an autonomous vehicle able to rove in an indoor, office-like environment with the potential presence of people, led to identify security patrolling, material transportation and floor cleaning as the most promising market niches. A few products are currently under test in operative conditions at the client site: TRC Helpmate is employed for transportation tasks in hospitals, Cybermotion SR2 is used in patrolling service for intrusion detection and environmental monitoring of large buildings. Each of the mentioned products relies upon a mobile platform, a navigation system based on external sensors with obstacle avoidance capabilities and an operator interface.

At Telerobot, we decided to set up a prototype not explicitly addressed to a particular application, but sufficiently representative of a class of tasks, whose commonalty is autonomous navigation in an indoor environment, such as an office, an industrial complex or a hospital. Our goal is to make the robot able to navigate in our offices among predefined points, in order to accomplish simple missions such as "go to Mr. X's office". The external requirements include the ability to cope with people wandering around, a minimal impact on the office environment, a friendly user interface to make it usable by a visitor too.

At the same time, some more constraints, derived by the will to shrink the gap between basic research and product development, have been added:

- Self-consistent system, with all the computational hardware and the power supply on-board, without any umbilical link;

- accurate cost/performance analysis, that implies to reduce the capabilities in order to limit the hardware costs;

- few functions carried out with high reliability;

- a fine exterior appearance.

In the next section an overview of the prototype will be given, together with a comparison with existing similar objects and the main technological innovation will be highlighted. In the following, some technical aspects of the prototype will be described in greater detail: The software architecture, the local navigation module based on the ultrasound sensors and the vision based global position maintenance. Finally a few information on carried out experiments and performance data will be given, together with plans for future developments.

2   DESCRIPTION OF THE PROTOTYPE

A key feature of the prototype is its functional architecture, that is, how the software modules are connected together in order to generate a coherent behaviour. Even if many self-standing navigation and perception modules have been developed, only a few are actually integrated in the prototype in order to match system design and hardware constraints: The obstacle avoidance capability is obtained through an ultrasonic sensor belt, whilst the correction of dead reckoning errors is demanded to an independent Computer Vision system.

The logic architecture, inspired by the Brooks' *behavioural control* concepts [2], is decomposed in almost independent layers of competencies each one in charge of single well defined task, such as obstacle avoidance or global position maintenance. The obstacle avoidance strategy is *reflexive*, that is the trajectory is heuristically determined on the basis of sensor readings rather than accurately planned starting from a reconstructed local environmental map. The suboptimality of the obtained trajectory is largely compensated by the fast response time which allows to navigate safely at an acceptable speed also in presence of cluttered environments.

The two most mature products in the field are the TRC HelpMate© and the Cybermotion Navmaster©. Both the systems are characterised by a great deal of operative reliability and a high level of engineering, but the technical approaches are typical of the time frame in which the development was carried out, that is the second half of the eighties, when ultrasonic mapping was the leading technology. In fact both the systems rely on sophisticated map making capabilities and environmental feature extraction in order to match salient features (mainly walls, doorways or special sonic reflectors) to global maps and, therefore, correct dead reckoning position estimations.

HelpMate is somehow more modern since many heterogeneous sensory modules, each expert in a particular task, are combined together through data fusion to achieve a more robust overall behaviour. Specifically, obstacle detection is made safer by means of a structured light system in addition to ultrasonics, and monocular vision is employed to solve some particular tasks, such as precision docking to elevators.

Our system relies much more on Computer Vision, that is nowadays a more economic technology in terms of hardware power consumption, volume and cost. Moreover map making and planning capabilities are limited as far as possible in order to spare computational power and software development costs. Our point is that it is better to deal efficiently with as greater number of real situations as possible, rather than to solve problems in general but at an higher cost in terms of system complexity. Figure 1.a shows the integrated prototype and 1.b depicts the system hardware components with the data paths.

3 THE ROBOT FUNCTIONAL ARCHITECTURE

The functional architecture is a mix between a behavioural control approach, or *subsumption architecture*, and the more classical hierarchical architecture.

The *behavioural* concepts that can be found are the following:

- it is possible to identify at least two layers of behaviours, characterised by different tasks, based on different world models, able to process diverse sensory data and able to generate motion commands.

- There is neither a unique central world representation nor a single actuation control module, in charge of issuing motion commands.

- there is not a detailed path planning algorithm: A reactive, sensor-based strategy is preferred for the intrinsic real time performance and safety guarantee, notwithstanding the non optimality and the risk of unrecoverable deadlocks.

However the intrinsic parallelism of the subsumption architecture is not achievable with a single process single CPU system, therefore, a hierarchical dependency among modules with vertical transmission of commands and status information is preserved and used as a synchronisation mean.

Figure 2 illustrates how the software modules are organised:

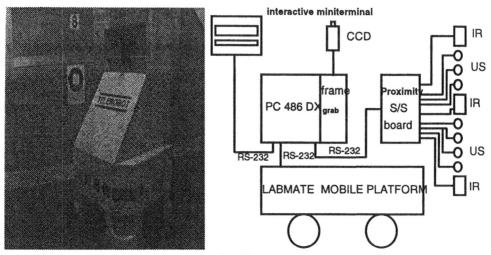

a. The mobile robot.   b. The system breakdown structure.

Figure 1: The exterior appearance of the robot and the main components: Low level motor control is assigned to a commercial platform, as well as the proximity sensors interface board. The central computer is an industrial PC 486 and a frame grabber on one of the five expansion slots. An interactive miniterminal provides an onboard user interface.

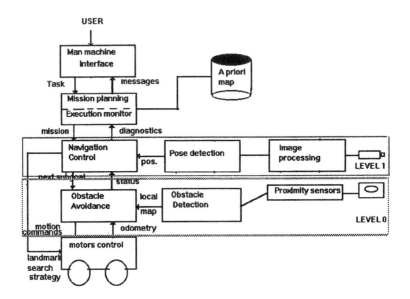

Figure 2: The software architecture of the navigation system.

- *layer 0* is responsible of the short term navigation and the avoidance of unexpected obstacles. This behaviour is achieved through the co-operation of a perceptual module *Obstacle Detection*, that acquires and elaborates proximity sensor read-outs, and a planning module that implements the obstacle avoidance strategy and generates motion commands to be issued to the low level motion controller.

- *layer 1* is in charge of the overall mission management and global position detection and tracing. At this level, Computer Vision algorithms are used to search for artificial or natural landmarks and to determine the robot pose with respect to a global map. The reconstructed position is used to correct odometry estimations. The navigation control module is able to issue motion commands while the robot is in the landmark search phase, therefore, it generates a *behaviour* exactly as the layer 0. While searching for an artificial landmark, the robot rotates on spot and acquires images until the expected pattern is actually spotted. Even if the subsumption paradigm requires for independent modules, there is a loose hierarchical dependency between the two layers, in fact the Navigation Control module controls the mission evolution by generating the sequence of navigation subgoals and verifying the mission status.

The other modules, that make up the system, are not considered behaviour layers, since they do not generate autonomously motion commands, but are auxiliary to the rest of the system. Actually the whole global map management and mission decomposition, that is the generation of a navigation subgoal list able to link the starting point to the given overall goal, and the user dialogue are demanded to special modules acting off-line with respect to the real navigation mission.

3.1 Reflexive Obstacle Avoidance based on Ultrasonic Sensors

Ultrasonic maps are typically used in navigation system described in the literature. A rather classical planning algorithm utilises the analogy with virtual potential field in order to model the repulsive force exerted by obstacles onto the robot, that is at the same time attracted towards a goal. That approach requires a good reconstruction of the shape and position of obstacles and a static, but possibly complete, knowledge of the environment around the robot location.

A more viable approach, described in [3], augments the potential field idea including a mechanism for integrating continuously fresh information coming from sensors. The world is modelled as a 2D histogram centred in the robot location, whose cells contain the probability of presence of an obstacle *(occupancy grid)*. This probability is updated at each positive sensory read-out, therefore, false readings are weighted accordingly. Finally, each cell exerts a repulsive force as in the classical potential fields algorithm.

Our approach differs quite radically since no map based planning is ever done, but motion is decided locally on the basis of heuristics and a simple 1-D histogram world model. The method is somehow *reflexive*, since accurate path planning is substituted by a frequent world sensing: The actual path is determined by changes in the sensory data readings, and therefore, by the environment itself. That implies sub optimal trajectories but high adaptability to a dynamically changing world.

158

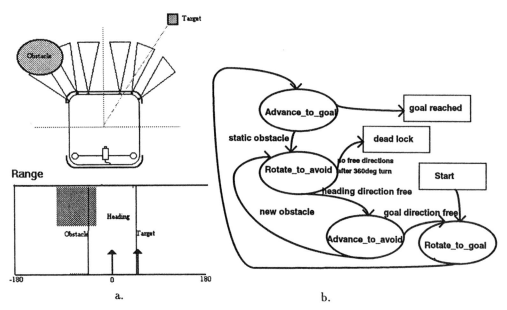

a.   b.

Figure 3: a. The polar histogram with superimposed the actual heading, with the projection of the robot's boundaries, and the goal direction. – b. The avoidance strategy expressed as a state machine.

The world representation, shown in Figure 3.a, is limited to a polar histogram, constructed around the robot current location, of the range readings coming from the ultrasonic sensors within a predetermined attention threshold. The histogram represents synthetically the occluded directions and their angular extension with respect to the vehicle centre of mass.

This representation is not maintained all along the navigation mission, but just when avoiding obstacles: In this way every sensory readings is stored in the histogram and taken into account even after robot motions, just updating this sort of local map centred in the robot location applying the necessary geometric transformations, according to the odometry estimate of motion. The short-term local memory approach allows to manoeuvre in constrained environments with only 8 sensors arranged in the front side of the robot. In fact, panoramic acquisition can be done combining robot's rotations on spot and the local memory mechanism. The use of just few sensors, not only allows a quick sensing loop but also, if interrogated in such a way that contiguous units are never polled in sequence, generates few false readings due to crosstalks and multiple echoes.

The avoidance strategy can be represented as the state machine shown in Figure 3. b, where state changes are determined by events shown up in the polar histogram and motion actions are associated to each state.

- **Advance_to_goal** is the state corresponding to obstacle free straight navigation segments. The change of state is determined either by the target reaching or by the detection of a static obstacle in the navigation path. An obstacle is considered static if at least one ultrasonic sensor gives a range value below a danger threshold, that persists for a given number of polling cycles. In this case the robot comes to a halt, otherwise the alarm is attributed to a spurious sensory read-out or to a moving obstacle crossing the robot's path and the robot just slows down without changing its orientation.

- **Rotate_to_avoid** is the beginning of the avoidance phase. The robot turns on spot and makes a panoramic acquisition of the environment until finds out a free corridor in a direction roughly tangential to the obstacle. This test is performed by verifying that a polar histogram sector centred around the current robot's heading direction and corresponding to the projection of the vehicle dimensions onto the polar map, does not intersect a histogram sector labelled as occluded.

- in the **Advance_to_avoid** state the robot moves following the avoiding direction until the goal is no more occluded by the obstacle, and that event corresponds to the end of the obstacle overcome phase. A new obstacle detection makes the robot stop and return to the previous state.

- **Rotate_to_goal** allows the robot to recover the original path towards the goal before moving in the Advance_to_goal state.

3.2 Computer Vision For Global Positioning

The navigation system needs a periodic position and orientation estimation coming from an external sensor in order to reset drifts of the odometry. This is provided through a Vision system able to detect, when required, the shape of artificial landmarks located in known positions and the robot's pose with respect to them. Each artificial landmark consists of a black annulus on a white background, visible in Figure 1.a; they are situated all along the predefined route of the robot in positions stored in the global map. The technique that allows to recover the 3D position and attitude of the camera from a single image is known as *model based perspective inversion* and is illustrated in other papers [4].

Moreover, in certain environments such as hallways, another Vision algorithm is able to recover the robot's absolute orientation from the analysis of the vanishing points generated on the image by parallel straight lines belonging to the 3D structure of the environment. Which algorithm is invoked is specified in the global map and therefore depends on the robot's current position.

The global map is not a complete 3D map but rather an abstract description based on *points of interest* located in strategic positions for the navigation. Neighbouring points of interest are connected by vectors, whose module represents the length of the path and the vector orientation is the robot's heading direction. A valid navigation plan is a sequence of points of interest, or subgoals, to be reached.

When the robot reckons to be on a point of interest, it can take either of the following actions:

- Stop and verify its position by observing the relative artificial landmark or vanishing point (this is specified in an attribute of the point of interest), reset odometric errors and turn towards the following subgoal;

- Execute a motion command such as an absolute turn.

4 EXPERIMENTS AND FUTURE WORKS

Experiments have concerned so far the determination of performance parameters relative to the single modules and the evaluation of the overall robot's behaviour in operative conditions. Specific tests have regarded the estimation of the accuracy and repeatability of the vision-based positioning technique, the evaluation of the ultrasonic and IR sensors performance, the analysis of the escape trajectories carried out in presence of various obstacles typologies.

It is quite difficult to quantify the performance of the obstacle avoidance strategy. The most correct evaluation measure is the time necessary to negotiate an obstacle, since encompasses both the computation time and the optimality of the escape manoeuvres. The prototype is currently able to traverse a 5.5 m path obstructed by an obstacle in about 40 to 50 s, depending on the obstacle dimensions. The normal speed is 38 cm/s and the avoidance speed is 12 cm/s, whereas the perception/control cycle runs at 17 Hz on the PC486.

Occasional deadlocks can arise in case of large concave obstacles, such as wall corners. The avoidance strategy is optimal for small obstructions arising along a normally free path, but is not able to cope with labyrinth-like situations that are unusual in normal working conditions, unless a major fault of the global navigation level has occurred.

Moreover, experiments demonstrated that it is possible to negotiate very constrained environments and to traverse doorways with exactly the same strategy, just changing dynamically software and system parameters such as the vehicle velocity and the security thresholds applied over the ultrasonic polar histogram.

Future works will follow two parallel threads: since the availability of a free ranging vehicle will allow to perform experiments in a real environment, a thorough test campaign will be carried out in order to validate the current navigation strategy; on the other hand, new sensing technologies, such as rotating laser scanners and solid state rate gyroscopes, and new perception algorithms, such as the function of following a line painted on the floor or recognising natural landmarks, will be experimented and added to the system.

Acknowledgements

The authors acknowledge the fundamental contribution of Prof. Sandini, and his research team in the initial phase of the project. Roberta Balbi contributed to the system development and integration.

References

[1] F. Ferrari, E. Grosso, G. Sandini, M. Magrassi, A Stereo Vision System for Real Time Obstacle Avoidancein Unknown Environment, IROS-90, Tokyo, July 1990.

[2] R. Brooks, A Robust Layered Control System for a Mobile Robot, IEEE Journal of Robotics and Automation, vol.RA-2, n.1, March 1986.

[3] J. Borenstein and Y. Koren, The Vector Field Histogram – Fast Obstacle Avoidance for Mobile Robots, IEEE Trans. oon Robotics and Automation, vol.7, No.3, June 1991.

[4] G.B. Garibotto and S. Masciangelo, 3D Computer Vision for Navigation/Control of Mobile Robots, in Machine Perception, AGARD Lecture Series 185, 1992.

F
IMAGE CODING
AND TRANSMISSION

Time-Varying Image Processing and
Moving Object Recognition, 3 – V. Cappellini (Ed.)
163

Tradeoffs on Motion Compensation/DCT Hybrid Coding:
an efficiency analysis

Carlos Pires
IST

Enrico Polese
CSELT

Mario Guglielmo
CSELT

Centro Studi Laboratori Telecomunicazioni SpA
Via Reiss Romoli 274 10148 Torino Italia
(Fax: +39 11 2286190 Tel:+39 11 2286198)

Abstract

After the development of an audio/video coding standard for 1 Mbit/s digital storage media applications, the Moving Picture Experts Group (MPEG) has started a second phase having as target the development of a coding standard for CCIR 601 images at bitrates up to 10 Mbit/s for all digital Satellite and Terrestrial TV broadcasting, among other applications.

In this paper a general overview of MPEG's approach regarding video coding at bitrates up to 20 Mbit/s is made including some extensions towards very low loss coding in a multi-layered coding scheme. Simulation results will focus on essential parts of the coder, comparing results with already established coding schemes like ISO/Joint Photographic Experts Group (ISO/JPEG) for still colour pictures and the CCITT H.261 scheme [8].

The tested algorithm is an hybrid Motion Compensation/DCT, maintaining a similar structure to MPEG1 [2]. The coding process is based on a succession of coding decisions to minimize the distortion for the available bitrate. The coder can adaptively select from a number of operation modes at picture level and, within a picture, at macroblock level. Macroblocks are processed sequentially from left to right starting from picture up left corner [3].

In these schemes motion detection is made using the best matching technique and assumes a relevant role as it exploits the temporal redundancy present in motion video sequences. Some important improvements were recently introduced in the "coder part" regarding motion vector search. New approaches, more adapted to the field interlaced format of CCIR 601 input sequences, were proposed and included in MPEG Test Model (FAMC, Dual Field MC, etc.) [3].

1. The MC process and Entropy coding

As the available bitrate increases, it is expected an improvement on the final picture quality meaning that the DCT high frequency components cannot be coarsely quantized like it is systematically done in H.261 and MPEG1 algorithms. This fact introduces a clear tradeoff

164

between the MC process and the traditional run-length/Huffman coding of DCT coefficients. The DCT may no longer concentrate the differential signal's energy in the low frequency coefficients since the MC process produce much less correlated data than the original pixels [4, 5]. Moreover, in these hybrid schemes, input images are divided in small bi-dimensional blocks introducing *artificial edges* (blocking noise) which result in high frequency components to be coded.

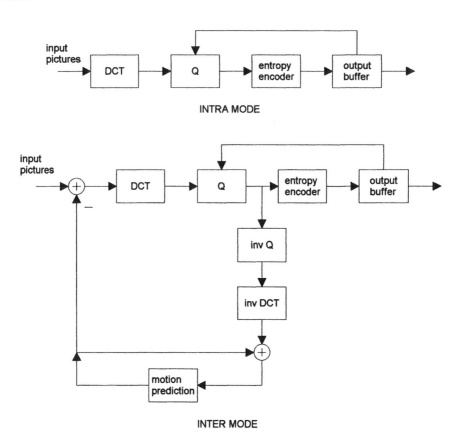

Figure 1 - Generic hybrid MC/DCT coder block diagram

In figure 2a and 2b results of typical DCT spectral shape for intra and differential MC blocks obtained averaging on tests sequences at bitrates of 4 and 9 Mbit/s are presented. It can be seen that for inter *error* blocks the statement of concentration of block's energy in the up-left corner DCT coefficients is no longer valid. The different histogram shape is considered for the intra and inter cases and new VLC tables are tested. The effectiveness of the entropy coding block adopted in MPEG is analyzed and compared with JPEG's approach. Universal

VLC entropy coding, referred in [7], is tested as an alternative to tune to the different intra/inter signal statistics.

78.1	2.22	0.71	0.33	0.20	0.13	0.06	0.02
5.10	0.76	0.25	0.15	0.09	0.05	0.03	0.01
2.63	0.61	0.25	0.11	0.08	0.07	0.04	0.01
1.52	0.43	0.18	0.09	0.06	0.04	0.03	0.00
0.94	0.32	0.16	0.07	0.14	0.14	0.04	0.00
0.77	0.23	0.13	0.07	0.15	0.19	0.05	0.00
0.66	0.18	0.10	0.05	0.06	0.07	0.03	0.01
0.61	0.14	0.08	0.05	0.04	0.02	0.01	0.01

(a)

7.86	4.28	2.58	1.66	1.20	0.95	0.69	0.33
5.65	3.01	1.91	1.27	1.11	1.47	0.57	0.28
4.45	2.56	1.67	1.10	0.82	0.67	0.52	0.25
4.30	2.43	1.56	1.01	0.81	0.65	0.50	0.22
4.10	2.33	1.47	0.94	0.70	0.60	0.45	0.20
4.10	2.30	1.44	0.90	0.79	0.75	0.49	0.20
3.94	2.23	1.41	0.88	0.92	0.90	0.50	0.18
3.67	2.18	1.35	0.83	0.70	0.60	0.40	0.18

(b)

Figure 2 - Average power distribution in a DCT block in (%):
(a) Intra block; (b) Inter block

2. The MC process and Quantization

Since the transform domain representation of intra and inter signals is quite different and MPEG uses as main bitrate control mechanism the quantization of DCT coefficients this step is quite relevant to the final picture quality. MPEG schemes use weighting matrices and a quantization step (mquant) updated with slice or macroblock granularity and that strictly depends on the fullness of the output buffer and on a measure of the local image energy. Comparison test results are presented based on MPEG core experiments (alternative block scanning patterns) and a new proposal based on the local macroblock activity based on the DC and first two DCT coefficients.

3. Filtering the Motion Vector Field

A critical point of the MC process is the production of coded random vectors which do not correspond to a real movement in the scene or do not represent overall gain regarding intra coding. An alternative bi-dimensional low pass filtering of the motion vector field is done using the Walsh Hadamard transform. This experiment is extended to the concept of multi-layer coding were a first, basic layer contain only a raw information of the vector field.

4. Layered coding

Multi-layer coding is a key concept of video coding considering either constant or variable bitrate coding. It assumes particular interest in transmission over noisy channels (like in radio broadcasting digital TV or mobile video phone applications) since it may code different resolution (spatial or frequency) layers with *strategic* error protection producing a guaranteed minimum picture quality output. From the simple two layer coding, dividing the available bitrate at the same spatio-temporal resolution (error requantization) to the more complex frequency scalability some tests were made on the efficiency and final quality performance as well as on graceful degradation on noisy environments. Layered coding is also used to try pseudo-lossless coding approach where a maximum error for pixel reconstruction is previously set and a *soft* quantization is implemented.

5. Lossless and pseudo-lossless coding

The area of lossless and very low loss coding (pseudo lossless) was found to be an important research topic by several of the national bodies in recent MPEG meetings. The applications identified so far are: archiving and communications of original picture sequences for studio applications which may require several code/decode iterations (with lossy algorithms this may become critical) and also medical, feature extraction and computer graphics and animation We partially tested an algorithm introducing only DPCM techniques within the general structure of MPEG syntax, using motion compensation. Among the many problems to be resolved the fact that lossless coding do not use any rate control mechanism (work in "open loop") is one of the most important. The first results report compression factors in the range of 2 to 3 (80 100 Mbit/s) for CCIR 601 4:2:2 pictures.

Other ideas under simulation try to introduce a new frame type, the L frame, in which some of the information is lossless coded. This hybrid scheme would keep compatibility with the MPEG rate control strategy.

References

[1] *A. Netravali, B. Haskell , "Digital Pictures: representation and compression", N.Y. (US) Plenum Press 1988*

[2] *CD Editorial Group, ISO/IEC JTC1/SC29/WG11 - MPEG 1 Committee Draft, March 1992*

[3] *MPEG TEST MODEL 2, Angra dos Reis - Brasil, August 1992*

[4] *C. Pires, S. Battista, "DCT/Motion Compensation for Broadcasting Applications", ITEC'92 Sapporo, July. 1992*

[5] *C. Pires, M. Quaglia, "Spectral Characteristics of Motion Compensated Image Signals", URSI'92 Malaga, Set. 1992*

[6] *JPEG Technical Specification, Rev 5 ISO/IEC JTC1/SC2/WG8 - JPEG 8-R5, Jan. 1990*

[7] *T. Fautier et al, VADIS contribution to MPEG: UVLD implementation, June 1992*

[8] *CCITT Recommendation H.261 "Video Codec for Audiovisual Services at px64 kbit/s", Geneva 1990*

Time-Varying Image Processing and
Moving Object Recognition, 3 – V. Cappellini (Ed.)
© 1994 Elsevier Science B.V. All rights reserved.

Neural Networks for Compression of Image Sequences

Stefano Marsi and Giovanni L. Sicuranza

DEEI, University of Trieste
via A. Valerio, 10 – 34127 Trieste – Italy *

The problem of the image compression via Neural Networks is considered. A parallel architecture is presented in which different parts of the image sequence are processed by different neural networks according to their complexity and elementary motion.

An algorithm for motion detection, applied on small blocks of data and used for classifying different kinds of motion is presented. Simulation results are also presented and discussed.

1. INTRODUCTION

The compression of images has recently become very attractive because of the wide range of applications. In particular for the compression of image sequences, which requires the computation of a lot of data in a relative short time, many efficient algorithms and high-throughtput architectures have been presented.

A large part of these methods is based on a two-dimensional transform able to reduce the spatial correlation between the data, while in the temporal direction the compression is based on prediction and interpolation methods[1][2].

Among the high-throughtput architectures, linear Neural Networks have recently been used with a good success[3][4][5], but still as a two-dimensional transform method. Usually a Two-Layer-Perceptron is used as a compander (compressor-expander) of the images. It is built with the same number of input and output nodes but with a reduced number of nodes in the hidden layer. During the training phase different images are fed into the input layer and the net is recursively adapted to reproduce the same images at the output nodes with the minimum error. After this phase, if the training set is sufficiently various, the net is able to operate the compression of the data: when the original image is fed into the input layer, a compressed version of it is available at the output of the hidden nodes and this compressed image can be successively expanded to the original dimensions through the output layer. In such a way the method works as a classical transform - filter - antitransform method[5], but with two important advantages:

1. the coefficients in the transform space do not need to be chosen with a particular algorithm because the dimension of this space is, by definition, smaller than the original space dimension.

---

*This work was supported by the MURST projects.

2. the transform can be made adaptable to the statistics of the input data.

Of course, to keep the size of the net into reasonable limits the images are partitioned into small blocks, or patterns, which are used as input data of the neural network.

This method is quite successful for the compression of still images; in the present work we consider the possibility of an extension in order to operate directly on sequences of images with a 3-D reduction of redundancy.

The principal idea is to build a few nets, which use as input and, of course, as output vectors, blocks of data read from all the three dimensions of the sequence (horizontal,vertical and temporal).

In such a way we are able to reduce the redundancy in the coded data taking advantage from the correlation both in the temporal and in the spatial dimensions.

Moreover, using a linear neural network instead of a classical 3-D transform, we can avoid the algorithm for the selection of the most important coefficients in the 3-D transform space.

The most evident difference between a 2-D net and the new networks, here proposed, is that the last ones are very sensible to the kind of the trained patters we use and in particular the nets show a great dependency from the type of motion present in the training-set. This dependency is very evident in the sense that if the patterns used as training set show a particular kind of motion, then, during the test phase, the net impresses the same motion to every compressed pattern. To take into account this characteristic, a structure based on a few NN's, each one specialized in the compression of a particular kind of patterns, has been studied (Sect. 2).

Of course to do this subdivison of the patterns a motion detection algorithm, able to distinguish what kind of motion is present in a block, has been studied and will be presented in Sect. 3.

At the end, in Sect. 4, experimental results will show the advantages and the drawbacks of the method.

2. PROPOSED STRUCTURE

In the proposed structure [Fig.1] we start subdividing the original sequence we want to code into a few groups of successive frames and then every group is cut into a given number of not-overlapped 3-D blocks that we shall call "macro-blocks". Now in every macro-block we can find pixels that may show a correlation both in the spatial and in the temporal dimensions.

Within the hypothesis that the motion is constant in the macro-blocks we subdivide them into several classes different for both the intensity and the direction of the motion. Now for each class we can choose a particular kind of nets that would be the best for coding the data of the macro-block.

These various NN's have been built with different size and different number of hidden nodes and each one was trained, of course, with a particular kind of patterns belonging to that particular class.

It is worth noting that if the motion is very slow we do not distinguish about its direction and we can use a very "deep" net in the temporal direction because there is

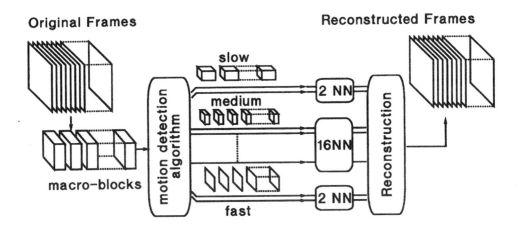

Figure 1: Proposed structure

a great temporal correlation of the data, while if the motion is very fast or irregular we must use a "thin" net because we cannot take advantage from the temporal correlation.

Moreover, for the medium motion (1 or 2 pixels / frame) we need at least eight nets for all the cardinal direction (2 horizontal, 2 vertical and 4 diagonal), but, for semplicity, we build only two of these i.e. one horizontal and one diagonal, while the others were obtained with elementary rotation of 90 degrees.

In particular we formed four different training sets in which all the patterns show the same elementary motion (i.e. horizontal, diagonal, slow-motion and fast/irregular-motion) and these ones were used to build four different kinds of NN's. For each kind we built several NN's with different number of hidden nodes and thus with a different compression degree, using a semplified version of the back-propagation algorithm[5].

Due to the different input dimensions of the various NN's we can notice that, before to compress the data, the macro-blocks need to be cut into smaller patterns having appropriate size according to the input dimensions of net chosen as coder-decoder.

In particular we choose a priori the dimensions shown in Table 1.

It is worth noting that more difficult is the motion in the macro-block to code, less deep is the net to use. At the same time if the net is thin and thus it is not possible to reduce the temporal correlation, then we extend its spatial dimensions to take advantage at least from the spatial correlation. The use of nets with spatial dimensions less then 4 pixels is avoided since in this case we are not able to take advantage from the spatial correlation and, in addition, the motion cannot be discovered inside this too little window. On the other hand, nets with spatial dimensions greater than 8 pixels should be avoided in order to limit the blocking effect.

A further discrimination can be made looking in the spatial domain: it is obvious that patterns extracted from smooth parts of the images are simpler to code than blocks which are heavily detailed[3], so we process simple patterns with nets having few hidden nodes,

kind of motion	spatial complexity	input dimension	hidden nodes
slow	low	4x4x8	2
	high	4x4x8	6
medium	low	4x4x4	4
	high	4x4x4	8
fast or	low	8x8x1	12
irregular	high	8x8x1	16

Table 1: Dimensions of the different nets used

while nets with a lower compression degree are required to code more complicated blocks (see also Reference [3][7] for a similar approach). It has to be noted that to simplify the architecture we use only two NN's for each class, one for simple patterns and one for heavily detailed patterns.

In such a way we are able to code the data with an adaptive transform which for every macro-block can find a good equilibrium between temporal and spatial compression thanks to the different input size and a good compromise between compression and reconstruction error thanks to the possible choice of nets with different number of hidden nodes.

To evaluate the spatial complexity of the pattern we look at the following parameters:

$$\sigma_k^2 = \sum_{i=0}^{d_x-1} \sum_{j=0}^{d_y-1} (x_{ijk} - \overline{x_k})^2 \tag{1}$$

$$\sigma_{max}^2 = max(\sigma_k^2) \quad \text{for any} \quad 0 < k \leq d_t - 1 \tag{2}$$

where: d_x, d_y, d_t are the spatial and respectively the temporal dimensions of the pattern, x_{ijk} is the value of the pixel in position i, j, k, and $\overline{x_k}$ is the mean over all the pixels of the pattern that belongs to frame k.

We use σ_{max}^2 as a measure of the spatial complexity: if the value calculated for a certain pattern is below a predefined threshold, the pattern is considered smooth and thus we code it with a net with a high compression degree; if this condition is not verified, we define the pattern as complex and we use a net able to preserve the details but with a lower compression degree.

3. THE MOTION DETECTION ALGORITHM

We propose to use, for the motion detection, a method based on two subsequent measures: the first one is used to indagate coarsely over the intensity and the perceptual sensibility, and the second one for a more improved motion estimation able also to find its

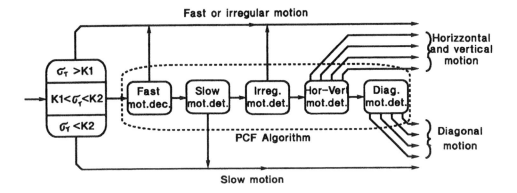

Figure 2: Motion Detection Algorithm

direction. To have a first discrimination of the temporal complexity of the macro-blocks we can calculate the variance:

$$\sigma_T^2 = \sum_{i=0}^{dx-1} \sum_{j=0}^{dy-1} \sum_{k=0}^{dt-2} \left(x_{ijk} - x_{ijk+1} \right)^2 \tag{3}$$

If the value of σ_T^2 is small we can say that or the motion is absent or it is hardly perceptible and in both cases we can code the macro-block using the nets for slow motion. On the contrary, if σ_T^2 is large, the motion is easily perceptible and thus, to have a good visual impression, we propose to compress it not so largely using the nets specialized on fast motion.

In this sense we have extabilished two tresholds as shown in Fig.2: if σ_T^2 is below the lowest one then we use the *'slow nets'*, while if it is above the highest one we use the *'fast nets'*; otherwise, we proceed with a better motion detection.

For the choice of a method able to distinguish the motion in intensity and direction we must take into account two important peculiarities: we want to classify the macro-blocks, so we need to detect the motion using only the data inside the macro-block, that is a quite small window to observe it; again, we need to compute the motion for the entire deepness of the macro-block.

We found that a two dimensional phase correlation function (2D-PCF) has a sufficient ability to recognize the motion inside quite small blocks; thus, we can use this algorithm to find the coordinates m_x and m_y of the motion estimation between two frames in the same block.[6]

The next step is to compute the motion inside the entire macro-block to classify it. Using the PCF we compute m_x and m_y between all the couples of consecutive frames in the macro-block. Let $\overline{m_x}$ and $\overline{m_y}$ be the means and M_x and M_y the maxima of m_x and m_y. Again we call Z the number of times that (m_x, m_y) was computed as $(0, 0)$. We then use the following algorithm to classify the macro-blocks:

1. if $M_x^2 + M_y^2 > 8$ then: or the motion is very fast (more than 2 pixels/frame) or something fast passed inside the macro-block during those frames; in both cases we need to use the "fast-nets"

2. else if $Z \geq 5$ then: there are not appreciable movements; we use "slow-nets".

3. else if $\overline{m_x}^2 + \overline{m_y}^2 < 0.25$ then: the motion is irregular; we need to use again the "fast-nets"

4. else there was a medium motion in a particular direction. Using the values of $\overline{m_x}$ and $\overline{m_y}$ we subdivide it into 8 classes with different angles for the motion direction. For each class we need to use a particular kind of nets.

The PCF algorithm has, generally, good performances in the motion recognition, but sometimes it can fail if the pattern is too much homogeneous so that just a little noise can compromise the result. These patterns, however, are included in the group of patterns in which the motion is hardly perceptible.

4. SIMULATION RESULTS

In the proposed method we use macro-blocks with dimensions of 16x16x8 pixels as a good compromise between accuracy in motion detection and subdivision of the sequence in parts of constant motion. In fact, the use of smaller macro-blocks can increase the subdivision of the sequence and thus the hypothesis that the motion will be constant in every macro-block becomes more realistic and the blocking effect is reduced, but on the other hand the motion detection algorithm becomes less accurate. In fact, looking at the motion through a smaller window the number of macro-blocks in which the motion detection fails increases. Moreover if the macro-blocks are too small we are not able to take advantage from the correlation of the data inside the macro-block.

Image degradation caused by compression is evaluated by the peak S/N ratio, defined as

$$PSNR = 10 \; log \; \frac{255^2}{e^2} \tag{4}$$

where e^2 is the mean square error between the original and the reconstructed image.

We have used as testing sequence the first 104 frames of the well known sequence "*Salesman*".

To have a good casuistry we have trained a few kinds of nets using different groups of patterns. In particular we have built 3 different groups of training patterns:

- one group using only patterns extracted from the first eight frames of the testing sequence;

- a second group using patterns extracted from the frames 8-16 and 56-64 that are particularly various (rich of motion) frames;

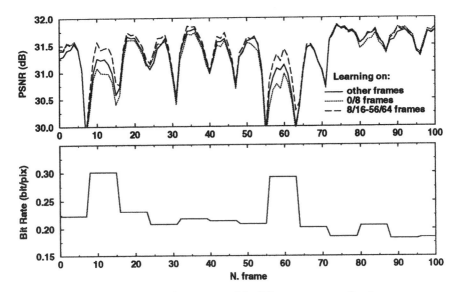

Figure 3: Performances with different groups of nets

- and a third group using patterns from other sequences completly different from the testing sequence.

As shown in Fig 3, we can note that all the groups of nets show good performances during the test phase. Of course the net performances are a little better when the test is done on the same frames that have been used as training set, but sometimes, as in the first case, we pay this improvement with worse quality on other parts of the sequence. This degradation is due to the fact that, probably, the training set has not been sufficiently various. This effect does not happen in the second case in which the frames used as training set have been selected to be various and significant for the sequence.

It is worth noting that the waves in the PSNR are probably due to the fact that the nets for the coding of the slow motion are not subdivided into different directions, so they code a "medium motion" for all the patterns. In such a way they code well the intermediate frames, but they make some mistakes at the beginning and at the end of the patterns. The performances of the system can be improved and this error removed using a finer subdivison of the nets both for motion and the spatial complexity.

In the third case we can see that good results, althought a little worse than in the second case, can be obtained also using nets adapted on patterns completely different from the testing sequence but sufficiently various.

To be able to compute the Bit-Rate in the channel transission we also studied the effect of the quantization of the data.

Various tests done on the sequence have shown that four–five bits are sufficient for a good reconstruction of the data. Althought four bits can be sufficient to reach a good PSNR value, in the decompressed frames a little noise effect is visibile. To reduce it and to have a better visual quality we choose to quantize the data with five bits.

Of course, to take into account also the side information we can notice that for every macro-block we need to say to the receiver what net has been choosen among the twenty possible nets. To do this we have to spend, without entropy coding, $log_2(20) = 4.32$ bits per macro-block that is a very little extra side information of $2.11 \cdot 10^{-3}$ bit per pixel.

The result obtained using these values is shown in Fig.3. Of course triggering in another way the various thresholds or using different nets with a different number of hidden nodes allows us to reach a different trade-off between compression rate (bit-rate) and quality (PSNR). Moreover the performances of this system can be improved using a finer subdivision of the nets for both motion detection and spatial complexity.

As final comments, two are the advantages of this method with respect to the methods based on prediction and interpolation of frames:
– a full refresh of the output can be made every few frames, according to the temporal dimension of the macro-blocks;
– the range of the variation of the PSNR is quite restricted (< 1.5 dB) with a consequent better visual quality.

References

[1] Atul Puri and R. Aravind "Motion - compensed Video Coding with adaptive Perceptual Quantization", *IEEE Trans. on Circuits and Systems for video technology* vol. 1, pp351 -361 , N. 4, December 1991.

[2] Moresco, Lavagetto and Cocurullo "Motion Adaptive Vector Quantization for Video Coding", Proc. EUSIPCO-92, Brussels, Belgium, August 24-27, 1992, vol 3, pp. 1357-1360, Elsevier Science Publisher B.V. 1992.

[3] S. Carrato and S. Marsi, "Parallel structure based on neural networks for image compression", *Electronics Letters*, vol. 28, pp. 1152-1153, June 1992.

[4] Mougeot M., Anzecott R. ,Angeniol B., "A study of image compression with backpropagation", in Heuralt J. (Eds.): NATO ASI Series, vol. F 68, Neurocomputing, (Springer-Verlag, Berlin, Heildelber 1990).

[5] S.Carrato, A.Premoli, G.L.Sicuranza,"Linear and nonlinear neural networks for image compression", *Proc. Int. Conf. on Digital Signal Processing*, Florence, Italy, Sept. 4-6, 1991.

[6] W. Meier and H.D. von Stein "Infrared Image Enhancement with Nonlinear Spatio–Temporal Filtering" Proc of EUSIPCO-92, Brussels, Belgium, August 24-27, 1992, vol 3, pp. 1397-1400, Elsevier Science Publisher B.V. 1992.

[7] Torres-Urgell, L. and Kirlin, R.L. "Adaptive image compression using Karhunen–Loeve transform", *Signal Processing*, December 1990, 21, pp. 303-313.

Time-Varying Image Processing and
Moving Object Recognition, 3 – V. Cappellini (Ed.)

Causal Estimation of the Displacement Field for Implicit Motion Compensation

F. Lavagetto and P. Traverso

DIST - Department of Communication, Computer and Systems Science
University of Genova, Via Opera Pia 11-a, 16145, Genova, Italy

Displacement field evaluation represents the basic component of any system for motion compensated (MC) video coding. In these systems frame prediction is performed using causal and non causal approaches, in real-time [1] and interactive applications [2], respectively. The encoded video signal typically consists of interleaved data streams encoding the estimated motion field jointly with the corresponding prediction error. Block based techniques are usually employed both for motion compensation and for error encoding, thus simplifying the co-decoding computational complexity. The basic underlying hypothesis usually assumed on the motion field is its smoothness, i.e. that nearby blocks are typically affected by similar motion except in correspondence of moving edges: in other words stating that feasible motion fields exhibit low entropy and can consequently be effectively predicted. In this paper we present a new approach to motion compensated video coding, based on block classification, where two different modalities are employed for coding the motion vector of each block. Specifically, blocks belonging to smoothly moving areas are motion compensated without requiring the transmission of their motion vectors, as they can be estimated implicitely at the receiver. Thanks to this mechanism, the rate-distortion performances of conventional H.261 coders have been improved, showing promising possibility of application for high-compression video coding. The results obtained with this technique are discussed and a performance comparison is presented with respect to conventional block-based techniques within very low bitrate applications.

1. INTRODUCTION

The block-based scheme employed by H.261 video coders has proven to be apparently unsatisfactory for services characterized by rates below 64 kbit/sec., when very high data compression is required. In fact, block-based techniques have inherent limits since they do not rely on any model of the scene structure. Therefore, present research is mainly focusing on unconventional object-oriented approaches capable to provide far higher compression. The improved rate-distortion performances are however paid, to a variable extent, in terms of increased co/decoding complexity and of more severe constraints on the class of applications. Several proposals, based on different scene models, have been

This work has been partially funded by the Italian National Research Council (Special Project on Telecommunication).

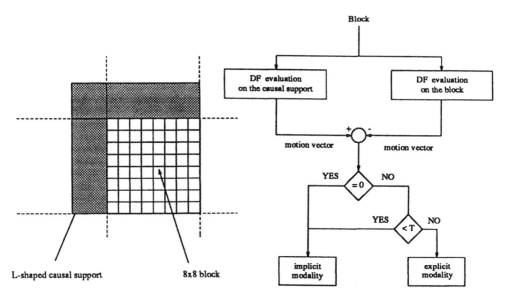

Figure 1: (Left) Causal support employed for the block motion estimation. (Right) Flow-chart of the algorithm for the estimation of the motion field.

recently presented showing very interesting results as far as the achievable compression is concerned, but also revealing some weakness when model failures occurr [3].

Indeed, block-based techniques are far from being obsolete. On the contrary, the short-term scenario is expected to be massively characterized by an increasing population of block-based video coders, though progressively adapted to new incoming applications and transport services. As an example, advanced research in image and video coding is presently issuing a variety of proposals for the adaptation of conventional block-based schemes to packet-switched networks [4-5].

Besides projects with long-term ambitious goals like that of pure object-oriented video coding, significant efforts have been also produced toward minor round-the-corner objectives like those of improved H.261-compatible schemes. Within this latter activity, very interesting goals have been reached, showing that some margin is still available to refine the algorithms and upgrade preexisting equipments [6]. Intuitively, greater improvements can be attained if the constraint on H.261-compatibility is relaxed, meaning that suitable modifications of both the algorithms and code syntax can be introduced while preserving the block-based DCT transform. Our proposal addresses exactly this area and show how it is possible to maintain the general block-based structure of the H.261 recommendation (and, consequently, of the involved co-decoding hardware), while achieving improved performances for very low bitrate applications.

The proposed algorithm, explained in detail in Section 2, exploits the spatial redundancy of video motion field by checking, in correspondence of each block of the image, if its displacement coincides or not with that of a surrounding neighborhood of reconstructed pixels. If the same motion affects both the block and its neighborhood, there is no need at all to transmit the block displacement since it can be implicitely estimated at the receiver. On the contrary, if the block lays on a moving edge, its displacement may significantly differ from that of the surrounding regions thus originating different motion estimates: in this case the block motion vector must be explicitely transmitted.

As described in the next Section, the H.261 syntax has been modified to classify

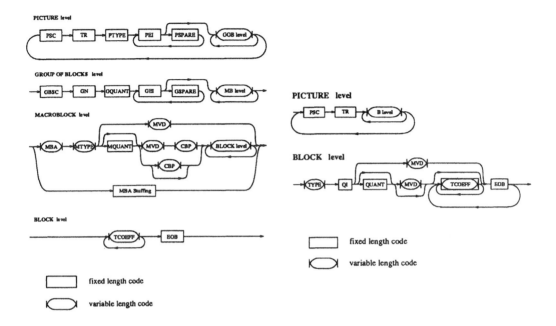

Figure 2: (Left) H.261 4-level hierarchy. (Right) Modified 2-level hierarchy.

the block motion as being predictable or unpredictable and to invoke explicit or implicit motion compensation, respectively. The achieved experimental results, reported and discussed in the last Section, concern the application of the proposed algorithm within a scheme for 8 kbit/sec. video coding.

2. IMPLICIT MOTION COMPENSATION

In case of implicit motion compensation the block displacement is predicted on a strictly causal support and its estimate can be performed at the decoder without requiring any side information, i.e. without requiring the transmission of the motion field. In case the estimated block motion should not correspond to the actual block displacement or, alternatively, should the prediction error exceed a given threshold, the actual motion vector is sent explicitly. As a consequence of this policy, the algorithm operates block by block according to two different modes, implicit or explicit, depending on the local smoothness of the displacement field. Specifically:

- **implicit mode:** only the prediction error is encoded and transmitted, while the block motion vector is computed at the decoder;

- **explicit mode:** both the actual motion vector of the block and the prediction error are encoded and transmitted.

The algorithm has been applied within a H.261-like system encoding blocks of 8x8 pixels and applying motion compensation to 16x16 pixel macroblocks. Motion is estimated on a "L-shaped" support with overlap with respect to the three nearby blocks, as shown on the left-hand side of Figure 1. The algorithm operates, as sketched on the right-hand side of the same Figure 1, through the following steps:

1. a block-matching technique is used to estimate the motion vector of the causal support with respect to a past reference frame;

2. the same technique is applied to estimate the actual motion vector of the block;

3. if the estimated motion vectors coincide, the implicit operating mode is employed;

4. in case the estimated motion vectors do not coincide:

   - if the block distortions corresponding to the two motion vectors differ more than a given threshold T, then the explicit operating mode is employed;

   - else, the implicit operating mode is employed.

Prediction			MQUANT	MVD	CBP	TCOEFF	VLC
Intra						x	0001
Intra			x			x	0000 001
	Inter				x	x	1
	Inter		x		x	x	0000 1
	Inter+MC			x			0000 0000 1
	Inter+MC			x	x	x	0000 0001
	Inter+MC		x	x	x	x	0000 0000 01
	Inter+MC+FIL			x			001
	Inter+MC+FIL			x	x	x	01
	Inter+MC+FIL		x	x	x	x	0000 01

Table 1: H.261 MTYPE field.

TYPE	CODE
Intraframe	0001
Interframe	001
Explicit estimation	01
Implicit estimation	1

Table 2: TYPE field in the modified syntax.

With respect to conventional MC schemes, the encoder and decoder complexity is increased. In fact, at the encoder, two different motion fields are always computed instead of one (the first with respect to the causal support and the second one with respect to the block), while at the decoder a further motion vector must be evaluated in case of implicit operating mode.

The above described algorithm has been inserted in a H.261 system, where some modifications in the coding/decoding procedures as well as in the code syntax have been introduced. Besides the management of the newly added information, necessary to switch between the implicit/explicit modality for motion estimation, also some syntax simplifications have been introduced. The H.261 4-level hierarchy has been reduced to a 2-level structure, as shown in Figure 2, where the MTYPE field has been replaced by the corresponding TYPE field shown in Tables 1 and 2.

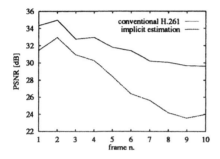 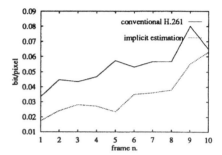

Figure 3: Performance comparison between the conventional H.261 coder and the modified scheme with implicit motion estimation. Experiments on the sequence "Miss America" at full frame frequency (30 frames/sec).

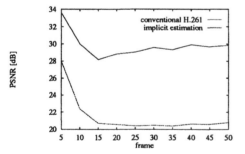 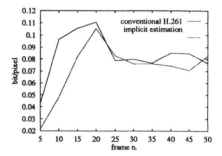

Figure 4: Performance comparison between the conventional H.261 coder and the modified scheme with implicit motion estimation. Experiments on the sequence "Miss America" with 1:5 temporal subsampling.

3. EXPERIMENTAL RESULTS

The proposed algorithm for motion estimation, together with the simplified code syntax, has been integrated in a non-standard scheme providing satisfactory results as far as compression is concerned.

Let us examine the improvements introduced by the new algorithm by discussing the results achieved on the test sequence "Miss America", QCIF format at full frame frequency (30 frames/sec.). In Figure 3 the PSNR and bitrate curves provided by a conventional H.261 coder are compared to those obtained by using the new scheme in implicit modality. In this experiment the motion field was always estimated with reference to the causal support, meaning that only the prediction error was encoded. A severe quality loss of more than 3dB is paid in exchange of significant bitrate reduction which is however assured only in case of smooth motion. In fact, as shown in Figure 4, in case of 1:5 temporal subsampling the bitrate saving is not guaranteed at all.

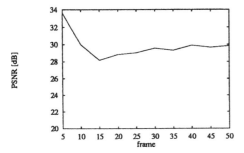 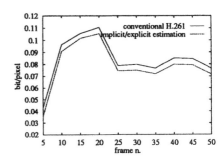

Figure 5: Performance comparison between the conventional H.261 coder and the novel scheme with implicit/explicit motion estimation. Experiments on the sequence "Miss America" at full frame frequency (30 frames/sec).

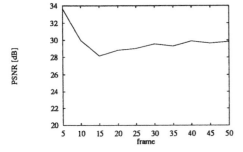 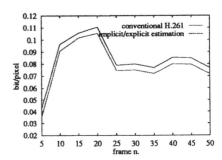

Figure 6: Performance comparison between the conventional H.261 coder and the novel scheme with implicit/explicit motion estimation. Experiments on the sequence "Miss America" with 1:5 temporal subsampling.

If the algorithm for motion estimation can switch between the implict and explicit modality, far better results are achieved as shown in Figures 5 and 6. In this experiment the threshold T has been set to zero, meaning that the implicit modality is applied only in case the block motion vector coincides with that estimated on the causal support. The percentage of occurrence for the explicit motion estimation is plotted in Figure 7, in case of full frame frequency and of 1:5 temporal subsampling. In Table 3 the rate-distortion performances of a conventional H.261 coder are compared to those achieved through the use of the new algorithm. Fixing the reconstruction quality, an appreciable reduction of the bitrate due to motion vector encoding is obtained, in contrast to a slight increment of the bitrate for error encoding. The transmission overhead is reduced, as far as the motion vector component is concerned, because many vectors are not transmitted at all but implicitly estimated at the receiver. On the other hand, the higher bitrate required for error encoding is due to the fact that it includes the overhead due to implicit/explicit block classification.

A final experiment has been performed by applying the new algorithm to small format color sequences (1/9 CIF) for very low bitrate coding. The results achieved on "Miss America" sequence, for 8 kbit/sec. transmission, are shown in Figure 8.

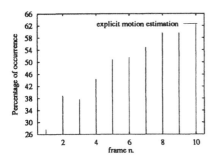
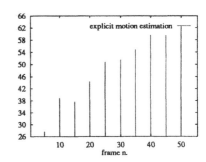

Figure 7: Percentage of occurrences of the explicit motion estimation. Experiments on the sequence "Miss America" at full frame frequency (left) and with 1:5 temporal subsampling (right).

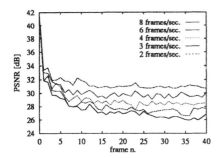
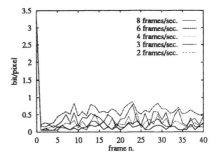

Figure 8: Sequence "Miss America" coded at 8 kbit/sec. by means of the new algorithm for implicit/explicit motion estimation. Depending on the temporal subsampling, different values of the PSNR (left) and of the bitrate (right) are obtained.

REFERENCES

1. CCITT, Video codec for audiovisual services at p*64 kbits/s, Recommendation H.261, CDM XV-R 37-E, 1990.
2. MPEG ISO-IEC JTC1 1/SC 29/WG11, Coding of moving pictures and associated audio for digital storage media up to about 1,5 Mbit/s, 1992.
3. H.G. Musmann, M. Hoetter and J. Ostermann, Object-Oriented Analysis - Synthesis Coding of Moving Images, Signal Processing: Image Communication, Vol. 1, pp. 117-138 (1989).
4. R. Ter Horst, F. Hoeksema, G. Heideman, H. Tattje, Efficient ATM Transmission with 1-Layer H.261 Videocodecs, 5-th Int. Workshop on Packet Video, Berlin, D4 (1993).
5. T. Houdoin, J. Cochennec and J. Le Hir, Short Term Video on ATM Networks of Existing H.261 Video Terminals, 5-th Int. Workshop on Packet Video, Berlin, H1 (1993).
6. F. Pereira, and L. Masera, Two-layers knowledge-based videophone coding, Proc. PCS-90, Cambridge Ma, 6-1 (1990).
7. E. Badique', Knowledge-based Facial Area Recognition and Improved Coding in a CCITT-Compatible Low-Bitrate Video-Codec, Proc. PCS-90, Cambridge Ma, 9-1 (1990).

Frame	Conventional H.261		New Algorithm		
	Motion Field $\frac{bits}{pixel}$	Error $\frac{bits}{pixel}$	Motion Field $\frac{bits}{pixel}$	Error $\frac{bits}{pixel}$	PSNR dB
1	0.057755	0.078506	0.023636	0.080917	36.178799
2	0.059601	0.078506	0.025833	0.080261	37.232360
3	0.063217	0.078507	0.028214	0.084671	34.759180
4	0.065643	0.078507	0.029663	0.087311	34.979913
5	0.071930	0.078506	0.034149	0.090149	33.141078
6	0.074295	0.078507	0.037827	0.091323	32.808315
7	0.082443	0.078507	0.044907	0.09436	31.912691
8	0.087006	0.078506	0.051926	0.090835	31.912691
9	0.093964	0.078506	0.055450	0.096451	31.072773
10	0.096542	0.078507	0.058990	0.097092	31.182099

Table 3: Performance comparison between a conventional H.261 coder and the proposed scheme with the implicit/explicit modality for motion estimation.

Time-Varying Image Processing and
Moving Object Recognition, 3 – V. Cappellini (Ed.)
© 1994 Elsevier Science B.V. All rights reserved.

IMAGE SEQUENCE DATA COMPRESSION THROUGH ADAPTIVE SPACE-TEMPORAL INTERPOLATION

Francesco G.B. De Natale, Giuseppe S. Desoli and Daniele D. Giusto

Signal Processing and Understanding Group
Dept. of Biophysical and Electronic Engineering, University of Genoa
via Opera Pia 11a, Genoa 16145 Italy

A novel approach to video coding at very low bitrates is presented, which uses a spline-like interpolation scheme in a spatiotemporal domain. This operator is applied to a non-uniform 3D grid (built on sets of consecutive frames) so as to allocate the information adaptively.
The proposed method allows a full exploitation of intra/inter-frame correlations and a good objective and visual quality of the reconstructed sequence.

1. INTRODUCTION

Image-sequence coding techniques capable to reach very low bitrates are useful in a number of applications, in particular, videoconferencing and videotelephony.
In these applications, the necessity for attaining extremely high compressions (so as to allow the transmission over narrow-band ISDN channels) imposes some constraints on quality preservation. Most of the techniques developed and implemented in the last few years produce evident errors, in particular, an annoying blocking effect due to the use of large blocks coarsely encoded. Even the MPEG standard coding strategy, which performs quite well at compression ratios not exceeding 1:30, yields far worse results when stressed at higher factors.
In our opinion, this intrinsic limitation of block-coding techniques can be overcome only by using adaptivity criteria able to better exploit local characteristics of an image. Accordingly, the two main purposes of this work are: first, the definition of a suitable strategy to allocate information non-uniformly over a frame sequence; then, the development of an algorithm able to reconstruct the sequence from such sparse data, without introducing visible artifacts and discontinuities.
The proposed algorithm consists of two steps:
1. generation of a 3D non-uniform grid over a set of consecutive frames, the nodes of which contain information for the reconstruction process;
2. interpolation of these data by a spline-like method.
The paper is organized as follows: in section 2, the procedure for designing the 3D grid is outlined and the encoding of grid topology information is described. In section 3, the algorithm adopted for the interpolation of grid vertices is detailed and some possible variants are proposed. Finally, in section 4, the results achieved by applying the new method to a standard test sequence are reported, and some examples of decoded frames are given.

2. GENERATION OF THE GRID

The basis for the proposed scheme is a simultaneous processing of groups of frames (frame-stacks); this operation aims to manage spatial and temporal correlations homogeneously. Such a goal is reached by creating a non-uniform 3D grid on the frame-stack to be coded. The grid is characterized by a higher density of nodes in each area containing an intense activity, spatial (edges, textures) or temporal (movement).

The frame-stack is first subdivided into a set of volumetric blocks of maximum size (2^s+1, 2^s+1, 2^t+1). The dimensions must be odd numbers of pixels in order that there may always be a superposition of vertices among neighboring blocks. Then, the algorithm for the grid generation is applied to each separate block. Starting from a single mesh of maximum dimensions, with only the eight vertices as nodes, the algorithm proceeds through a recursive dicotomic subdivision in the spatial or temporal domain, guided by the evaluation of an error measure. A spatial split involves the generation of four equally-sized sub-blocks, with dimensions ($2^{s-1}+1$, $2^{s-1}+1$, 2^t+1), and consequently the generation of ten new nodes; a split in the temporal direction generates two equally-sized sub-blocks (2^s+1, 2^s+1, $2^{t-1}+1$) and consequently four new nodes.

In practice, several different configurations may result, more stretched in the spatial or temporal domain, depending on the necessity for matching rapid changes in one or the other domain. The structure described above can be regarded as a tree (see Figure 1), in which each node can produce four sons (after spatial splitting, it behaves like a quaternary tree) or two sons (after temporal splitting, it is simply a binary tree).

In addition to flexibility and ease of implementation, this structure presents two more advantages. First, it is very inexpensive to encode: using a Huffman-like code, it is possible to assign only one bit (configuration 0) to the most probable case (i.e., *no split*) and two bits (configurations 10 and 11) to the other cases (i.e., *temporal split* and *spatial split*), thus obtaining a very compact code. Second, starting from a 3D block with odd dimensions, it is possible to always have a superposition of the grid vertices of adjacent sub-blocks; this results in a better interpolation continuity (as explained in the following) and in saving in information (fewer vertices to be transmitted).

The splitting process goes on until each sub-block fulfils the error condition in both directions, or reaches the user-defined minimum dimensions. If a block satisfies a fixed quality criterion, it becomes a leaf of the tree, and a 0 bit is transmitted to indicate the end of the splitting process; otherwise, if a block reaches the minimum dimensions, the 0 bit can be saved, as the decoder knows that this block cannot be split any further.

3. VERTEX INTERPOLATION

During the splitting process, it is necessary to evaluate the error caused by the approximation in order to decide if the current block must be further subdivided and in which direction; to measure this error, the interpolation of the grid data must first be performed. The mathematical function chosen, for both this step and the final reconstruction, is a separable interpolator, linear in the three directions (x,y,t). The use of a similar function in a 2D domain (bilinear interpolation) is not new, and has been applied in still-picture coding with good results [1].

As shown in Figure 2, the computation of this function for a single block can be performed by calculating, in sequence:
1. the 1D linear interpolation of block edges in the x direction;
2. on the basis of these data, the bilinear interpolation of two faces of the 3D block in the (x,y) domain (i.e., the 2D spatial blocks corresponding to the first and last temporal frames considered);
3. finally, the linear interpolation of each pair of corresponding pixels in the temporal direction (trilinear interpolation).

It should be stressed that a trilinear interpolation, although very simple to compute, is not a 1st-order interpolator. In fact, from the equations below, it can be deduced that it is a spline function of the 3rd order of the variables (x,y,t).

Given a 3D vector Z of dimensions $n*m*l$, the interpolation of a generic point Z_{ijk} can be obtained from the values of the eight vertices of Z as follows:

i)   compute the value of the point Z_{i00} (linear interpolation of Z_{000} and Z_{n00}):

$$Z_{i00} = \frac{Z_{n00} - Z_{000}}{n} i + Z_{000}$$

ii)   compute the value of the point Z_{im0} (linear interpolation of Z_{0m0} and Z_{nm0}):

$$Z_{im0} = \frac{Z_{nm0} - Z_{0m0}}{n} i + Z_{0m0}$$

iii) compute the value of the point Z_{ij0} (linear interpolation of Z_{i00} and Z_{im0}):

$$Z_{ij0} = \frac{Z_{im0} - Z_{i00}}{m} j + Z_{i00} =$$

$$= \frac{(Z_{n00} + Z_{0m0}) - (Z_{000} + Z_{nm0})}{nm} ij + \frac{(Z_{000} - Z_{0m0})}{m} j + \frac{(Z_{n00} - Z_{000})}{n} i + Z_{000} = A_1 ij + A_2 j + A_3 i + A_4$$

iv)   repeat the operations from (i) to (iii) to calculate point Z_{ij1}:

$$Z_{ij1} = \frac{Z_{im1} - Z_{i01}}{m} j + Z_{i01} =$$

$$= \frac{(Z_{n01} + Z_{0m1}) - (Z_{001} - Z_{nm1})}{nm} ij + \frac{(Z_{001} - Z_{0m1})}{m} j + \frac{(Z_{001} - Z_{0m1})}{n} i + Z_{001} = B_1 ij + B_2 j + B_3 i + B_4$$

v)   compute the value of the point Z_{ijk} (linear interpolation of Z_{ij0} and Z_{ij1}):

$$Z_{ijk} = \frac{Z_{ij1} - Z_{ij0}}{l} k + Z_{ij0} =$$

$$= \frac{B_1 - A_1}{l} ijk + \frac{B_2 - A_2}{l} jk + \frac{B_3 - A_3}{l} ik + \frac{B_4 - A_4}{l} k + A_1 ij + A_2 j + A_3 i + A_4$$

The above computations, however, require a considerable use of floating-point operations and are therefore unsuitable for fast implementations. To overcome this drawback, the interpolation can be computed by using only integer numbers, and by applying an iterative (or recursive) top-down process to the tree layers. Therefore, given the eight vertices of a 3D block, the central points of each segment, of each face, and of the block are computed sequentially, as shown in Figure 3.

To speed up the error evaluation function, a complete approximation of the block is not required. To decide if spatial splitting has to be performed, only the first frame of the block is examined, whereas, to decide on temporal splitting, only the linear interpolations of the four pairs of corresponding vertices in the time direction are calculated. This simplification is justified by the following considerations:

- if there is no movement in a block, the first face can be considered *representative* of the entire block, so spatial splitting can be performed on it;
- if a block is smooth (i.e., it does not contain high frequencies), the four vertices of each face can be considered *representative* of the entire face, hence the temporal splitting condition depends only on these vertices.

When the interpolator has to be calculated on the global tree structure (for the final reconstruction), some additional problems arise, due to the necessity for maintaining the interpolation continuity. It is worth noting that a tree structure may produce configurations in which some vertices do not match with those of the neighboring blocks, thus causing discontinuities if such information is neglected. This drawback may strongly affect the visual quality of the reconstructed image. Figure 4 shows two typical examples of possible discontinuity points.

To face such situations, a different method can be adopted for the global interpolation. The algorithm implemented is of the recursive type, based on the tree structure of the grid. At each recursion, the points belonging to a lower level of the tree hierarchy are considered; the temporal interpolation is computed first (in order to solve discontinuities of the (b) type), then the spatial one (to handle problems of the (a) type).

Starting from a maximum spatial level s and a maximum temporal level t, i.e., a block with dimensions $(2^s+1, 2^s+1, 2^t+1)$, the sequences of interpolated points observe the following order:

approximate:	*for levels:*	*block size:*
t-level.........(t-1)	s-level..............(s)	$(2^s+1, \quad 2^s+1, \quad 2^{t-1}+1)$
s-level.........(s-1)	t-level..............(t, t-1)	$(2^{s-1}+1, \quad 2^{s-1}+1, \quad 2^{t-1}+1)$
t-level.........(t-2)	s-level..............(s, s-1)	$(2^{s-1}+1, \quad 2^{s-1}+1, \quad 2^{t-2}+1)$
s-level.........(s-2)	t-level..............(t, t-1, t-2)	$(2^{s-2}+1, \quad 2^{s-2}+1, \quad 2^{t-2}+1)$
. . . . .		

As the maximum t-level is in general lower than the maximum s-level (to reduce storage requirements), the t-level will reach its minimum value after a few iterations. From this point on, s-level approximations only will be performed at each recursion.

4. EXPERIMENTAL RESULTS

The results achieved by the proposed interpolation strategy are quite satisfactory. The technique can be regarded as the basis for a sequence-coding method characterized by a wide compression range. At the lowest bitrates, sequences with good visual qualities were reconstructed, starting from a very small number of samples; high fidelity was achieved for mid-range compressions. Some tests were also performed where the interpolation was used as the first part of a two-source coder, and the residual was scalar quantized and entropy coded, to obtain very high quality, even if at lower compression rates.

In Figure 5, four consecutive frames of the classical test sequence *Miss America* are shown; the spatial resolution is 256x256 pixels, and only the intensity component (8 bpp) was extracted from the original RGB data. In Figure 6, the same frames are shown after the co-decoding process; the bitrate achieved is 0.13 bpp (the compression factor is about 1:60), with an average PSNR of 33.8 dB.

In this test, a frame-stack made up of 16 frames was used; the average size of the tree was about 69100 nodes, 36900 of which were leaves. The corresponding grid vertices used for the interpolation were 53200. Table 1 gives the average distributions of the blocks in the spatiotemporal domain: the minimum allowed dimensions were 3x3x3, and the maximum dimensions were 65x65 in the spatial domain (only square blocks were considered) and 17 (i.e., all the input frames) in the temporal domain.

The information related to the tree structure was coded by means of a Lempel-Ziv entropy encoder, so achieving a bitrate lower than 0.4 bit per tree-node (for instance, the tree in the example required 3480 bytes). To encode the vertex values, a DPCM algorithm was used, followed by a Huffman encoder; the resulting rate was about 2.4 bits per sample.

Figure 7 presents the frame-by-frame differences between original and interpolated data (Figures 5 and 6, respectively). It can be noticed that the residual is distributed quite uniformly over the sequence, and that it is in general concentrated near the image edges. These characteristics make it possible to discard the residual when an extremely low bitrate is needed, or to encode it through a coarse quantization if it is necessary to improve the final reconstruction quality (the residual contains highly decorrelated high-frequency information). Finally, Figure 8 shows the same decoded sequence as in Figure 6. The encoded residual was added to it in order to achieve a more accurate reconstruction: the bitrate was 0.25 bpp, with a PSNR of 35.8 dB.

ACKNOWLEDGMENTS

This research work was carried out within the framework of the EUREKA Prometheus Project (Pro-Com sub-project), and partially funded by the National Research Council (CNR) of Italy.

REFERENCES

[1]   P.M.Farrelle, *Recursive Block Coding for Image Data Compression*, Springer-Verlag, New York, 1990.

Table 1. Average number of blocks generated by 3D splitting for each block size.

tdim\sdim	3x3	5x5	9x9	17x17	33x33	65x65
3	28656	5266	1335	263	54	8
5	500	123	31	5	0	0
9	254	85	33	6	1	0
17	198	74	23	3	1	0

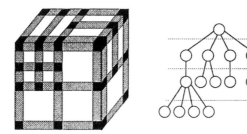

Figure 1. An example of 3D subdivision and the corresponding tree.

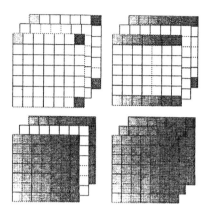

Figure 2. Trilinear interpolation of a 3D block.

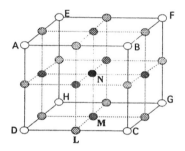

○ starting vertices (A, B, C, D, E, F, G, H)

⊛ interpolation of segments' central points (e.g.: $L = \frac{D+C}{2}$)

◉ interpolation of faces' central points (e.g.: $M = \frac{D+C+H+G}{4}$)

● interpolation of block's central point ($N = \frac{A+B+C+D+E+F+G+H}{8}$)

Figure 3. Integer computation of the interpolator.

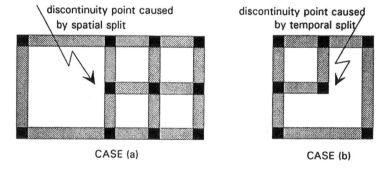

discontinuity point caused
by spatial split

discontinuity point caused
by temporal split

CASE (a)

CASE (b)

Figure 4. Examples of possible discontinuities in grid interpolation.

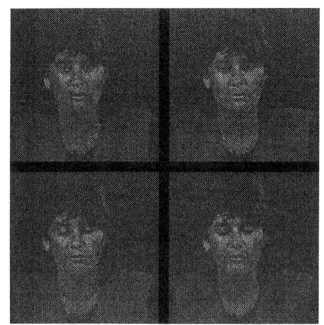

Figure 5. Four frames of the original test sequence *Miss America:*
256x256 pixels in size, 8 bpp (b/w).

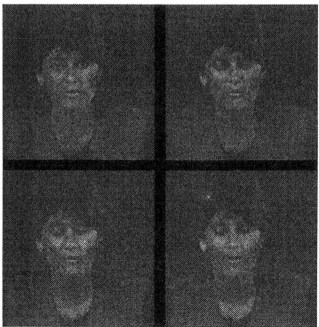

Figure 6. Interpolation of the four frames in Figure 5:
bitrate: 0.13 bpp (CR 1:60), PSNR: 33.8 dB.

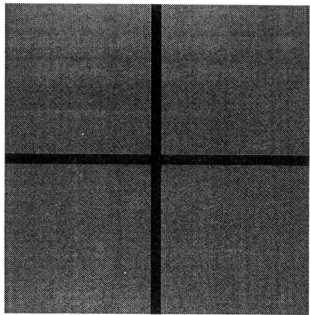

Figure 7. Residual images (differences between the frames in Figures 5 and 6)

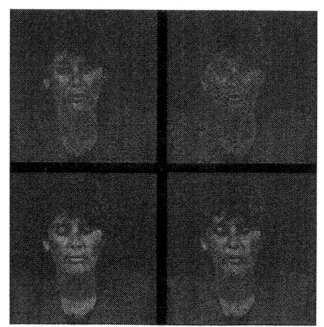

Figure 8. Final reconstruction, with the addition of residual information;
bitrate: 0.25 bpp (CR 1:32); PSNR: 35.8 dB.

Time-Varying Image Processing and
Moving Object Recognition, 3 – V. Cappellini (Ed.)

Linear Phase IIR QMF Banks Based on Complex Allpass Sections for Image Compression

F. Argenti[a], V. Cappellini[a], A. Sciorpes[a] and A.N.Venetsanopoulos[b]

[a]Dipartimento di Ingegneria Elettronica, Universita' di Firenze
Via di Santa Marta, 3 - 50139 Firenze, Italy

[b]Department of Electrical Engineering, University of Toronto
10 King's College Road, Toronto, Ontario, Canada

In this work the problem of linear phase QMF banks for subband coding applications is addressed. A new method for designing and implementing two-channel analysis/synthesis banks of IIR filters with whole sample symmetric impulse responses is presented. The design procedure is based on the representation of the filters by means of complex allpass functions. The importance of such structures has been already highlighted in the literature for the property of being structurally lossless and for the low sensitivity to coefficient quantization. The results of the application of the new family of filter banks to image compression are also presented.

1. INTRODUCTION

During the last few years several properties and applications of subband coding systems have been pointed out [1]. In this work we investigate on the design of perfect reconstruction (PR) Quadrature Mirror Filter (QMF) IIR banks having the properties of orthogonality and linear phase, i.e. having a symmetrical impulse response. These properties are achieved at the price that the filters of the analysis and synthesis banks are no more causal, i.e. they involve direct and inverse direction of recursion to achieve stability: this fact may not pose a problem when finite extent signals are dealt with, for example in subband coding of images. Different types of symmetries can be defined [2]: a symmetrical or antisymmetrical impulse response having a central term is referred to as a whole sample symmetric (WSS) or as a whole sample antisymmetric (WSA) filter, while a symmetrical or antisymmetrical impulse response which does not have a central term is referred to as a half sample symmetric (HSS) or as a half sample antisymmetric (HSA) filter. In [3] the conditions that IIR banks of filters must satisfy to achieve, simultaneously, the properties of perfect reconstruction, orthogonality and linear phase are stated and a constructive method given for the case of half sample symmetry. In this work we present a design method for a class of WSS filters which is based on the use of complex allpass sections. Such structures can be used for producing pairs of doubly-complementary filters, as presented in [4]: here, new constraints on the filters are imposed to obtain linear phase and perfect reconstruction. Moreover, it is well-known from the literature [5][6] that the filters composed by allpass sections, as the doubly complementary filters, exhibit also the advantage of a low sensitivity with respect to coefficient quantization. In the final part of this work the results of the application of the new family of filter banks to image compression are presented.

In the following a superscript * will denote the conjugate of a complex number, a subscript * will denote that the conjugate of the coefficients of a function must be taken and $\tilde{F}(z)$ will denote $F_*(z^{-1})$. Finally, a function is said to be allpass if $F(z)\tilde{F}(z)=1$.

2. POWER COMPLEMENTARY, LINEAR PHASE IIR BANKS COMPOSED BY COMPLEX ALLPASS SECTIONS

The aim of this work is the study and design of filter banks for two-channel subband coding systems with the following characteristics: structural perfect reconstruction; linear phase; low computational cost; stable frequency response with respect to coefficient quantization; orthogonality. A general scheme of a two-channel subband decomposition system is shown in Fig. 1. The reconstructed signal $\hat{X}(z)$ is retrieved from

$$\hat{X}(z) = \frac{1}{2} [F_0(z)\ F_1(z)] \begin{bmatrix} H_0(z) & H_0(-z) \\ H_1(z) & H_1(-z) \end{bmatrix} \begin{bmatrix} X(z) \\ X(-z) \end{bmatrix} \tag{1}$$

The filter bank is said to satisfy the perfect reconstruction condition if $\hat{X}(z) = cz^k X(z)$ for some real c and some integer k.

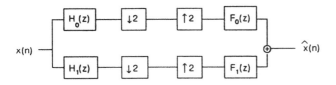

Figure 1. Subband coding system.

Let $h_0(n)$ and $h_1(n)$ denote the impulse responses of the filters $H_0(z)$ and $H_1(z)$, respectively. A bank of filters is said to be orthogonal if the following relationships hold:

$$\sum_n h_i(n)h_i^*(n-2k) = \delta_k \qquad i=0,1,\ \forall\ k\in Z$$

$$\sum_n h_0(n)h_1^*(n-2k) = 0 \qquad \forall\ k\in Z \tag{2}$$

As can be seen from (2) the impulse responses of the filters are orthogonal with respect to even shifts, that is the autocorrelation function of the impulse responses is zero when computed with lags ± 2, ± 4, ± 6, ... This fact implies that the transfer functions $\tilde{H}_i(z)$, $i=0,1$, satisfy

$$H_i(z)H_{i*}(z^{-1}) + H_i(-z)H_{i*}(-z^{-1}) = 1 \qquad i=0,1$$

$$H_0(z)H_{1*}(z^{-1}) + H_0(-z)H_{1*}(-z^{-1}) = 0 \tag{3}$$

If a filter has a linear phase then it satisfies one of the following conditions:

$$R_s(z^{-1}) = z^{-k}R_s(z)$$
$$R_a(z^{-1}) = -z^{-k}R_a(z)$$

(4)

where k is an integer. $R_s(z)$ and $R_a(z)$ are symmetric and antisymmetric filters, respectively. If k is even, the impulse responses of the filters have a central sample (WSS and WSA), otherwise they do not have a central sample (HSS and HSA). In [3] it is shown how the properties of linear phase and orthogonality do not exclude each other in perfect reconstruction filter banks if the filters are IIR, at the price that the filters are non-causal.

Consider now a pair of filters H(z) and G(z) having real coefficients and composed by the sum and the difference of complex allpass sections, i.e. having the form

$$H(z) = \frac{1}{2}\left[A_{c1}(z) + A_{c2}(z)\right]$$
$$G(z) = \frac{1}{2j}\left[A_{c1}(z) - A_{c2}(z)\right]$$

(5)

where $A_{c1}(z)$ and $A_{c2}(z)$ are allpass functions with complex coefficients.

In [6] it is demonstrated that given a filter H(z)=P(z)/D(z) with P(z) a symmetric polynomial of odd degree another filter G(z)=Q(z)/D(z) can be found, with Q(z) an antisymmetric polynomial, such that the filters H(z) and G(z) can be rewritten as sum and difference of real allpass sections and form a power complementary pair, i.e.

$$H(z)\tilde{H}(z) + G(z)\tilde{G}(z) = 1$$

(6)

Instead, if the numerator of H(z) is a symmetric polynomial with even degree then a filter G(z) can be found having a symmetric polynomial as numerator, such that the pair of filters H(z) e G(z) satisfy the power complementary property (6) and can be expressed in the form (5). Since the sum H(z)+jG(z) is an allpass function the name *doubly-complementary* pair has been given to such structures [4].

The doubly-complementary filters previously described do not have linear phase: let us impose this new constraint and see how this property affects the structure of the filters. If the pair of filters H(z) and G(z), expressed as in (5), comprises a low-pass and a high-pass filter then the symmetries are of the WSS type [7], that is

$$H(z^{-1}) = z^{-m}H(z)$$
$$G(z^{-1}) = z^{-m}G(z)$$ m *even*

(7)

The symmetry condition imposes constraints on the allpass functions $A_{c1}(z)$ e $A_{c2}(z)$. In fact, by substituting (5) in (7) we get

$$A_{c1}(z^{-1}) = z^{-m}A_{c1}(z)$$
$$A_{c2}(z^{-1}) = z^{-m}A_{c2}(z)$$

(8)

2.1 Application to subband coding systems

Consider now the case of a two-channel subband coding system and let us impose, as in [4], $G(z)=H(-z)$. In this case we have to use a modified QMF bank [8][4], with the analysis and synthesis banks given by:

$$H_0(z) = H(z)$$
$$H_1(z) = z^{-1}G(z) = z^{-1}H(-z) \tag{9}$$

$$F_0(z) = z^{-1}H(z^{-1})$$
$$F_1(z) = G(z^{-1}) \tag{10}$$

The scheme of the modified QMF system is shown in Fig. 2.

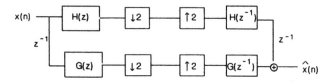

Figure 2. Modified QMF bank.

The banks of filters which have been described satisfy the orthogonality property (3), as can be easily verified by direct substitution of (9) in (3) and by considering the decomposition of $H(z)$ and $G(z)$ given by (5).

By using some of the previous results, we will adopt a simplified notation. Eq. (5) can be rewritten by considering an unique allpass function $A_c(z)=A_{c1}(z)=A_{c2*}(z)$:

$$H(z) = \frac{1}{2}\left[A_c(z) + A_{c*}(z)\right]$$
$$G(z) = \frac{1}{2j}\left[A_c(z) - A_{c*}(z)\right] \tag{11}$$

If between the filters $H(z)$ and $G(z)$ the relationship $G(z)=H(-z)$ exists, then the allpass function $A_c(z)$ must satisfy the condition [4]

$$A_c(z) = jA_{c*}(-z) \tag{12}$$

Moreover the hypothesis of symmetry on the filters implies

$$A_c(z^{-1}) = z^{-m}A_c(z) \tag{13}$$

with m an even integer. Eq. (12) implies that if z_p is a pole of $A_c(z)$, then also $-z_p^*$ is a pole of $A_c(z)$. Eq. (13), instead, implies that if z_p is a pole of $A_c(z)$, then also $1/z_p$ is a pole of $A_c(z)$. From these considerations we can conclude that the allpass function $A_c(z)$, in the case of WSS perfect reconstruction QMF banks, can be factorized as follows:

$$A_c(z) = \eta z^\gamma \prod_{i=1}^{N_1} \frac{j\beta_i + z^{-1}}{1 - j\beta_i z^{-1}} \frac{j\beta_i + z}{1 - j\beta_i z} \prod_{k=1}^{N_2} \frac{-\alpha_k^* + z^{-1}}{1 - \alpha_k z^{-1}} \frac{\alpha_k + z^{-1}}{1 + \alpha_k^* z^{-1}} \frac{-\alpha_k^* + z}{1 - \alpha_k z} \frac{\alpha_k + z}{1 + \alpha_k^* + z} \tag{14}$$

where η is a constant such that $|\eta| = 1$, γ is an integer, β_i and α_k are real and complex values, respectively. Then, eq. (14) can be rewritten as

$$A_c(z) = \eta z^\gamma B(z) B(z^{-1}) \tag{15}$$

where B(z) is a stable allpass function with complex coefficients given by

$$B(z) = \prod_{i=1}^{N_1} \frac{j\beta_i + z^{-1}}{1 - j\beta_i z^{-1}} \prod_{k=1}^{N_2} \frac{-\alpha_k^* + z^{-1}}{1 - \alpha_k z^{-1}} \frac{\alpha_k + z^{-1}}{1 + \alpha_k^* z^{-1}} \tag{16}$$

It is easy to verify that $B(z) = \pm B_*(-z)$.

The conditions (12) and (13) impose some constraints on the choice of the values η and γ. From (12) it derives $\eta^2 = j(-1)^\gamma$, while (13) yields $\gamma = m/2$.

Fig. 3-a   Fig. 3-b

Figure 3. Pole position for H(z) produced by a first order (a) and a second order (b) allpass function B(z).

It can be noted that B(z) is factorized in allpass sections having poles on the imaginary axis or pairs of poles of the form $(\alpha, -\alpha^*)$. In [4] these conditions were verified by the whole function $A_c(z)$, while now they hold only for a factor of $A_c(z)$: hence, the new filter banks have less degrees of freedom, but they have the properties of linear phase and perfect reconstruction. In our case the poles of $A_c(z)$ occur in pairs on the imaginary axis ($j\beta$, $1/j\beta$ with $\beta \in R$) or in quadruplets (α, $-\alpha^*$, $1/\alpha$, $-1/\alpha^*$ with $\alpha \in C$). However, the poles of the transfer functions H(z) and G(z) include also the conjugates of the poles of $A_c(z)$. In fact, as shown in [6], since the denominators of $A_c(z)$ and $A_{c*}(z)$ do not have common factors, then cancellation between pole-zero in the sum $A_c(z) + A_{c*}(z)$ or in the difference $A_c(z) - A_{c*}(z)$ do not occur. Possible configurations of poles of the transfer function H(z) produced by a function B(z) composed by a single first order section (i.e. with only one pole on the imaginary axis in $j\beta$) or by a single second order section (i.e. with a pair of complex poles in α and $-\alpha^*$) are shown in Fig. 3-a and Fig. 3-b, respectively.

The problem of implementing the banks with a minimum computational cost is not addressed here for brevity's sake. However, in [7] it is shown how the subbands obtained from a cascade of N_1 first order cells can be computed with $2N_1$ multiplications.

3. EXAMPLES OF FILTER DESIGN

As shown in the previous Section only one or two degrees of freedom are left for the design of a four or an eight poles transfer function $H(z)$ in the two different cases of having a first order or a second order complex allpass section, respectively. The position of the poles is constrained by the structure of the filters: experimental design tests have shown that better results can be obtained by using a cascade of only first order sections.

Table I
Results of filter numerical design

Nominal ω_s	N_1	β_i	Gain at $\omega=\omega_s$ (Db)	First ripple gain (Db)	Gain at $\omega=\pi$ (Db)
0.54 π	5	+5.683484 E-01 -9.040434 E-02 -6.855018 E-01 -9.503068 E-01 +9.079501 E-01	-37.05	-34.52	-51.67
	4	+2.627580 E-01 -4.257299 E-01 -8.888370 E-01 +8.024073 E-01	-23.67	-38.18	-55.83
0.60 π	4	+2.523451 E-01 -4.073094 E-01 -8.653298 E-01 +7.714136 E-01	-63.58	-59.11	-54.31
	3	+6.295464 E-01 -9.696486 E-02 -7.797596 E-01	-29.45	-43.08	-50.72
0.70 π	2	+3.523630 E-01 -5.909896 E-01	-29.70	-42.83	-45.29

In Table I some results of numerical design of orthogonal WSS digital filter are shown. These filters were obtained by using classical numerical optimization routines for minimizing the following function:

$$\Phi(\beta_1,\beta_2,...,\beta_{N_1}) = \int_{\omega_s}^{\pi} |H(\omega)|d\omega \tag{17}$$

where ω_s denotes the lower bound of the stopband. In Table I the number of first order allpass sections composing B(z), i.e. N_1, and the relative coefficients are given; some characteristics of the filters, that is the gain at $\omega=\omega_s$, $\omega=\pi$ and the height of the first ripple in the stopband, are also shown. The magnitude of the frequency responses is presented in Fig. 4, where the filters of Table I are denoted by (a)-(e), in the order.

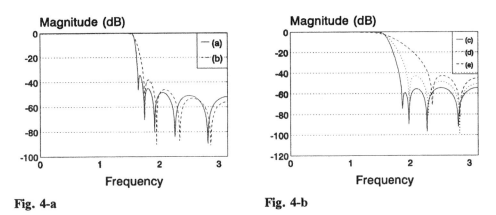

Fig. 4-a   Fig. 4-b

Figure 4. Frequency responses of the filters presented in Table I.

Other techniques can be used: in [7] a design method based on a flatness constraint, in which the coefficients of the allpass sections of the filters are found in closed form, is presented.

4. APPLICATION TO IMAGE COMPRESSION

Monodimensional filter banks can be applied to the decomposition of bidimensional signals, such as images, by using a separable filtering scheme which operates along the rows and the columns. In so doing, an input image is decomposed in four subbands: the first represents a lower resolution version of the image, while the others contain horizontal, vertical and diagonal details. In our coding scheme the lower resolution subband has been further decomposed so that a seven subband decomposition is obtained.

The effectiveness of the new filter banks when applied to image coding has been analysed by considering the compression ratios obtained by using the fixed distortion/variable bit rate compression algorithm described in [9]: efficient data reduction is achieved by using a run-length encoding of high-frequency subbands, while DPCM is used for the lowest resolution subband. The test image "Lenna", 512x512, 8 bit/pixel was used. Fig. 5 shows the results of the compression as diagrams of the PSNR vs the bit rate. As can be seen an optimum value of the selectivity of the filters exists. In fact, a highly oscillating impulse response corresponds to very selective filters, so that *ringing* effects are produced around the edges present in the image: in these areas coding can be very expensive in terms of number of bits. Best results are obtained for filters (c) and (d): in these cases the subbands are computed with 8 and 6 multiplications per input pixel.

198

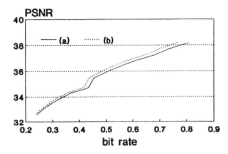

Fig. 5-a

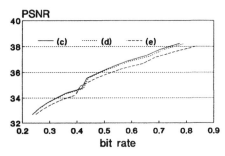

Fig. 5-b

Figure 5. Results of compression on the test image "Lenna", 512x512 pixel, 8 bit/pixel.

5. CONCLUSIONS

In this work a new family of IIR perfect reconstruction filter banks having the properties of linear phase and orthogonality is presented. Some filter banks, designed by means of numerical approximation routines, have been applied to image coding and compared in terms of the compression ratio. Low order filters seem very interesting from the point of view of both the computational cost and the performance.

REFERENCES

1. P.P.Vaidyanathan, "Multirate digital filters, filter banks, poliphase networks, and applications: a tutorial", *Proc. of the IEEE*, vol. 78, no. 1, pp. 56-93, Jan. 1990.
2. C.Herley, M.Vetterli, "Wavelets and recursive filter banks", tech. rep. CU/CTR/TR 299-92-9, Columbia University.
3. C.Herley, M.Vetterli, "Linear phase wavelets: theory and design", in *Proc. IEEE Int. Conf. ASSP*, Toronto, Canada, May 1991, pp.2017-2020.
4. P.P.Vaidyanathan, P.A.Regalia, S.K.Mitra, "Design of doubly-complementary IIR digital filters using a single complex allpass filter, with multirate applications", *IEEE Trans. Circuits Syst.*, vol. CAS-34, n. 4, pp. 378-388, Apr. 1987.
5. T.Saramaki, T. Yu, S.K.Mitra, "Very low sensitivity realization of IIR digital filters using a cascade of complex all-pass structures", *IEEE Trans. Circuits Syst.*, vol. CAS-34, n. 8, pp. 876-886, Aug. 1987.
6. P.P.Vaidyanathan, S.K.Mitra, Y.Neuvo, "A new approach to the realization of low-sensitivity IIR digital filters", *IEEE Trans. Acoust., Speech, Signal Processing*, vol. ASSP-34, n. 2, pp. 350-361, Apr. 1986.
7. F.Argenti, A.Sciorpes, "Design of IIR Linear Phase QMF Banks based on Complex Allpass Sections" tech. rep. n. 930101, Dipartimento di Ingegneria Elettronica, University of Florence.
8. C.R. Galand, H.J.Nussbaumer, "New quadrature mirror filter structures", *IEEE Trans. on Acoustics, Speech, Signal Processing*, vol. ASSP-32, pp. 522-531, June 1984.
9. J.C.Darragh, R.L.Baker, "Fixed distortion subband coding of images for packet switched networks", *IEEE Journal Selected Areas Commun.*, vol. JSAC-7, pp. 789-800, June 1989.

Time-Varying Image Processing and
Moving Object Recognition, 3 – V. Cappellini (Ed.)
© 1994 Elsevier Science B.V. All rights reserved.

Somatic features extraction for facial mimics synthesis

W. Bellotto, S. Curinga, P. Ghelfi and F. Lavagetto

DIST - Department of Communication, Computer and Systems Science
University of Genova, Via Opera Pia 11-a, 16145, Genova, Italy

The approach described in this paper addresses the problem of modeling a typical videophone scene by means of wire-frame flexible structures and suitable analysis/synthesis algorithms. Relying on a large basis of a priori knowledge, an adaptive model has been devised, capable to reproduce the somatic characteristics of human faces. The developed algorithms have been tested on simple videophone sequences whose frames have been synthesized from the model by means of few estimated parameters. The achieved experimental results show concrete possibilities of application in interpersonal video communications where very low bitrate is required.

1. INTRODUCTION

Within interpersonal visual communication, a significant part of information is conveyed through the expression of the speaker face which, together with speech, can be reasonably considered the most powerful means of subjective interaction. Thanks to the integration with facial mimics, the acoustic message associated to speech gains appreciable richness and allows an improved impact on the human perceiver. In order to preserve as much as possible this critical information, many algorithms have been proposed [1-2] oriented to image segmentation with increased resolution in correspondence of highly informative face subregions like eyes and mouth.

These approaches are generally referred to as model-based, meaning that a basis of a priori knowledge is exploited to drive both the image segmentation and coding. By means of this knowledge very high performaces can be attained, with the consequent drawback however of restricting, somehow severily, the class of applications: the more detailed is the model the less generic is its use. As a consequence, suitable adaptation mechanisms are to be variously employed in order to provide the required characteristics of robustness and flexibility. Among the many proposals, those based on flexible structures for modeling the video scene have proven quite satisfactory when employed within videophone applications. More specifically, with reference to simplified "head-and-shoulder" sequences, many attempts have been done during the last few years aimed at the adaptive modeling and animation of human faces by means of anatomical models [3-5]. The introduced modeling complexity and the involved sophisticated algorithms lead to very

This work has been partially funded by the Italian National Research Council (Special Project on Telecommunication).

high quality performances, partially paid in terms of reduced applicabilitY, expecially if also the synchronous speech signal is analyzed and used as an integration for the dynamic adaptation of the model [6-7].

The approach we have followed and which is here described addresses the problem of facial animation from a similar point of view, namely that of using a flexible model and knowledge-based adaptation procedures. The developed system can be characterized as "analysis-synthesis video coding", meaning that some form of analysis is performed on the incoming real frames to drive the synthesis from the model. All these kinds of systems can be generally described as based on the fundamental procedures of model adaptation and animation. The former procedures rely on simplifying hypotheses over the typology of faces to be modeled, so that only minor adaptations are expected for fitting a prede-fined generalized model on any specific face. The latter procedures, on the other hand, exploit simplifying hypotheses on the motion affecting the scene, mainly due to head movements and facial mimics. Model adaptation typically requires significant computa-tion and transmission overhead but is usually performed only once at the beginning of the communication. Conversely, model animation must be performed nearly at frame rate to preserve the visual information associated to the face expression, so that its computing and transmission burdens are strictly constrained by the real-time requirement.

In Section 2 we provide a brief description of the problems related to the repre-sentation of human faces as far as their role in interpersonal video communication is concerned. In Section 3 the general structure of the model is sketched out while the analy- sis procedures are introduced and discussed in Section 4. Experimental results and final conclusions are then reported in Sections 5.

2. REPRESENTATION OF HUMAN FACES

Human faces represent a formidable source of interpersonal information deserving extreme accuracy of representation as they can be considered to reflect the internal mood of the speaker. The muscle structures responsible of facial mimics have been shown to be activated and controlled through mechanisms closely related to complex psycological agents according to rather comprehensive and general rules.

A multitude of muscles can variously be activated to originate more than 60 distin-guishable facial actions. However, as it results from Ekman and Friesen studies [8], these many muscle configurations can be reduced to a minimum set of twenty basic expressions without noticeable loss of information. This complexity reduction is explained by conside-ring the fact that facial muscles are not independent from each other. In contrast, they are closely correlated. Let us think of this simple experiment: once the shape of mouth and eyebrows is fixed no further significant facial muscle activity is possible.

The anatomical structure described in Figure 1 is evidently too complex and unjusti-fied from the point of view of our application purposes. During an interpersonal visual communication, images mostly consist of 2D frontal projections of the speaker face so that only a far reduced subset of all possible muscle actions can be perceived. Moreover, as the animation of the model for the synthesis of the facial expressions should be driven by the analysis of the incoming real images, a too complex though anatomically correct structure looks prohibitive. As a final consideration, the simplicity of the model sounds reasonably in favour of its genericity and adaptivity.

Looking at a talking face, most of the attention is focused on the eyes region where

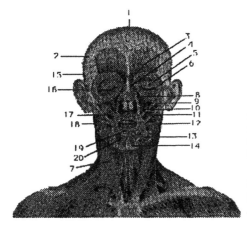

1) Galea aponeurotica   2) Aponeurosis temporalis
3) Frontalis   4) Procerus   5) Pars palpebralis
6 Pars orbitalis   7) Pellicciaius   8) Pars angularis
labii superioris 9) Pars orbitalis 10) Pars zygomatica
11) Zygomatic   12) Risorius   13) Triangularis
14)Mentalis 15) Auricularis superioris 16) Auricularis
inferioris 17) Buccinator 18) Masseter 19) Orbicularis
oris 20) Levator labii inferioris

Figure 1: Main muscles responsible of human facial mimics

most of the perceivable information is concentrated. Significant attention is also paid by the observer to the mouth region where lips movements are tracked to check synchronism with speech. Besides eyes and mouth, a third very critical element is represented by the eyebrows whose shape represents the final result of many muscle actions affecting the forehead region. Thus, the representation of these regions, referred in the following as primary somatic features, requires high fidelity while lower quality can be generally accepted in correspondence of less important areas of the face.

Different solutions have been prospected as far as the representation problem is concerned, ranging from 2D to $2\frac{1}{2}$D and 3D structures, each of them exhibiting pros and cons in variable amount. Starting from the Parke model [9], the idea of employing wire-frame structures has gained more and more consensus thanks to their intrinsec flexibility which easily allows the reproduction of facial mimics by means of geometrical deformations [10-12].

Among the proposed wire-frame structures, a great variability exists as far as their geometry and adaptation mechanisms are concerned. Some of them rely on a predefined structure whose parameters are suitably set from face to face for the best fitting. Other models, instead, grow directly on the specific faces adapting themselves to their somatics. Whatever adaptation procedure is employed, primary somatic features must be given anyway a very accurate representation, as even a small approximation error can easily originate unacceptable artifacts at animation time. Because of this fact, the model we have developed is tailored directly on the face by means of suitable operators exploiting different kinds of a priori knowledge, as described in the next Section.

3. THE FLEXIBLE FACIAL MODEL

From the various considerations presented in the previous Section it is apparent that the model must adapt itself as close as possible to the speaker's face. In addition, the

Figure 2: Wire-frame model adapted on the face of "Miss America". Bidimensional (left) and threedimensional structure (right).

adaptation process must be automatic, fast and precise despite the large variability of facial somatics which can be encountered. Our approach has been based on a few basic hypotheses:

- the face is affected by motion;
- the scene background has no significant motion;
- the face exhibits strong axial symmetry;
- the primary somatic features are completely visible;
- no occlusion is present (like eye-glasses, beart or mustaches).

The face symmetry axis is expected to be approximately vertical as only slight rotations of the face are accepted. No variations of the scale factor are managed as the speaker is supposed to be at constant distance from the camera. In these hypotheses, it is apparent the basic role played by the face symmetry axis, as the success of the subsequent analysis heavily depends on its correct estimation. A priori knowledge is in fact available on the somatic topology of a human face but some form of normalization with respect to 2D roto-translation on the image plane must still be performed: the face symmetry axis and the head silhouette are actually used for this purpose.

The speaker silhouette is obtained by segmenting the most relevant moving region of the scene by means of interframe analysis relying on the hypothesis of fixed or slowly changing background. The shape of the segmented region is then suitably refined by a priori knowledge and the head component is separated from the shoulder. The head region is then subdivided into four different areas, namely that of hair, hair-and-eyes, eyes-and-nose and mouth. As described in the next Section, the face symmetry axis is finally evaluated by analyzing the image inside the eyes-and-nose region where maximum is the symmetry information.

Taking as a reference the so obtained symmetry axis, all the other somatic features are extracted and a wire-frame model is adapted to the speaker face. The modeling structure is composed of a net of triangles with variable resolution covering the face region. Specific subnets are used to model each somatic feature like eyes, mouth and eyebrows, in order to reproduce as closely as possible their geometrical and pictorial characteristics. Some triangles are given more importance than others as they correspond to mimically active parts of the face. They are accordingly grouped in sets, generically called muscles, and are affected by predefined deformation rules: by shrinking and stretching these triangles the resulting visible effect is very similar to that originated by the actual facial expressions.

The construction of the triangle net is obtained through sequential operations performed on a face-independent uniform grid which is progressively adapted to fit at best the somatic characteristics of the speaker. Features like eyes, eyebrows and nose cannot be modeled through the uniform grid but, conversely, require much more fidelity in terms of pictorial resolution and geometrical approximation. To achieve this goal, as shown in Figure 2, rectangular patches of triangles are erased in correspondence of the primary somatic features and new adapted wire-frames are inserted in place of them.

The number of triangles in the wire-frame structure is variable, depending on the extension of the face area, with typical values ranging from 100 to 300, for 2D models, up to 1000 for 3D models. The color parameters associated to each triangle are coded at the beginning of the communication, together with the model parameters.

As no real 3D information is assumed to be available, the face can be represented either bidimensionally as a planar mosaic of triangular patches (see Figure 2, left) or threedimensionally by means of a priori knowledge. The 3D model, differently from the 2D solution, requires much more computation but also allows more sophisticated rendering effects. On the right-hand side of Figure 2 an example is shown where a 3D model is adapted to the well-known videophone sequence "Miss America". Observing the triangle net, the synthetic depth information is rather visible expecially in correspondence of the nose, as well as the presence of adapted wire-frames in correspondence of the mouth and the eyes.

4. THE ANALYSIS ALGORITHMS

The face symmetry axis is estimated by minimizing a suitable functional within an image window internal to the head silhouette. The face is assumed to be rather frontal and vertical (no significant rotation) so that the symmetry axis is approximately aligned with the column direction. The symmetry functional is computed, for each column k of the window, as follows:

$$S(k) = \frac{2}{XY} \sum_{j=0}^{Y} \sum_{i=1}^{X/2} ||x(j, k - i) - x(j, k + i)|| \tag{1}$$

where $x(j, i)$ is the luminance value of the pixel corresponding to the j-th raw and i-th column within the $X \cdot Y$ pixel window.

The estimation of the face symmetry axis is rather sensitive to the particular image window used for the computation of the functional $S(k)$: the eyes-and-nose region usually exhibits more axial symmetry than other face regions where less somatic morphology is available. An example is reported in Figure 3 showing the symmetry axis estimated on a frontal frame extracted from the standard sequence "Miss America".

The eye-and-nose region is divided by the symmetry axis into a left-hand side window, assumed to contain the left eye, and a right-hand side window, containing the right eye. The luminance distribution $x(j, i)$ is analyzed, separately inside each window, along a direction orthogonal to the symmetry axis in order to detect the presence of the eye-pattern, simply modeled as five consecutive alternate dark-bright segments, i.e. a dark-bright-dark-bright-dark configuration. The central bright-dark-bright subpattern models

204

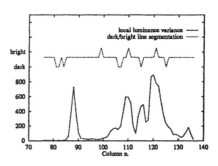

Figure 3: Estimation of the eye position: the eyes-through line is segmented into dark-and-bright patterns (dotted curve). Each matching eye-pattern is weighted by means of a reliability coefficient proportional to the local luminance variance (dashed curve). Estimation performed on the sequence "Miss America", frame n.20

the iris region while the peripheral dark-bright and bright-dark subpatterns model the regions of the eye corners.

Luminance is converted to a dark/bright binary representation $B(j, i)$ according to:

$$B(j, i) = \begin{cases} dark & if \ x(j, i) \ < \ E \\ bright & if \ x(j, i) \ > \ E \end{cases} \tag{2}$$

where E represents the local average luminance evaluated inside the window.

The binarized luminance $B(j, i)$ is then convolved with the eye pattern and the local maxima of the filter response are detected. These maxima points are subsequently weighted by means of a reliability coefficient, proportional to the local luminance variance, in order to enhance the filter response in correspondence of structured texture, typical of the eye region. The most reliable matching pattern is consequently chosen as an estimation of the iris position, as shown on the righthand-side of Figure 3.

Similar processing is then applied for the extraction of eyebrows, nose and chin parameters. The described analysis algorithms are employed both for the model adaptation and animation; however, while the model must be adapted to the speaker face with very high precision, less accurate estimates are allowed for its real-time animation. Errors occurring during the model adaptation can originate coarse approximations which become critical or even unacceptable in correspondence of somatic features. Let us consider a significant error in the adaptation of the eye submask: this results in wrong geometric modeling of the eye components and wrong association of the color parameters. The effects of this kind of error are very evident at animation time, when non-natural eye mimics is synthesized on the model.

In contrast, errors affecting the estimation of the parameters when they are employed to animate a previously adapted model, are far less important. This kind of errors leads to unprecise but always acceptable mimics synthesis. The wrong estimate of the eye or mouth aperture introduces a sligth distortion which is only temporary and most of the time invisible to the human perceiver. Somatic features parameters therefore, can be estimated by means of simplified analysis algorithms exploiting the video interframe correlation.

5. EXPERIMENTAL RESULTS

The analysis/synthesis algorithms previously described have been applied to different test sequences showing interesting possibility of application. All the simplifying hypotheses we made on the videophone scene were satisfied by the test sequences, where no occlusion occurred as well as no 3D motion of the speaker head. All the test sequences have the QCIF format with variable frame frequency ranging from 25 to 30 frames/sec. but with appreciable differences as far as the facial somatics of the speaker, background structures and lighting conditions are concerned. The curves plotted on the left-hand side of Figure 4 describe the temporal tracking of the face symmetry axis during the first frames of the sequence "Miss America". The picture on the right-hand side of the same Figure shows the result provided by the analysis algorithms for the extraction of the somatic features. In Figure 5 a frame drawn from the same sequence "Miss America" is shown together with the synthesized frame. The synthesized facial expression looks slightly different from the original one but the subjective information has been preserved. A rough evaluation of the required transmission overhead can be figured out in terms of less than 100 bit/frame for coding the animation parameters.

Further experiments are presently carried out to check whether some form of 3D information can be inferred from the analysis of multiple views of the speaker to improve the head model [13]. Finally, the integrated analysis of speech is foreseen to provide helpful information for the animation of the mouth and for lips synchronization.

REFERENCES

1. F. Pereira and L. Masera, Two-layers knowledge-based videophone coding, Proc. PCS-90, Cambridge Ma, 6-1 (1990).
2. E. Badique', Knowledge-based Facial Area Recognition and Improved Coding in a CCITT-Compatible Low-Bitrate Video-Codec, Proc. PCS-90, Cambridge Ma, 9-1 (1990).
3. K. Waters, A Muscle Model for Animating Threedimensional Facial Expression, Computer Graphics, Vol. 22, pp. 17-24 (1987).
4. N. Magnenat-Thalmann, E. Primeau and D. Thalmann, Abstract Muscle Action Procedures for Face Animation, The Visual Computer, Vol. 3, pp. 290-297, (1988).
5. D. Terzopoulos and K. Waters, Analysis of Facial Images Using Physical and Anatomical Models, Proc. IEEE 3rd Int. Conf. on Computer Vision, Osaka, pp. 727-732 (1990).
6. S. Morishima, K. Aizawa and H. Harashima, Model-based facial image coding controlled by the speech parameter, Proc. PCS-88, Turin, 4-4 (1988).
7. B.P. Yuhas, M.H. Goldstein Jr., and T.J. Sejnowski, Integration of Acoustic and Visual Speech Signals Using Neural Networks, IEEE Communications Magazine, pp. 65-71, (1989).
8. P. Ekman and W.V. Friesen, Facial Action Coding System. Consulting Psychologists Press, Stanford University, Palo Alto (1977).
9. F.I. Parke, Parameterized Models for Facial Animation, IEEE Computer Graphics and Applications, Vol.2, pp. 61-68 (1982).

206

10. K. Aizawa, H. Harashima and T. Saito, Model-Based Analysis-Synthesis Image Coding (MBASIC) System for Person's Face, Image Communication, Vol. 1, pp. 139-152 (1989).
11. R. Forchheimer and T. Kronander, Image Coding - from Waveforms to Animation, IEEE Trans. on ASSP, Vol. 37, pp. 2008-2023 (1989).
12. F. Lavagetto, A. Grattarola, S. Curinga and C. Braccini, Muscle Modeling for Facial Animation in Videophone Coding, Proc. IEEE Int. Workshop on Robot and Human Communication, Tokyo, pp. 369-375 (1992).
13. C. Braccini, S. Curinga, A. Grattarola and F. Lavagetto, "3D Modeling of Human Heads from Multiple Views", Proc. Int. Conf. on Image Processing: Theory and Application, SanRemo, pp. 131-136 (1993).

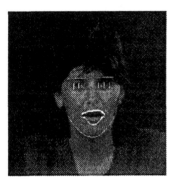 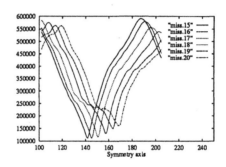

Figure 4: (Left) The face symmetry axis is tracked from frame to frame. The symmetry functional always exhibits a single sharp minimum leading to good estimations. The simulation performed on the sequence "Miss America", from frame n. 15 to frame n. 20, has yielded the results shown in the plot. (Right) Chin and jaw contours have been extracted by means of the procedure described in the text.

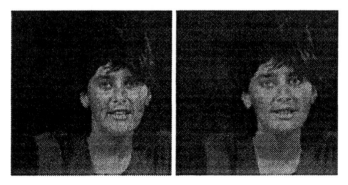

Figure 5: Example of facial expressions synthesized from "Miss America" sequence: original (left) and synthesized (right).

Time-Varying Image Processing and
Moving Object Recognition, 3 – V. Cappellini (Ed.)
© 1994 Elsevier Science B.V. All rights reserved.

HDTV coding based on edge detection and two-channel decomposition

Roberto Cusani, Stefano Corti

INFOCOM Dpt, University of Rome "La Sapienza", Via Eudossiana 18, 00184 Rome, Italy

A new procedure for HDTV coding is presented and its performance analysed. After the motion compensation, based on the classical block-matching technique, the error residual frame is split into low-pass (LP) and high-pass (HP) components. From the latter, a complex filter extracts the edge points and compute their strength and orientation. At the receiver side, these are employed to recover the HP component. The choice of a suitable (complex) edge extractor filter, maximally covering the high-frequency regions in the frequency plane, allows to preserve sharpness in the final reconstruction without enhancing the quantization noise.

1. INTRODUCTION

Many techniques have been proposed in the past years for HDTV coding. A common solution is to employ motion compensation (MC) between successive fields or frames, and then to transmit the motion vectors and the residual error sequence, the latter being coded by Discrete Cosine Transform (DCT) or Subband Coding (SBC). In particular, some coding schemes split the residual error sequence into four subbands, then code the low-frequency subband by DCT and the high-frequency subbands by PCM or DPCM [1]. This should exploit the ability of the subband decomposition to separate different image features, i.e. the discontinuities in the horizontal, vertical and diagonal directions, that exhibit low spatial correlation and for this reason are coded by simple PCM or DPCM; and the ability of DCT to efficiently code smooth intensity variations, characterizing the LP subband. Because of the subsampling property, the number of pels after the subband decomposition is the same of the original image. It is observed, however, that a single spatial event (for example, an edge point) is present in all the HP subbands, and this could limit the resulting coding efficiency. Moreover, the analysis filters are designed so as to give a perfect or nearly-perfect reconstruction in the absence of quantization noise or channel errors, thus excluding many filter choices.

In the past years some coding schemes have been designed based on the decomposition of the (still) image into low-pass (LP) and high-pass (HP) components by means of a narrow-band LP filter. The HP component, constituted mainly by edges, is coded by edge detection and contour tracing, eventually after a further decomposition in directional sub-images as in [2]. Some recent techniques analyse the HP component by a complex edge enhancer filter, its complex output representing pixel-by-pixel the edge strength, by its magnitude, and the edge orientation, by its phase [3]. The basic idea is that the transformation in the 'edge domain' allows to code the edge information in its 'natural' domain, i.e. edge strength and orientation, such as it is (probably) perceived by the human observer. This feature should overcome the drawback, with respect to the SBC strategy, of the lack of a subsampling theorem. The use of complex filters in [3] avoids the introduction of a filter bank with different directional responses, as in [2], thus reducing the amount of information to be coded and allowing a flexible coding of edge orientation.

The above concepts are exploited in the coding scheme proposed in this paper, with reference to HDTV sequences. It is based on an intraframe/interframe mode, both modes employing a two-channel decomposition (LP+HP) of the image. The HP component is further processed by an edge extractor filter, with complex impulse response, and then the position, the magnitude and the orientation of the detected edges are coded. In particular, interframe mode employs forward or backword/forward motion compensation (MC) and transmits the motion vectors as well as the motion-compensation prediction residual,frame.

The effectiveness of the procedure has been evaluated on the CCIR test sequence 'Mobile & Calendar'.

2 - THE CODING PROCEDURE

2.1. Motion compensation

Motion compensation is performed with a classic block-matching algorithm, with blocks of dimension 8x16 pels and a maximum displacement is ±7 pels. In order to achieve a better encoding, two consecutive interlaced fields are merged to generate a complete frame. This allows to halve the motion vectors transmission cost, without considerably deteriorate the quality of the estimated frame. The motion vectors, which exhibit a strong spatial correlation (contiguous blocks should have about the same displacement), are coded after an horizontal scansion with differential encoding, followed by run-length coding (RLC) of zero values. The original frame and the residual error are shown in Fig.1.

2.2. Two channel decomposition

The residual error frame is encoded after a 2-channel decomposition, employing a low pass filter and an edge extractor filter with complex impulse response (see Fig.2, left).

The low pass component is obtained by a Gaussian low-pass filter and is downsampled by a factor 2; the frequency profile $H_0(\omega_1,\omega_2)$ (see Fig.2, right) is chosen so as to avoid appreciable aliasing.

The high frequency component is obtained by the difference between the original image and the quantized, upsampled LP image. In this way a (partial) recovery of errors due to undersampling and quantization is obtained. The up-sampling operation, here as well as at the receiver side, is performed through bilinear interpolation.

The HP component is passed through a (complex) edge extractor filter $H_1(\omega_1,\omega_2)$. The output is a complex image where the magnitude represents the edge intensity, and the phase represents the edge orientation with respect to the coordinate axes. Magnitude and phase of the edge points are then quantized and transmitted.

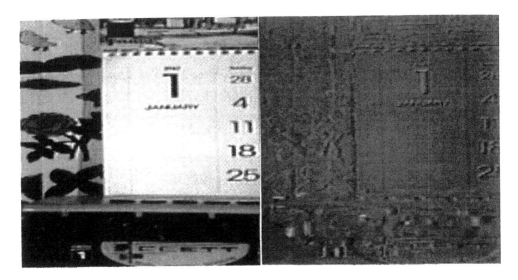

Fig.1 - Detail (zoom x4) of the original frame (left); MC prediction residual, (right)

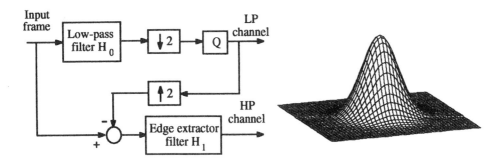

Fig.2 - Transmitter block diagram (left); frequency response of the LP filter (right)

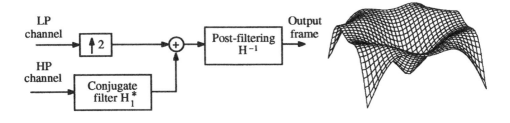

Fig.3 - Block diagram of the receiver (left); frequency response of filter MFILT (right)

The reconstruction process follows the procedure of Fig.3 (left), where the received LP image is upsampled and the edge image is filtered by the conjugate filter $H_1^*(\omega_1,\omega_2)$, so as to reconstruct (apart from the quantization) the original edge image.

After adding these two images the reconstruction is not perfect, even in the absence of quantizations and channel errors; this is because at this point the global transfer function is not unitary, but is given by the expression:

$$H(\omega_1,\omega_2) = H_0(\omega_1,\omega_2) + [1 - H_1(\omega_1,\omega_2)] \, |H_1(\omega_1,\omega_2)|^2$$

$H(\omega_1,\omega_2)$ is non-zero in the whole frequency plane, so that it is possible to invert it everywhere. This is performed in the post-filtering block, having transfer function $1/H(\omega_1,\omega_2)$.

The edge extractor filter $H_1(\omega_1,\omega_2)$ exhibits a band-pass behaviour, with a zero in the origin and a decay at very high frequencies. So, in the post-filtering step the quantization noise (introduced in both channels) is strongly amplified in correspondence of the high frequency regions, thus severely degrading the quality of the reconstructed image. Therefore a particular filter (denoted by MFILT) was designed, with maximal occupation in the frequency plane (see Fig.3, right). This reduces the effects of the quantization noise, at the same time preserving the sharpness of the reconstruction. The MFILT filter is a separable, complex filter with short impulse response, defined as:

$$h_1\,(i,j) = h_R\,(i,j) + j\,h_I\,(i,j)$$

with:

$$h_R\,(i,j) = \begin{bmatrix} 1 \\ -1 \end{bmatrix}\begin{bmatrix} -1 & 5 & 5 & -1 \end{bmatrix} = \begin{bmatrix} -1 & 5 & 5 & -1 \\ 1 & -5 & -5 & 1 \end{bmatrix}; \qquad h_I\,(i,j) = \begin{bmatrix} -1 \\ 5 \\ 5 \\ -1 \end{bmatrix}\begin{bmatrix} -1 & 1 \end{bmatrix} = \begin{bmatrix} 1 & -1 \\ -5 & 5 \\ -5 & 5 \\ 1 & -1 \end{bmatrix}$$

3. QUANTIZATION AND CODING

3.1. LP component
In the intraframe mode the LP component is the subsampled version of the original image, and is coded by a simple DPCM, with RLC encoding of the zero values. In the interframe mode the low frequency component is extracted from the motion compensation residual, and coded by DPCM with RLC coding of the zero values. The quantized LP component is shown in Fig.4, left.

3.2. HP (edge) component
The (complex) edge image at the output of the $H_1(\omega_1,\omega_2)$ filter is represented in magnitude and phase, thus directly referring to edge intensity and orientation.

The magnitude is normalized at the maximum value Vmax, and then thresholded, the threshold value being the basic parameter for reconstruction quality. Isolated pels (i.e., those surrounded exclusively by pels under threshold), which doesn't represent adequately an edge structure, are set to zero. A four level uniform quantization (over the threshold) is then performed on the edge magnitude.

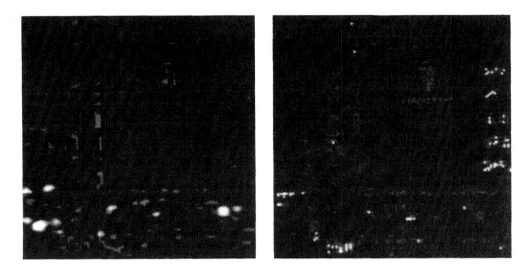

Fig.4 - Quantized LP image(left); Quantized magnitude of the HP (edge) image (right)

The quantized magnitude is coded following a 'tree-partitioning algorithm' (quad-tree). It is based on a progressive and recursive splitting of the original frame into four square sub-regions, thus localizing every significant pixel position (see Fig.5). In this way only one (or a few) null symbols are assigned to large, empty regions. The output bit stream is then RLC-coded, as such as the magnitude of the visited pels. The quantized edge component (in magnitude) is shown in Fig.4, right.

The phase is quantized by an eight level uniform quantizer, with $\pi/4$ steps; this value is a good compromise between the resulting entropy and an the accuracy in the reconstruction of edge orientation. The phase values (transmitted only for pels with magnitude over the threshold) are encoded with a 'scansion & tracing' algorithm, taking advantage of the correlation between pels belonging to the same edge: if the edge keeps its orientation in the space, all these pels exhibit the same quantized phase value. A starting edge point is at first located through a line-by-line scansion; then, on the basis of its phase value the next edge point is searched.

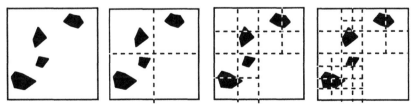

Fig.5 - Tree-partitioning algorithm for edge position coding

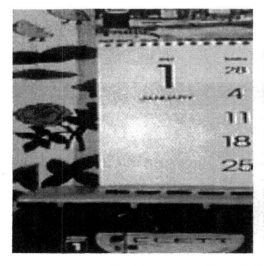

Fig.6 - Reconstructed frame (left); final residual error (right)

If the pel localised in this way is not an edge point (i.e., its magnitude is zero), the scansion phase restarts, searching for the next edge starting point. The resulting output stream, which contains long sequences of equal values, is then differentially encoded before transmission. The reconstructed frame and the final error are shown in Fig.6.

4. RESULTS ON THE TEST SEQUENCE

In our tests, we referred to HDTV transmission over 140 Mbit/sec capacity channels (CCIR 3rd hierarchical level), with a maximum available data bit/rate of about 120 Mbit/sec.

The effectiveness of the procedure has been evaluated on the CCIR test sequence 'Mobile & Calendar', with frame resolution 720x576 picture elements (pels). The numerical results, reported in Tab.I, are obtained referring to the luminance (Y) component.

The basic scheme consists in the transmission of one frame in intraframe mode (' I ' frame), followed by the transmission of NI frames encoded in interframe mode (' MC ' frames), as in Fig.7. This structure allows a periodic refresh (by means of the ' I ' frames) of the frame used for the motion-compensation, thus avoiding an excessive propagation of transmission errors and dealing with the problems coming from a change of scene. On the other hand, ' I ' frames need an higher bit/rates than ' MC ' frames at the same fixed SNR, so that NI must be chosen in accordance with the available channel capacity. In our tests we assumed NI = 6.

In Fig.8 the temporal behavior of SNR and bit rate is reported for the first seven frames of the test sequence.

Mobile & Calendar	Average SNR	Average entropy (global)	Corresponding bit/rate*
Forward MC	33,99 dB	1,02 bit/pel	~112 Mbit/sec
Forward/backward MC	34,02 dB	0,9 bit/pel	~99,5 Mbit/sec

Tab.I Performance of the proposed coding procedure (Mobile & Calendar test sequence)

Fig.7 - Intra/interframe scheme, with forward MC

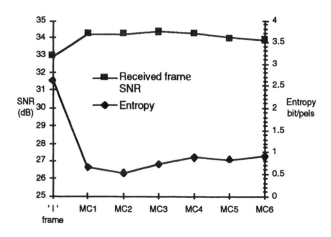

Fig.8 - Temporal evolution of SNR & Entropy (Mobile & Calendar test sequence)

---

* The bit/rate corresponding to an HDTV sequence is calculated with the formula $F_b = E \cdot (Q \cdot P_{tot} \cdot 2)$, with E the total average entropy, Q is the number of frames/sec (25), P_{tot} the total number of pels for every frame (1920 x1152) and 2 the factor due to the necessity of assigning at chrominance component an equal bit number of the luminance component.

It is also possible to assume an alternative scheme, with forward and backward motion-compensation of the 'I' frame as in Fig.9. In this way the bit rate reduces considerably (see Tab.I), even though the receiver needs a very large frame memory.

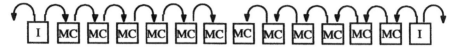

Fig.9 - Intra/interframe scheme, with forward and backward MC

5. CONCLUSIONS

The coding technique presented in this paper is based on the use of an edge extractor filter in the HP channel, which allows an encoding strategy adapted to human vision characteristics, and results in a good subjective quality of the reconstructed image. It seems to be well adapted to the structure of the MC error sequence, mainly constituted by edges. The two channel decomposition does not require a peculiar frequency profile of the analysis filters; in particular, the impulse response of the LP filter is not critical and an arbitrary FIR filter can be choosen, the only limitation being the aliasing due to the successive undersampling of this component.

In the proposed procedure the same basic coding technique has been applied to 'I' and 'MC' frames, apart from a different quantization of the LP component. It seems possible to reduce the cost of 'I' frames by a more appropriate coding, for example by using DCT. Moreover, a "post-filtering" is required; in our tests it has been performed by multiplication in the frequency domain, but in a practical application a finite response filter approximation should be necessary.

REFERENCES

1. K. Irie, R. Kishimoto, 'Adaptive sub-band DCT coding for HDTV signal transmission', Image Communication, Vol.2, No.3, pp.333-342, October 1990.

2. A. Ikonomopulos, M. Kunt, 'High compression image coding via directional filtering', Signal Processing, Vol.8, pp.179-204, 1985.

3. R. Cusani, G. Jacovitti, 'Adaptive Image Coding with Harmonic Angular Filters', Signal Processing V: Theories and Applications, Elsevier Service pub. B. V. (North Holland), pp.1377-1380, Aug. 1992.

4. 'Coding and transmission of High-definition TV signals', COST 206 Final Report, April 1992.

5. D. Le Gall, H. Gaggioni, C.T. Chen, 'Transmission of HDTV signals under 140 Mbit/s using a subband decomposition and discrete cosine transform', in Signal Processing for HDTV (L. Chiariglione ed.), pp.287-293, North-Holland, 1988.

G
REMOTE SENSING IMAGE PROCESSING

*Time-Varying Image Processing and
Moving Object Recognition, 3* – V. Cappellini (Ed.)
© 1994 Elsevier Science B.V. All rights reserved.

The concept of kinematical image and its practical use in radar target studies.

J.Bertrand [a], P.Bertrand [b] and L.Vignaud [c].

[a] CNRS and University Paris VII, 75251 Paris, France

[b] ONERA / DES, BP 72, 92322 Chatillon, France

[c] ONERA / DES and Ecole Normale Supérieure de LYON, France

1. INTRODUCTION

Capturing an image basically requires remote sensors. Unlike cameras or any imaging device, radars are usually not built to provide us with "eye ready" images, and radar imaging is always some processing on digital datas. Far from being a constraint, it enables us to freely reformulate both the notion of image and the kind of data to be collected to adapt them to the particular situation we are aiming at imaging. The following study concerns the generalisation of radar imaging [1,2] in order to allow the description of non stationary radar targets, i.e. any assembly of moving parts which may not be reflecting evenly at all time and at all frequencies, taking into account scintillating and dispersive properties. Of course, such a target requires a more complicated response to be processed as it will be detailed further on, but meanwhile it permits to access to an enlarged notion of image. This image makes use of densities in a four dimensional space : (x, t, v, f). In this space x holds for the distance of an elementary reflector, v for its radial velocity, f for its operating frequency and t for its instant of existence. A distribution $I(x, t, v, f)$ of such moving, non permanent and "coloured" bright points represents what we call a "kinematical image" to underline that it depends on t and v: the velocity of the bright points is determined by the constructive process of the kinematical image and does not result of some post-processing applied to time-dependent pictures [3].

In principle, the kinematical image of a radar target can be obtained by processing, via a coherent wavelet analysis, the values of its (inelastic) backscattering coefficient. The solvability of this problem has been established in a previous work [4] which was only dealing with theoretical considerations; it used a general approach which had been previously applied to reformulate the static radar imaging [5]. The object of the paper is to discuss the "practical" side of the subject. The two following sections respectively give a pattern of the method and a presentation of the illustrative applications which have been developed.

2. ANALYTICAL FRAME

Radar measurements consist in recording the changes undergone by an electromagnetic wave when interacting with a target. In static situations, the comparison between the incoming and outgoing waves can be performed for each frequency separately and the effect of the scattering is simply described by a complex factor depending on the frequency ("backscattering coefficient" or "radar hologram").

2.1. Inelastic backscattering coefficient

Things are however more complicated with non-stationary targets : the scattering process can only be represented by a linear operator acting on the radar signal, with a wide band of frequencies required for both emission and reception. If $S_{in}(f_1)$ is the incoming signal impinging upon the target, the outgoing signal $S_{out}(f_2)$ is given by :

$$S_{out}(f_2) = \int_0^\infty \sqrt{f_1/f_2}\, H(f_1,f_2)\, S_{in}(f_1)\, df_1 \tag{1}$$

where the kernel $H(f_1,f_2)$ is what we call "inelastic backscattering coefficient". We find it convenient to define f, the working frequency of an elementary reflector, as the frequency seen in the reflector system :

$$f = f_1 \frac{1-v/c}{\sqrt{1-v^2/c^2}} \quad \text{and} \quad f_2 = f \frac{1-v/c}{\sqrt{1-v^2/c^2}} \tag{2}$$

Thus $f = \sqrt{f_1 f_2}$ and $v = c\dfrac{f_1 - f_2}{f_1 + f_2}$ $\qquad(3)$

We will see in 2. how the function $H(f_1,f_2)$ can be calculated for some theoretical targets as well as simulated and computed through real static measurements obtained in an anechoic chamber. In this step stands the "direct problem".

2.2 Kinematical image and covariance group

Solving the "inverse problem" is a matter of finding a method to get the kinematical image $I(x, t, v, f)$, by processing the values of the backscattering coefficient $H(f_1, f_2)$. The postulate of physical relevance of the model of bright points in the (x, t, v, f) space implies a covariance constraint between the backscattering coefficient and the kinematical image. The main point is that the image has to remain coherent whatever the observer, and thus have a definite transformation law in a change of reference frame. The phase space (x, t, v, f) can be entirely explored by a particular group of point transformations $G_{\xi, \tau, \mu, \alpha}$: space-time translations $(\xi, c\tau)$, Lorentz transformations μ and dilatations α.

The isomorphism between this group and a direct product of two *affine groups* (groups of dilatations and translations) has been pointed out in [4] : the theory takes advantage of this remark, especially for the wavelet analysis presented next section. However, even if calculations have to be conducted *in the affine form*, the results will here be directly translated to the form *using the physical variables* .

The group (often named Weyl - Poincaré) acts on $H(f_1, f_2)$ and $I(x, t, v, f)$ in the following way :

$$G_{\xi, \tau, \mu, \alpha}[H(f_1, f_2)] = \alpha \, e^{-2i\pi\left[\left(f_1 + f_2\right)\xi / c - \left(f_1 - f_2\right)\tau\right]}$$

$$\times H\left(\alpha \frac{1 + \mu / c}{\sqrt{1 - \mu^2 / c^2}} f_1, \alpha \frac{1 - \mu / c}{\sqrt{1 - \mu^2 / c^2}} f_2\right) \tag{4}$$

$$G_{\xi, \tau, \mu, \alpha}[I(x, t, v, f)] = I(\alpha^{-1} \frac{1}{\sqrt{1 - \mu^2 / c^2}}((x - \xi) + \mu(t - \tau)),$$

$$\alpha^{-1} \frac{1}{\sqrt{1 - \mu^2 / c^2}}((t - \tau) + \mu / c^2(x - \xi)), \frac{\mu + v}{1 + \mu v / c^2}, \alpha f) \tag{5}$$

2.3 Wavelet analysis

The results of wavelet analysis associated with the affine group suggest, in this context, a very natural and convenient way to proceed. Suppose we know a complex function $\Phi_{x_0, t_0, v_0, f_0}(f_1, f_2)$, called "mother-wavelet", *attached* to the particular point P_0 with coordinates $(x_0 = 0, t_0 = 0, v_0 = 0, f_0 = 1)$: this means that it represents the inelastic backscattering coefficient of an elementary reflector localised around the origin of space-time, with velocity approximately equal to zero and a working frequency in the vicinity of 1. Since P_0 can be brought to an arbitrary point P thanks to a transformation (ξ, τ, μ, α) of the group, the points of phase space can be labelled by the parameters of the group : formula (5) leads to $(x, t, v, f) = (\xi, \tau, -\mu, \alpha^{-1})$

Then, using the transformation group on the mother-wavelet, we are able to build a displaced wavelet $\Phi_{x,t,v,f}(f_1, f_2)$ *attached* to any point (x, t, v, f).

$$\Phi_{x,t,v,f}(f_1, f_2) = G_{\xi, \tau, \mu, \alpha}\left[\Phi_{x_0, t_0, v_0, f_0}(f_1, f_2)\right]$$

$$= f^{-1}e^{-2i\pi\left[(f_1 + f_2)x/c - (f_1 - f_2)t\right]}$$

$$\times \Phi_{x_0, t_0, v_0, f_0}\left(\frac{f_1}{f}\frac{1 - v/c}{\sqrt{1 - v^2/c^2}}, \frac{f_2}{f}\frac{1 + v/c}{\sqrt{1 - v^2/c^2}}\right) \tag{6}$$

With $\{\Phi x, t, v, f\}$, we have formed an overcomplete basis which is stable by the action of the covariance group. The components of $H(f_1, f_2)$ in this basis are called the wavelet coefficients :

$$C(x, t, v, f) = \langle H, \Phi_{x,t,v,f}\rangle \tag{7}$$

where $\langle, \rangle$ represents the invariant scalar product for $G_{\xi, \tau, \mu, \alpha}$:

$$\langle H, H' \rangle = \int_{\Re^{+2}} H(f_1, f_2) H'^*(f_1, f_2) df_1 df_2 \tag{8}$$

We are also provided with an isometry relation and a reconstruction formula which allow us to finally define the kinematical image as :

$$I(x, t, v, f) = \frac{1}{K_\Phi}|C(x, t, v, f)|^2 \tag{9}$$

where K_Φ is the wavelet admissibility coefficient :

$$K_\Phi = \int_{\Re^{+2}} \Phi_{x_0, t_0, v_0, f_0}(f_1, f_2)\Phi^*_{x_0, t_0, v_0, f_0}(f_1, f_2)\frac{df_1}{f_1}\frac{df_2}{f_2} \tag{10}$$

This image only depends on the choice of the "mother wavelet". Next section will discuss about the choice of an optimal wavelet through the study of the uncertainty relations.

2.3 Optimal wavelet and uncertainty relations

The size of the elementary reflectors in the (x,t,v,f) space obeys uncertainty relations. These relations come from the non commutation of some of the operations that can be performed on the analysed signal. They are non trivial because of coupled terms.

Here it is possible to realise all the transformations on $H(f_1,f_2)$ by the infinitesimal operators F, V, β_X, β_T :

$$F=\sqrt{f_1 f_2} \; ; \; V=c\frac{f_1-f_2}{f_1+f_2} \; ; \; \beta_X=-\frac{1}{2i\pi}\left(f_1\frac{d}{df_1}+f_2\frac{d}{df_2}+1\right) \; ; \; \beta_T=-\frac{1}{2i\pi}\left(f_1\frac{d}{df_1}-f_2\frac{d}{df_2}\right) \quad (11)$$

The commutation relations between these operators are :

$$[\beta_X,V]=[\beta_T,F]=0 \; ; \; [\beta_X,F]=-\frac{1}{2i\pi}F \; ; \; [\beta_T,V]=-\frac{c}{2i\pi}\left(1-\frac{V^2}{c^2}\right) \quad (12)$$

Thus the uncertainty relations are given using the standard deviations corresponding to the operators :

$$\sigma_{\beta_X}^2 \sigma_F^2 \geq \frac{1}{16\pi^2}\langle F\rangle \quad \sigma_{\beta_T}^2 \sigma_V^2 \geq \frac{c^2}{16\pi^2}\left(1-\frac{\langle V^2\rangle}{c^2}\right) \quad (13)$$

This leads us to propose an optimal mother-wavelet Φ_0 which corresponds to the minimum of the uncertainty relations, i.e the state for which equality holds in (13). It represents a trade-off between localisation and spreading in the phase space.

$$\Phi_0(F,V)=\frac{1}{F}e^{-2\pi\lambda_F(\ln(F)-F)}\times(1-V^2/c^2)^{-\pi c\lambda_V} \quad (14)$$

In this formula λ_F and λ_V are adjustable parameters which characterise the spreading of the states in variables F, V, thus allowing to control the resulting image "blur" and resolution accordingly.

With $\Phi_{x_0,t_0,v_0,f_0}(f_1,f_2)=\Phi_0\left(\sqrt{f_1 f_2},c\frac{f_1-f_2}{f_1+f_2}\right)$ formula (6) can now be rewritten :

$$\Phi_{x,t,v,f}(f_1,f_2)=f^{-1}e^{-2i\pi\left[(f_1+f_2)x/c-(f_1-f_2)t\right]}$$

$$\times\Phi_0\left(\frac{\sqrt{f_1 f_2}}{f},\frac{(1-v/c)f_1-(1+v/c)f_2}{(1-v/c)f_1+(1+v/c)f_2}\right) \quad (15)$$

3. SELECTED APPLICATIONS :

A convenient object to present our results is a digital colour movie made of a sequence (t) of images, whose pixel are situated at range-velocity (x,v) with colour or grey scale intensity coding $I(x,t,v,f)$. Frequency information may be added via colour (with R.G.B. coding 3 frequencies for instance): a "bright point" is then said to be "white" if all its frequency components are reflected, and "coloured" if it does not reflect every of them.

3.1 Some theoretical simulations :

In order to experiment our image construction frame, we need to know $H(f_1,f_2)$ for some simple theoretical cases.

A similar method to the one developed by Salopek and Razavy [6] consists in expressing the E.M. field at the reflection boundary. For a permanent mirror situated at $x(t)$ it leads to :

$$H(f_1,f_2) = \frac{2f_1}{f_1+f_2} \int e^{-2i\pi\left((f_1-f_2)t+(f_1+f_2)x(t)/c\right)} dt \tag{16}$$

For a scintillating mirror situated at x_0 with a reflectivity modulation $R(t)$:

$$H(f_1,f_2) = \int R(t) e^{-2i\pi\left((f_1-f_2)t+(f_1+f_2)x_0/c\right)} dt \tag{17}$$

On this model, we may simulate $H(f_1,f_2)$ for several moving bright points with various non stationary behaviour including scintillating and dispersive features. Perhaps one of the best candidate for these simulations is to put for $H(f_1,f_2)$ the wavelet itself, as it can be easily localised and stretched. The special case $H(f_1,f_2)=1/\sqrt{f_1f_2}$ can be interpreted as a flashing "white" point localised at $x=t=0$ with a total uncertainty concerning its velocity.

Computations have shown a good accuracy with forecast uncertainty relations and resolution power.

3.2 Application to Range-Doppler radar imaging:

Laboratory studies of radar targets make a general use of a monostatic coherent radar. In a typical application the target rotates around a fixed axis perpendicular to the radar line of sight and polarisation dependence is disregarded. The acquired data is a sampling of the complex backscattering coefficient for different frequencies of irradiation and different angular orientations of the target: $H(f_1,\theta)$.

If we simulate a virtual rotation of the target by letting $\theta \rightarrow \theta(t) = \Omega t$, we may express

$$H(f_1, f_2) = \int H(f_1, \Omega t) e^{-2i\pi f_2 t} dt = FT\left(H(f_1, \Omega t)\right) \tag{18}$$

Wavelet analysis is used on $H(f_1, f_2)$ to get $I(x, t, v, f)$. For a fixed frequency, it gives Range-Velocity images of the target through time (or angular orientation, equally). But furthermore, the velocity is proportional to the cross-range resolution (in far field approximation), which enables us to see $I(x, t, v, f)$ as a classical 2-dimensional image $I(x, \theta, y, f)$ of the target reflecting f in the θ-direction. An illustration is presented in Figure 1, where we present a reduced-scale missile as seen via range velocity imaging at different moments of its rotation and for an average frequency of the analysing band.

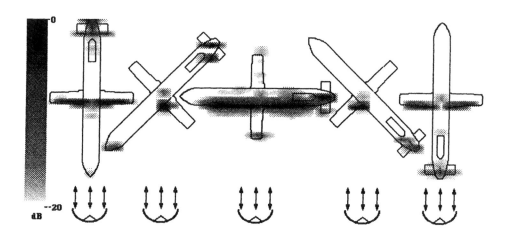

Figure 1 : Range-Velocity Radar Images of a Rotating Target (measurements by ONERA DES/SMS).

4. CONCLUDING REMARKS :

This method is a natural extension of the "bright point" analysis of the static radar imaging. It allows to display the information contained in the inelastic backscattering coefficient of a target as a kinematical image, which maps the density of bright points in a 4-dimensional space (position, time, velocity, and frequency). In some sense, it generalizes works on the range-velocity analysis [7] in relaxing the hypothesis of bright point reflecting uniformly for all frequencies and all times. It differs from the traditional movie-like image in explicitly including the process of velocity in its frame.

Numerical simulations on theoretical coefficients, modeling various dynamic situations such as reflectivity modulations in time, "colour" modulations, and accelerations, have illustrated the flexibility of the representation. The method has also been applied successfully to the analysis of a bidimensional static target by simulating a time-dependant rotation. Finally, it is worth stressing that the method yields a complete control of the uncertainty relations, with respect to the four variables and this has been verified in the numerical applications.

From a more general point of view, we try to solve an "inverse problem" and show that the wavelets transformation, when combined with a physical approach, may bring a solution to it.

REFERENCES :

1. D.A.Ausherman, A.Kozma, J.L.Walker, H.M.Jones and E.C.Poggio, "Developments in Radar Imaging", IEEE Trans. on Aerospace and Electronic Systems, **20**, pp.363-398, (1984).

2. D.L.Mensa *High resolution radar imaging* Artech House, USA, 1981.

3. See articles on "optical flow" etc... For instance, L.Jacobson and H.Wechsler. "Derivation of optical flow using a spatio-temporal-frequency approach". Computer Vision, Graphics, and Image processing 38, 29-65 (1987).

4. J.Bertrand and P.Bertrand, "Reflectivity study of time-varying radar targets", Proc. 6th European Signal Processing Conference. EUSIPCO-92 pp.1813-1816.

5. J.Bertrand, P.Bertrand and J.Ovarlez, "Dimensionalized wavelet transform with application to Radar Imaging", Proc. IEEE Int. Conf. Acoust., Speech, Signal Processing ICASSP-91 pp.2909-2912.

6. M.Razavy and D.Salopek, "On the inverse problem of reflection from a moving boundary", Europhys.Lett., 2 (3), pp.161-165, (1986).

7. E. Feig and F. A. Grünbaum, "Tomographic methods in range-Doppler radar", Inverse Problems 2 (1986), pp 185-195.
    H. Naparst, "Dense target signal processing", IEEE-IT 37 (1991), pp 317-327.

Time-Varying Image Processing and
Moving Object Recognition, 3 – V. Cappellini (Ed.)
© 1994 Elsevier Science B.V. All rights reserved.

Change Detection in Remote-Sensing Images by the Region-Overlapping Technique

P.Pellegretti, M.Salvatico, S.B.Serpico

Dept. of Biophysical and Electronic Eng.
Univ. of Genoa - ITALY

The problem of detecting changes in multitemporal images is considered to the purpose of developing a method that reduces the effects of registration, segmentation and classification errors on detection accuracy. To this end, a previously developed data fusion technique is used, which creates a spatial correspondence between the segmentation regions extracted from couples of images. Neural Networks are used to perform single-image classification. Results obtained on couples of images are compared to detect real changes.

1. INTRODUCTION

Detection of changes in remote-sensing images [1] is an important task for territory control in various applications, such as the monitoring of forests, coast erosion, desertification, and so on.

Let us consider the case in which the third dimension is negligible, that is, the data to be analyzed are 2D images of objects which are flat or, anyway, whose perspective analysis is not relevant. In addition, let us consider objects that may be described by region primitives (extracted by image segmentation). In this situation, the detection of the changes occurred during the time between the acquisition of two images may be obtained by classifying the regions of the two images separately, then comparing the classification results. However, this procedure implies that a spatial or a structural correspondence between the two images be created. In order to obtain a spatial correspondence, the two images must be registered, so that each couple of pixels at the same co-ordinates in the two images corresponds to the same ground point. A problem with this approach is that registration inaccuracies cannot be avoided, so the correspondence is never perfect. The structural approach tries to find a correspondence between arrangements of primitives (regions, in our case) extracted from the two images. This approach does not require registration; on the other hand, it involves a quite complex structural matching process. In addition, it is not sensitive to small changes of the primitives (e.g., a small change in the boundary of a region), which is an important information in several applications. Finally, both approaches may suffer for segmentation and classification errors.

Purpose of this paper is to present the basic concept of a method for detecting changes in multitemporal images, which is based on the "spatial-correspondence" approach, and reduces the effects of registration, segmentation and classification errors on detection accuracy. A classification of changes into categories is defined, which is used to characterize detected changes.

2. TYPES OF CHANGES

Various types of changes may be detected by comparing the classification results of two images. If this results are given by segmentation maps, where a cover type (i.e. a class label) is assigned to each region, then we may define the following four types of changes:

1. change in the boundary of a region, but not in its cover type ;
2. change in the cover type of a region but not in its boundary;
3. merging of regions with different cover types into a region with a single cover type;
4. splitting of a region into subregions with different cover types.

A couple of images, with two instances for each type of change, is shown in Fig.1. The image at left represents the first component, extracted by the Karhunen-Loeve transform, of a set of multiband images acquired by an Airborne Thematic Mapper (ATM) sensor. The image at right has been generated from the previous one by introducing a rotation of 0.5 degrees, centred in the lower left corner, and simulating some changes. Fig.2 shows the results of the segmentation of the images in Fig.1. In the left image we have indicated, by the corresponding number in the list of the types of changes, the regions in which simulated changes have been introduced.

3. EFFECTS OF PROCESSING ERRORS ON DETECTION ACCURACY

Three kinds of errors may occur in the processing of a couple of images, that affect change detection accuracy: registration, segmentation and classification errors.

3.1 Registration Inaccuracies
We hypothesize that the two images to be compared have already been registered to a few pixels of accuracy, so that the error is smaller than the minimum size of objects of interest. This kind of inaccuracy gives rise to a displacement of the boundaries of corresponding regions in the two images, which may be detected as a change. For example, a change of the first type (change in the boundary of a region) may be erroneously detected.

Fig.1. A couple of images with two instances for each type of change. Left: first principal component of a set of multiband Airborne Thematic Mapper images of an agricultural area (Feltwell, UK). Right: the same image after a rotation of 0.5 degrees with the centre in the lower-left corner and the insertion of simulated changes.

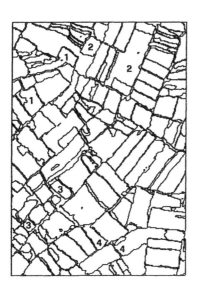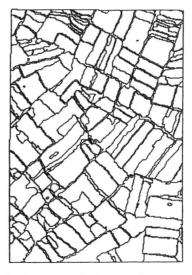

Fig.2. Segmentation results obtained on the images of Fig.1. Left: the regions including simulated changes are indicated by the corresponding numbers in the list of changes.

3.2 Segmentation Errors

The most serious segmentation error is the merging of regions with different cover types in a single region. This may lead to different errors. If the error is in the first of the two images, then the splitting of a region into more regions is detected (change of the fourth type). If the error occurs in the second image, then the merging of more regions into one region is detected (change of the third type). In addition, this kind of segmentation error causes also a classification error: a single cover type is assigned to more regions with different cover types. The single cover type may also be different from all the cover types of the regions involved.

3.3 Classification Errors

In addition to those induced by segmentation errors, other classification errors may occur. Mainly, they can be due to poor separability of classes in presence of noise and class variability, to an inadequate training set (in the case of supervised classification), to intrinsic limitations of the adopted classifier. If an identical error occurs in the two images to be compared, then no effect would be observed on change detection. More frequently, different classification errors may appear in the two images. In this case a false change of cover type (second type) would usually be detected.

4. CHANGE DETECTION BY THE "REGION OVERLAPPING" TECHNIQUE

4.1 The Region Overlapping Technique

"Region-Overlapping" (ROVL) [2] is a technique for fusing multisensorial or multitemporal images. The basic idea is to superimpose the region boundaries derived from the segmentation maps of two images. In this way, a new segmented map, with more regions than each of the two original maps, is obtained. In such a map, each region is labelled as a "kernel", that is, a significant region, or a "sliver", that is, a spurious region generated by registration or segmentation inaccuracies. In the multitemporal case, slivers may also be generated by small changes in region boundaries.

4.2 Change Detection

The map produced by ROVL is used in our approach as a common segmentation map for the two images. All changes are revealed by regions with two different cover types in the two images. On the other hand, false changes may be detected due to processing errors. It is less frequent that a change is masked by a processing error, so that it is not detected.

For classification purposes, we have adopted multilayer feed-forward neural networks (NNs), with sigmoidal nonlinearities [3]. We use such networks as supervised classifiers; to this end, we train them with the backpropagation algorithm. In particular, we give as input to the NN the feature vector of a region and we obtain as output a vector whose components represent the probabilities that such a region belongs to the different classes:

$$out_i(\underline{X}_j^k) = p(r_j^k \in \omega_i / \underline{X}_j^k) \tag{1}$$

where out_i is the i-th component of the NN output vector ; r_j^k is the j-th region of the segmentation map produced by ROVL, as it appears in the k-th image, and $\underline{X}_j^k$ is its feature vector (computed in the k-th image); ω_i is the i-th class.

A classification map for each of the two image may be obtained by taking, for each region, the component of the NN output vector with the maximum value and assigning the region to the corresponding class.

The information obtained in this way may be used to detect changes. By consulting also the segmentation maps of the two images separately and the class label of neighbouring regions, the type of each change may be identified. To this end, we are developing a fuzzy rule-based module.

5. REDUCING THE EFFECTS OF PROCESSING ERRORS

5.1 Detection of Registration Inaccuracies

Registration inaccuracies may be revealed by the presence of typical arrangements of slivers created by the displacement between the boundaries of corresponding regions in the two images. In the image areas where inaccuracies occur, the number of slivers is also higher than usual. A simulated example of registration inaccuracy is given by Fig.1 (as already said, the right image has been obtained by rotating the left image). The output map produced by ROVL is given in the left part of Fig.3, and the related slivers are shown in its right part. As one can note, many slivers appear, especially in the higher part of the image, where the effects of the rotation are stronger. Also, the orientation of slivers is not uniformly distributed, but it is influenced by the parameters of the rotation. The results of this kind of analysis may be used for improving the registration, or just to identify the reason of slivers in a registration inaccuracy rather than in a change.

5.2 Correction of Segmentation Errors

Let us consider the above mentioned segmentation errors, consisting in the merging of neighbouring regions with different cover types. If such errors occur in one image but not in the other, then they are automatically corrected by ROVL. In fact, thanks to the overlapping process, it is enough that the separation among regions be present in one of the two image, to make it take place in the output map. One can observe this kind of correction by comparing the left parts of Figs.2 and 3. The latter shows a more detailed segmentation map, due to the fact that the segmentation in the right part of Fig.2 (which has been overlapped to that in the left part to obtain the left map in Fig.3) is more detailed. For instance, a segmentation error in the upper left part of the image has been corrected.

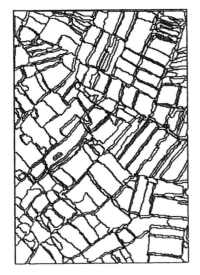 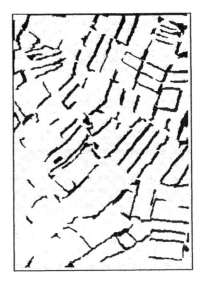

Fig.3. Results obtained by applying the Region Overlapping technique to the segmented images in Fig.2. Left: output segmentation map. Right: display of the "slivers" (in black).

5.3 Reduction of Classification Errors by a Probabilistic Approach

Depending on the application, the occurrence of changes in a time interval between two images may be more or less probable. This is equivalent to say that there is a more or less strong correlation between the two images. If sufficient a priori knowledge (or training images to extract it) is available, one may define (or compute) the probability of change and, possibly, the probability of each type of change. This may be used for classification purposes. In particular, the above mentioned NNs give us, for each image separately, an estimation of the probability that every region belongs to every possible class. Then, we can use these estimations to classify each region in the two images contemporaneously, and to decide if a change occurred or not. We assign a region to the class ω_i in the first image, and to the class ω_l in the second image, where ω_i and ω_l is the couple of classes that gives the following maximum:

$$\max_{\omega_i, \omega_l} \quad p(r_j^1 \in \omega_i, r_j^2 \in \omega_l / \underline{X}_j^1 \underline{X}_j^2). \tag{2}$$

Let us consider the case in which an estimation of the probability of change, and of no change is available. If one hypothesizes that: i) the feature vectors of the samples of the class ω_m are determined by a constant feature vector $(\underline{X}_m)$ plus an additive random vector $(\underline{n})$ due to noise:

$$p(\underline{X}_j^k / r_j^k \in \omega_m) = p(\underline{X}_j^k = \underline{X}_m + \underline{n}^k); \tag{3}$$

ii) $\underline{n}^1$ is independent from $\underline{n}^2$; iii) the probability of change and of no change for a given region is independent from the class it belongs to; then one may look for the classes ω_i and ω_l that give the maximum of the following expressions:

$$\text{for } i=l \quad p(r_j^1 \in \omega_i / \underline{X}_j^1) \, p(r_j^2 \in \omega_i / \underline{X}_j^2) \, p(\text{no change}) / p(\omega_i) \tag{4}$$

$$\text{for } i=/l \quad p(r_j^1 \in \omega_i / \underline{X}_j^1) \, p(r_j^2 \in \omega_l / \underline{X}_j^2) \, p(\text{change}) / (1 - p(\omega_i)). \tag{5}$$

The couple of classes $(\omega_i; \omega_l)$ can be different from that found by looking for the maximum of the two probabilities:

$$p(r_j^1 \in \omega_i / \underline{X}_j^1) \tag{6}$$

$$p(r_j^2 \in \omega_l / \underline{X}_j^2) \tag{7}$$

separately. In particular, this is useful to solve ambiguous situations in which, among the components of the NN output vector of a region, the maximum value is just a little larger than at least one of the remaining values.

Obviously, in addition to the time correlation between two images, one can improve classification results by utilizing also the spatial correlation between the subparts of the same image. The correlation between regions may be more or less high depending on the application and on segmentation results. For example, if most ground objects are fragmented in many regions, then a high correlation can be observed.

6. CONCLUSIONS

In this paper, a new method of change detection has been proposed, which is based on a previously developed technique for the fusion of couples of images. Different types of changes have been defined; a couple of images with simulated instances of each change type have been shown. The effects of image processing errors on change-detection accuracy have been considered. In this view, the proposed change detection method seems to be able to reduce such effects remarkably. Simulated images of the type shown in this paper will be the first data on which we will experiment our change detection system, which is currently under development.

REFERENCES

1. A.Singh, "Digital Change Detection Techniques Using Remotely-Sensed Data" Int. J. Remote Sensing, vol.10 (6), pp.989-1003.

2. A.Agostinelli, C.Dambra, S.B.Serpico, G.Vernazza, "Multisensor Data Fusion by a Region-Based Approach", Int. Conf. on Acoustics Speech and Signal Processing, ICASSP 91, Toronto (Canada), May 14-17, 1991, vol.4, pp.2617-2620.

3. J.L. McClelland, D.E.Rumelhart and PDP Research Group, Exploration in Parallel Distributed Processing: A Handbook of Models, Programs, and Exercises, 1988.

Time-Varying Image Processing and
Moving Object Recognition, 3 – V. Cappellini (Ed.)
© 1994 Elsevier Science B.V. All rights reserved.

ASSESSMENT OF TEMPORAL VARIATIONS IN INFORMATION CONTENT OF TM SCENES FOR DIFFERENT VEGETATION ASSOCIATIONS

C. CONESE and F. MASELLI

I.A.T.A.-C.N.R., P.le delle Cascine 18, 50144 Firenze, Italy.

ABSTRACT

The evaluation of the information content of satellite data about different cover types is important for ecological studies. In particular, it is fundamental for the multispectral scenes acquired on vegetated surfaces by the Landsat Thematic Mapper, which was specifically planned for terrestrial applications. Such information content depends on several factors which can not be established "a priori" and, above all, it can strongly vary during a growing season due to phenological variations of plants. A nonparametric method based on Mutual Information Analysis was recently proposed by the authors to estimate the information expressed by TM images about digitized cover categories. In the present work this property is employed to measure the information content of three TM acquisitions from a rural area of Tuscany in different seasons (winter, spring and summer). Six vegetation cover categories were considered representative for four main plant associations (coniferous and deciduous woods, olive groves and vine-yards). The information content of the TM scenes was very different for these associations and strongly variable during the growing season. These results are of high interest for their ecological implications and should be kept in mind when dealing with multitemporal TM imagery.

1. INTRODUCTION

After the advent of third generation satellites such as Landsat 5 and Spot, remotely sensed data are routinely used for monitoring earth resources on different spatial scales. In particular, the seven channels of Landsat Thematic Mapper were specifically planned to assess vegetation status, and several examples of this application are available (Horler and Ahern, 1986, Conese et al., 1988, Franklin et al., 1990, Cohen and Spies, 1992). In these cases, a point of primary importance concerns the informative value of TM imagery taken over different landscapes and in different seasons for vegetation discrimination. Clearly, the information content of TM images depends other than on the sensor's characteristics on the spectral features of the imaged scenes and, particularly, for vegetated areas, on the type, quantity and status of plants and on their phenological stages. As such, information content is expected to be extremely variable in space and time, and the study of these variations deserves particular attention.

Up to now, only few investigations have dealt with the information content of TM bands regarding vegetation cover types (Horler and Ahern, 1986, Mausel et al., 1990). Most of these studies were based on the use of parametric procedures, such as eigenvector analysis and divergence measurement. These procedures require some assumptions on the distributions of the data which are often questionable. Thus, it can be supposed that nonparametric methods would be more appropriate for estimating the information content of TM data about specific themes.

Such a nonparametric method was recently proposed by the authors (Conese and Maselli, 1993). That method is based on the concept of common entropy and is particularly suited to evaluating the informative value of TM scenes for vegetation discrimination. In the present work, the method is applied to TM scenes taken in three seasons (winter, spring and summer) on a rural area in Tuscany (Central Italy) mainly covered by coniferous and deciduous woods, olive groves and vine-yards. The information content of each band relative to these cover types is first estimated; then, some advanced classifications are performed to assess the actual value of that information. The results show some inportant aspects connected with the utilization of multitemporal TM imagery.

2. STUDY AREA AND SATELLITE DATA

The study site is located south of Florence at approximately 43° 40' North latitude, 11° 15' East longitude. This rural area belongs to the so called "Chiantishire" and corresponds to 4 geographic maps 1:25000 each of about 10x10 km. The terrain is gently rugged, with elevations ranging from 200 to 800 m. Its main cover types, fully described in Conese and Maselli (1991), are oak and chestnut woods, pine forest, olive groves and vine-yards.

The ground references were collected in summer 1988 by direct ground survey and the aid of aerial photographs. Six vegetation cover types were chosen as representative for the area (Table 1). Small plots of about 1 ha were considered and digitized in the reference system of the 4 geographic maps. To mantain compatibility with the TM imagery, a pixel size of 30x30 m was selected, so that each plot corresponded to a square of 3x3 pixels, with different grey-values for each cover type.

The TM images were extracted from three frames 192/30 acquired on 20 February, (scene A), 25 May (scene B), and 14 August (scene C), 1988. For each passage 4 sub-scenes were geometrically corrected and superimposed on the 4 maps by a bilinear interpolation algorithm trained on ground control points. The geometric accuracy of the process was below 1 pixel for both X and Y axes.

Table 1 - Vegetation cover classes considered and relevant numbers of reference pixels.

Class	Number of Reference Pixels
1) Pine forest	423
2) Chestnut wood	207
3) Oak wood 1	459
4) Oak wood 2	216
5) Vine-yard	387
6) Olive grove	405

3. DATA PROCESSING

3. 1. Mutual Information Evaluation

A method based on Mutual Information Analysis (MIA) was recently proposed by the authors (Conese and Maselli, 1993) to evaluate the information content of TM scenes with respect to a particular theme. The procedure uses the concept of common entropy between two variables which vary jointly (Kullback, 1959, Davis and Dozier, 1990). In our case, the two variables are the TM image of a certain channel and the superimposed digitized map with the ground references. For the reference image, the entropy H(Y) can be computed as:

$$H(Y) = - \sum P_{(i)} \ln P_{(i)} \tag{1}$$

where
$P_{(i)}$ = relative frequency of grey-level i.

From this definition, the global information is:

$$I(Y) = N \, H(Y) \tag{2}$$

where
N = number of elements (pixels) in the image.

The joint information shared between the two images can be computed as:

$$I(X,Y) = N \ln N + \sum \sum F_{(i,j)} \ln F_{(i,j)} - \sum F_{(i,+)} \ln F_{(i,+)} - \sum F_{(+,j)} \ln F_{(+,j)} \tag{3}$$

where
$F_{(i,j)}$ = frequency of X and Y with levels i and j;
$F_{(i,+)}$ = marginal summary of level i;
$F_{(+,j)}$ = marginal summary of level j.

Followingly, the mutual information (MI) expressed by the TM image with respect to the digitized map can be computed as:

$$MI(X,Y) = I(X,Y)/I(Y) \tag{4}$$

The methodology was applied to the 7 bands of the 3 TM passages. MI was evaluated both globally for the 6 cover classes (Figure 1) and for each single class against all the others; in Figures 2a-c the results for four representative cover types are shown (classes 1, 3, 5, 6).

3.2. Maximum Likelihood Classifications

In order to quantify the importance of the estimated mutual information content for discriminating the cover types of the area, the values previously obtained were compared with the classification accuracies derived from each TM band. An advanced maximum likelihood classifier was adopted to exploit its mixed parametric/nonparametric properties (Skidmore and Turner, 1988, Maselli et al., 1992).

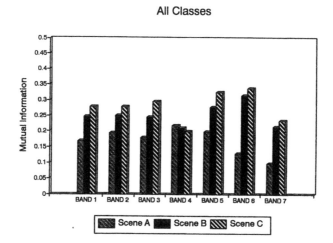

Figure 1 - MI of the 3 study scenes estimated for all classes.

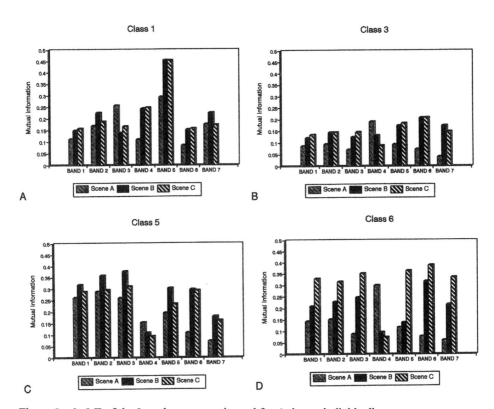

Figure 2a-d - MI of the 3 study scenes estimated for 4 classes individually.

As demonstrated in the previous work (Maselli et al., 1992), that classifier is quite efficient and robust and can be used with success also in heterogeneous, complex landscapes. In essence, the classifier works so as to minimize the modified discriminant function D for each class:

$$D = (X-M)' \, C^{-1} \, (X-M) + \ln |C| - 2 \ln Pr \tag{5}$$

where
X = pixel vector;
M = mean vector of the class considered;
C = variance-covariance matrix of the class considered;
Pr = prior probability of the class considered derived from the frequency histograms of the training sets.

The classification process was trained on all the ground references and applied separately to each of the 21 TM bands. The classification accuracies were estimated as Kappa coefficients of agreement compared to the reference pixels. The Kappa coefficient takes into account all the elements of a confusion matrix and was recommended as a standard measurement in remote sensing studies by several authors (Congalton et al., 1983, Hudson and Ramm, 1986, Rosenfield and Fitzpatrick-Lins, 1986). Mathematically, the Kappa coefficient can be computed from an error matrix as:

$$K = (N \sum x_{(i,i)} - \sum x_{(i,+)} x_{(+,i)}) / (N^2 - \sum x_{(i,+)} x_{(+,i)}) \tag{6}$$

where + represents summation over the index of the matrix.

The Kappa coefficients found for the 21 classifications were plotted and regressed against the relevant mutual information levels in Figure 3.

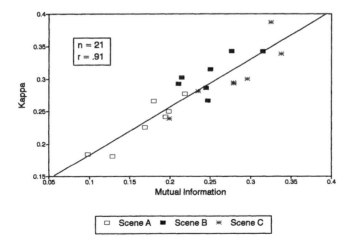

Figure 3 - Relationship between MI and Kappa for the 21 study images.

4. RESULTS

4.1. Mutual Information Analysis

The results of MIA show several interesting features of the data set examined. For all the six vegetation classes there is a well definite pattern of MI content which characterizes the four main spectral intervals of the TM data. The three visible bands (1-3) vary almost simultaneously, with a marked increase in information content from winter to summer (scenes A-C). In the near infrared channel (band 4), MI is always approximately constant, so that its relative contribution is the highest in winter and decreases in the other seasons. The middle infrared bands (5 and 7) show a steady increase similar to that of the visible channels. The thermal infared band (6) has low mutual information in winter but the highest levels in spring and summer.

As regards the single classes, the most noticeable feature of class 1 (pine forest) is the high informative value of band 5 and the low value of band 6 (Figure 2a). Class 2 (chestnut wood), has a more irregular trend which is not shown here. Class 3 (dense oak wood) behaves almost in the same manner as class 4 (thin oak wood); they show a low information content in all bands, with a trend which is very similar to the global one (Figure 2b). More peculiar characteristics can be noted for the two agricultural cover types. Class 5 (olive grove) has always high information content in the visible bands and low in band 4 (Figure 2c); a peak in spring is present for the middle infrared bands, while band 6 is highly informative only in spring and summer. For class 6 (vine-yards), a high information content in band 4 in winter can be seen, with a strong decrease in the other seasons; the opposite trend is present for the other bands, with notable values in the visible, middle infrared and thermal infrared intervals.

To comment on the present results, it can be noted that, in general, the increase in information content from winter to summer for the visible bands is probably due to the improved contrast during the growing season between cover types with low and high leaf coverage (LAI = leaf area index). The same increase in the middle and thermal infrared channels can be attributed to a similar higher contrast in apparent moisture content and evapotranspirative power, respectively. Accordingly, the different behaviours in these three spectral intervals for each class are mainly connected with the relevant dynamics of leaf expansion and evapotranspiration during the season. Also, the moderate information content of band 4 in spring and summer is probably due to the high, almost uniform reflectance of both vegetation and soils in this band.

4.2. Image Classifications

As can be seen from Figure 3, the use of single bands, each with limited discriminative power, leads do relatively low Kappa accuracies (0.15-0.40). Concerning the 3 seasons, the winter acquisition has the lowest accuracies, while higher scattering of Kappa coefficients is present in summer with respect to spring.

In general, the linear relationship between mutual information and Kappa holds quite well, with a high correlation coefficient ($r = 0.91$). In effect, the classifier used has partly nonparametric properties which exploit the same information measured by MIA. Accordingly, the scattering found in the relationship can be partially attributed to the parametric component of the classifier, which is presumebly in lower accordance with mutual information measures (Maselli et al., 1992).

5. CONCLUSIONS

The importance of multitemporal information for the interpretation of remotely sensed data has been highlighted in numerous investigations (Buttner and Csillag, 1989, Conese and Maselli, 1991). Particularly, multitemporal image data are fundamental for studying vegetation associations, both natural and agricultural, due to the complex dynamics of plant species during a growing season.

The present work has focussed on the evalutation of TM data from different seasons to discriminate plant associations in a complex rural area in Tuscany. The methodology, based on mutual information analysis, is specifically suited to evaluating the nonparametric information content of each TM band for the identification of the cover types examined.

The results obtained have highlighted the importance of different band subsets for discriminating the various classes in the 3 study seasons. Globally, the winter passage has been shown to be the least informative, while the spring and summer passages have almost the same information content. Also, the mutual information estimates show marked variations for the different classes.

These results should be kept in mind when choosing TM data for applications in similar landscapes. Actually, the methodology proposed is quite flexible and can be efficiently applied to other areas and imagery data.

REFERENCES

Buttner, G., and Csillag, F., 1989. Comparative study of crop and soil mapping using multitemporal and multispectral SPOT and LANDSAT TM data. Rem. Sens. of Env., 29:241-249.

Cohen, W.B,. and Spies, T.A., 1992. Estimating structural attributes of Douglas-fir western hemlock forest stands from Landsat and Spot imagery. Rem. Sens. of Env., 41:1-17.

Conese, C., Maracchi, G., Miglietta, F., Maselli, F., Sacco, V.M., 1988. Forest classification by Principal Component Analyses of TM Data. Int. Journal of Rem. Sensing, 9:1597-1612.

Conese, C., and Maselli, F., 1991. Use of multitemporal information to improve classification performance of TM scenes in complex terrain. ISPRS Journal of Phot. and Rem. Sensing, 46:187-197.

Conese, C., and Maselli, F., 1993. Selection of optimum bands from TM scenes through mutual information analysis. ISPRS Journal of Photogr. and Remote Sens., forthcoming.

Congalton, R.G., Mead, R.A., Oderwald, R.G., 1983. Assessing Landsat classification accuracy using discrete multivariate analysis statistical techniques. Phot. Engeen. and Rem. Sensing, 49:1671-1678.

Davis, F.W., and Dozier, J., 1990. Information analysis of a spatial database for ecological land classification. Phot. Engeen. and Rem. Sens., 56:605-613.

Franklin, J., Logan, T.L., Woodcock, C.E., and Strahler, A.H., 1986. Coniferous forest classification and inventory using Landsat and digital terrain data. IEEE Trans. Geosci. Remote Sens., GE-24:139-149.

Horler, D.N.H., and Ahern, F.J., 1986. Forestry information content of Thematic Mapper data. Int. Journal of Rem. Sensing, 8:1785-1796.

Hudson, W.D., and Ramm, C.W., 1987. Correct formulation of the Kappa coefficient of agreement. Phot. Engeen. and Rem. Sensing, 53:421-422.

Kullback, S., 1959. Information Theory and Statistics. John Wiley, New York.

Maselli, F., Conese, C., Zipoli, G., Pittau, M.A., 1990. Use of error probabilities to improve area estimates base on maximum likelihood classifications. Rem. Sensing of Env., 31:155-160.

Maselli, F., Conese, C., Petkov, L., Resti, R., 1992. Inclusion of prior probabilities derived from a nonparametric process into the maximum likelihood classifier. Phot. Engeen. and Rem. Sensing., 58:201-207.

Mausel, P.W., Krambler, W.J., Lee, J.K., 1990. Optimum band selection for supervised classification of multispectral data. Phot. Engeen. and Rem. Sensing, 56:55-60.

Rosenfield, G.H., and Fitzpatrick-Lins, K., 1986. A coefficient of agreement as a measure fo thematic classification accuracy. Phot. Engeen. and Rem. Sensing, 52:223-227.

Skidmore, A.K., and Turner, B.J., 1988. Forest mapping accuracies are improved using a supervised nonparametric classifier with Spot data. Phot. Engeen. and Rem. Sensing, 54:1415-1421.

Time-Varying Image Processing and
Moving Object Recognition, 3 – V. Cappellini (Ed.)
© 1994 Elsevier Science B.V. All rights reserved.

CLUTTER MAPS WITH ADAPTIVE UPDATING FOR THE DETECTION OF SEA TARGETS

M. Naldi

University of Rome "Tor Vergata", Department of Electronic Engineering, Rome, Italy

1. INTRODUCTION

Clutter maps have traditionally been employed as detection devices, thanks to their better performance in a spiky clutter environment with respect to cell averaging CFAR [1],[2]. Their simplicity and ease of implementation make desirable their application to target/environment couples other than those used in [2] (Rayleigh target embedded in Rayleigh clutter).

In particular, when confronted with the task of detecting ships, the scheme proposed in [1] and [2] is affected by the following deficiencies :

- the available dynamic range is poor to cope with the wide variations in radar cross section values exhibited by sea targets : a light boat can be represented as a 25 square meter target, but a fregate may be a 200000 m^2 one (see pp. 11.16-11.19 of [3]);
- slow targets go largely undetected due to the self-masking phenomenon : a target entering a resolution cell and staying in it for a few scans, contributes to raise the threshold and therefore progressively cancels itself;
- the behaviour of the map in a non-Rayleigh environment, and sea clutter is typically non Rayleigh, is not CFAR.

To overcome these limitations a new scheme has been proposed [4], which incorporates a log detector, to achieve a wider dynamic range, and an updating inhibition mechanism to reduce self masking.

In this paper a refined scheme, based on the same concepts illustrated in [4], is analysed, relying only partiallly on the simulation approach chosen in [4]. After an explanation of the map working, with an emphasis on the inhibition mechanism which allows adaptive updating, the expression for the false alarm rate is derived and both false alarm and detection curves are given for a number of working conditions.

2. THE UPDATING INHIBITION MECHANISM

A schematic of the proposed clutter map is given in Fig. 1. As aforementioned, it makes use of a logarithmic detector as a means of achieving a higher dynamic range. A device selects the maximum among M adjacent range cells, thus adding robustness against locally nonhomogeneous clutter. The maximum selector output constitutes then the input to the heart of the clutter map, a β-loop integrator whose behaviour is governed by the simple difference equation :

$$y_n = \beta \cdot y_{n-1} + (1-\beta) \cdot x_{n-1} \qquad (1)$$

where x_n and y_n are the integrator input and output at the n-th scan, β is the feedback gain, controlled by the difference between the input signal and the threshold at the previous scan.

If β is held constant, as in traditional clutter maps, eq. (1) can be solved iteratively leading to the following expression :

$$y_n = \beta^n \cdot y_0 + \sum_{i=0}^{n-1} (1-\beta) \cdot \beta^i \cdot x_{n-i-1} \qquad (2)$$

In the end this exponential smoother provides an asymptotically unbiased estimate of the input mean value, supposing of course that all the x_n are i.i.d. variables.

Setting the feedback gain value establishes a trade-off between the steady state accuracy of the estimate and the readiness to adapt to changing clutter conditions. Concisely said, the higher the feedback gain, the better the steady state accuracy, but the slower the map dynamics.

Controlling the amount of feedback allows therefore to slow down the updating process. As fast updating is responsible for the threshold build-up when a target is present and for the resulting self masking, increasing feedback after a detection is tantamount to freezing the threshold and therefore helps in reducing the target cancellation.

The control signal for the feedback gain is the difference between the signal and the threshold. It is to be noted that the variation of feedback is performed in the scan following the detection and lasts a single scan.

Two different control laws can be adopted : a) hard inhibition in the case of a step β vs. control signal curve; b) soft inhibition if there is a gradual transition towards map locking (see fig. 2).

All the results given in the following refer to the hard inhibition case. The analysis for the soft inhibition map would lead to performances which are somewhat midway between those presented here and those exhibited by traditional maps.

In order to see how performances are affected by the introduction of the updating inhibition, a heuristic reasoning can be applied. Basically the effect of the inhibition mechanism is to discard from the threshold setting process those clutter samples which trespass the threshold itself (which is supposed to have been initially determined letting the feedback gain constant). Given the Pfa values for which the map is designed, the number of clutter samples not taken into account represent a very minor portion of the underlying population and therefore we can assume that the false alarm rate is not significantly affected by the variable feedback introduction. This reasoning can be further enforced by considering that map locking concerns just the scan following a false alarm (or a detection). Assuming that clutter has produced a false alarm at the n-th scan (the threshold having been set on the basis of the previous n-1 clutter samples) with a p probability, the probability of a false alarm at the (n+1)-th scan is again p, as the threshold has not changed and clutter samples are assumed to be independent scan by scan.

As to the probability of detection, it is certainly higher, on an intuitive basis, than in the traditional scheme, because the threshold after a detection is kept fixed instead of being raised due to the presence of a target. Indicating by $T_1 = T$ the steady state threshold (after ∞ clutter samples) at the first appearance of the target, by $T_2 \neq T_1$ the threshold at the second staying of the target (determined by ∞ clutter samples and one target sample), by X_1 and X_2 the target+clutter signal at the first and second staying of the target, the probability of detection P_{D2} at the second staying is :

$$P_{D2} = P(X_2 > T_2) = P(X_2 > T_2 / X_1 > T_1) \cdot P(X_1 > T_1) + P(X_2 > T_2 / X_1 < T_1) \cdot P(X_1 < T_1) =$$
$$= P(X_1 > T_1) \cdot P(X_1 > T_1) + P(X_2 > T_2 / X_1 < T_1) \cdot P(X_1 < T_1) = \tag{3}$$
$$= P_{D1} \cdot P_{D1} + P(X_2 > T_2 / X_1 < T_1) \cdot (1 - P_{D1})$$

having considered that, if $X_1 > T_1$, the threshold stays unchanged.

In the absence of an explicit expression for $P(X_2 > T_2 / X_1 > T_1)$, unfortunately the only information that eq. (3) can supply is the lower bound for P_{D2}, given by P_{D1}^2.

3. FALSE ALARM RATE

As established in the previous section, the false alarm rate can be computed in the simple case of constant β, the results being approximately valid for the variable feedback as well.

An additional assumption we make is to focus on a clutter-dominated environment, i.e. with noise representing a negligibile portion of background unwanted signals with respect to clutter. In this case we need a model for sea clutter returns only. A reasonable one for the clutter envelope is a Weibull pdf, expressed by :

$$f_V(v) = \left(\frac{v}{q}\right)^{\alpha-1} \cdot \frac{\alpha}{q} \cdot \exp\left[-\left(\frac{v}{q}\right)^{\alpha}\right] \qquad v>0 \tag{4}$$

where V is the clutter envelope, α is a shape parameter typically ranging from 1.4 to 2 (see Table 2.7 of [5]) and q is a scale parameter related to the average radar cross section $\overline{\sigma_0}$ of clutter and to the gamma function $\Gamma(\bullet)$ by eq. (5) :

$$q = \sqrt{\frac{\overline{\sigma_0}}{\Gamma\left(1 + \frac{2}{\alpha}\right)}} \tag{5}$$

The parameter α governs the skewness of the pdf curve : lower values of α correspond to heavier tails. The Rayleigh distribution can be obtained as a particular form of the Weibull family setting $\alpha=2$: the case of a noise-dominated environment is thus included in the more general case of a Weibull background.

After passing through the log detector, the clutter's pdf assumes the following expression :

$$f_U(u) = \alpha \cdot e^{\alpha \cdot (u-\mu)} \cdot e^{-e^{\alpha \cdot (u-\mu)}} \tag{6}$$

where $\mu = \ln(q)$.

The maximum selector gathers these returns in M by M batches, each batch pertaining to M adjacent range cells, and extracts the maximum in each batch :

$$X = \max(U_1, U_2, ..., U_M) \qquad (7)$$

The number M is generally limited to 4, to avoid the effects of non homogeneous clutter.

As the U_i's are i.i.d. random variables, whose pdf is given by eq. (6), the pdf of X (clutter at the input of the β-loop is (see App. A) :

$$f_X(x) = M \cdot \alpha \cdot e^{\alpha(x-\mu)} \cdot e^{-e^{\alpha(x-\mu)}} \sum_{k=0}^{M-1} (-1)^k \cdot \binom{M-1}{k} \cdot e^{-k \cdot e^{\alpha(x-\mu)}} \qquad (8)$$

Its mean value is (see App. B) :

$$E[X] = \mu - \frac{C}{\alpha} + \frac{1}{\alpha} \sum_{k=2}^{M} (-1)^k \cdot \binom{M}{k} \cdot \ln(k) \qquad (9)$$

where C is Euler's constant (approx. 0.577).

The variance of X can be computed making use of the relation :

$$\text{Var}[X] = E[X^2] - \{E[X]\}^2 \qquad (10)$$

and using the expression of the mean square value of X given in App. C.

The corresponding moments of the output Y of the β-loop are given asymptotically by (see [2])

$$E[Y] = E[X]$$
$$\text{Var}[Y] = \frac{1-\beta}{1+\beta} \cdot \text{Var}[X] \qquad (11),(12)$$

A complete characterization of Y would require the knowledge of its pdf, which is rather difficult to determine. In the following we assume it to be Gaussian.

The threshold T is finally given by :

$$T = \gamma + Y \qquad (13)$$

It has then the same pdf as Y, with the mean value just shifted by γ.

The detection process is performed comparing each U_i sample with the threshold T. The probability of false alarms is therefore given by :

$$P_{fa} = \int_{-\infty}^{+\infty} \left[\int_t^{\infty} f_U(u) du \right] f_T(t) dt = \frac{1}{\sqrt{2\pi}\sigma_T} \int_{-\infty}^{+\infty} e^{-e^{\alpha(t-\mu)}} \cdot e^{-\frac{(t-E[T])^2}{2\sigma_T^2}} dt \qquad (14)$$

With the change of variable $z = \dfrac{t - E[T]}{\sigma_T \sqrt{2}}$ eq. (14) can be put in the form :

$$P_{fa} = \frac{1}{\sqrt{\pi}} \int\limits_{-\infty}^{+\infty} e^{-e^{\alpha\{E[T]+z\cdot\sigma_T\sqrt{2}-\mu\}}} \cdot e^{-z^2} dz \qquad (15)$$

and computed using Hermite's integration formula (see [7]).

Though not apparent from eq. (15), the insertion in this equation of the proper expressions for E[T] and σ_T reveals that the false alarm rate doesn't depend on μ, i.e. on the clutter power, but does however depend on α. We have therefore a CFAR behaviour, but not on the whole range of pdf's belonging to the Weibull family.

The two sets of curves shown in Figs. 3 and 4 (relative to the M=1 case) illustrate the strong dependence of Pfa on the threshold step-up γ. These pictures also show that increasing the feedback gain gives a better estimate of clutter parameters, allowing to lower γ.

4. DETECTION CURVES

Assuming a Swerling 0 model (non fluctuating) with amplitude A for ships, the pdf of the composite target+clutter signal after the log detector follows a Weibull-Rice law (see [9]) :

$$f_S(s) = \int\limits_0^{2\pi} \frac{\alpha \cdot \ln(2) \cdot e^{2s}}{2\pi} \cdot e^{-\ln(2)\left[e^{2s}-2A\cdot e^s\cdot\cos\theta+A^2\right]^{\alpha/2}} \cdot \frac{1}{\left[e^{2s}-2A\cdot e^s\cdot\cos\theta+A^2\right]^{(1-\alpha/2)}} d\theta$$
$$(16)$$

Due to the rather unwieldy form of this pdf, the detection probability has been evaluated by a MonteCarlo simulation. The number of runs performed (1000) guarantees an accurate estimate for the range of P_D values of interest ($P_D>0.8$), as the probability of falling within ±3% of the true P_D value is higher than 95%. Hereafter some of the curves obtained are shown. all of them refer to the M=1 case, with Pfa=10^{-6}. They show that even after 10 stayings the fall in the target visibility is quite low.

5. CONCLUSIONS

A new scheme of clutter map, incorporating a mechanism to inhibit updating in the presence of a target, to be used in a marine environment, has been proposed. Its CFAR property has been demonstrated for each clutter pdf belonging to the Weibull family. though the false alarm rate will depend on the Weibull shape parameter. Self masking, deeply affecting traditional clutter maps, has been reduced to negligible values.

ACKNOWLEDGEMENT

I wish to thank Dr. Ing. Di Lazzaro of Alenia S.p.A. for his valuable suggestions and his encouragement to publish this paper.

Appendix A - Pdf of clutter at the input of the integrator

Given a batch of M i.i.d. random variables U_i, whose common pdf is given by eq. (6), the maximum X among them has a pdf given by :

$$f_X(x) = \frac{dF_U^M(x)}{dx^m} = M \cdot F_U^{M-1}(x) \cdot f_U(x) \qquad (A.1)$$

The cumulative distribution function of U can be found by integrating eq. (6) :

$$F_U(x) = 1 - e^{-e^{\alpha(x-\mu)}} \qquad (A.2)$$

Its (M-1)-th power can be computed recalling the binomial power expansion, obtaining :

$$F_U^{M-1}(x) = \sum_{k=0}^{M-1} (-1)^k \cdot \binom{M-1}{k} \cdot e^{-k \cdot e^{\alpha(x-\mu)}} \qquad (A.3)$$

The insertion of eqs. (6) and (A.3) in eq. (A.1) leads to the final form :

$$f_X(x) = M \cdot \alpha \cdot e^{\alpha(x-\mu)} \cdot e^{-e^{\alpha(x-\mu)}} \sum_{k=0}^{M-1} (-1)^k \cdot \binom{M-1}{k} \cdot e^{-k \cdot e^{\alpha(x-\mu)}} \qquad (A.4)$$

Appendix B - Expected value of clutter at the input of the integrator

Inserting The expression of X's pdf in the definition of the expected value of X, we obtain :

$$E[X] = M \cdot \alpha \sum_{k=0}^{M-1} (-1)^k \cdot \binom{M-1}{k} \int_{-\infty}^{+\infty} x \cdot e^{\alpha(x-\mu)} \cdot e^{-(1+k) \cdot e^{\alpha(x-\mu)}} dx \qquad (B.1)$$

After the change of variable $t = e^{\alpha(x-\mu)}$ we have :

$$E[X] = M \sum_{k=0}^{M-1} (-1)^k \cdot \binom{M-1}{k} \left[\frac{\mu}{1+k} + \frac{1}{\alpha} \int_0^\infty \ln(t) \cdot e^{-(1+k)t} dt \right] \qquad (B.2)$$

Using form. (4.331/1) of [6], and noting that $\sum_{k=0}^{M-1} (-1)^k \cdot \binom{M}{k+1} = 1$, eq. (B.2) can be finally rewritten as :

$$E[X] = \mu - \frac{C}{\alpha} + \frac{1}{\alpha} \sum_{k=2}^{M} (-1)^k \cdot \binom{M}{k} \cdot \ln(k) \qquad (B.3)$$

Appendix C - Mean square value of clutter at the input of the integrator

Defining a variable $Z = X^2$, its pdf is given by the expression (see Ex. 5-11 of [8]) :

$$f_Z(z) = \frac{f_X(\sqrt{z}) + f_X(-\sqrt{z})}{2\sqrt{z}} \tag{C.1}$$

Using eq. (A.4) for $f_X(x)$, after some manipulations we have :

$$E\left[X^2\right] = E[Z] = M \sum_{k=0}^{M-1} (-1)^k \cdot \binom{M-1}{k} \cdot \int_0^{\infty} \left[\mu + \frac{\ln(t)}{\alpha}\right]^2 \cdot e^{-(1+k)t} dt \tag{C.2}$$

Expanding the squared expression inside the integral and recalling form. (4.331/1) of [6], we have :

$$E\left[X^2\right] = M \sum_{k=0}^{M-1} (-1)^k \cdot \binom{M-1}{k} \cdot \left\{ \mu^2 \cdot \left[\frac{1}{1+k}\right] - 2\frac{\mu[C + \ln(1+k)]}{\alpha(1+k)} + \frac{1}{\alpha^2} \int_0^{\infty} [\ln(t)]^2 \cdot e^{-(1+k)t} dt \right\} \tag{C.3}$$

Recalling now form. (4.335/1) of [6], we obtain in the end :

$$E\left[X^2\right] = \mu^2 + \frac{\pi^2}{6\alpha^2} - \frac{2\mu \cdot C}{\alpha} + \frac{M \cdot C^2}{\alpha^2} - \frac{1}{\alpha^2} \sum_{k=2}^{M} (-1)^k \cdot \binom{M}{k} \cdot [C + \ln(k)]^2 +$$

$$+ 2\frac{\mu}{\alpha} \sum_{k=2}^{M} (-1)^k \cdot \binom{M}{k} \cdot \ln(k) \tag{C.4}$$

REFERENCES

1. E.N. Khoury, J.S. Hoyle : "Clutter maps : design and performance", IEEE National Radar Conference, Atlanta, March 1984, pp. 1-7
2. M. Lops, M. Orsini : "Scan-by-scan averaging CFAR", IEE Proceedings Pt. F, no. 6, December 1989, pp. 249-254
3. M. Skolnik (ed.) : *Radar Handbook*, McGraw-Hill, 1990 (2nd ed.)
4. M. Di Lazzaro, M. Naldi, R. Pangrazi : "A detection technique for sea targets in coastal radars", IEE International Conference Radar 92, Brighton, 12-13 October 1992, pp. 106-109
5. G. Picardi : *Elaborazione del segnale radar* (in Italian), Franco Angeli, 1988
6. I.S. Gradshteyn, I.M. Ryzhik : *Table of integrals, series, and products*, Academic Press, 1980 (2nd ed.)
7. M. Abramowitz, I.A. Stegun : *Handbook of mathematical functions*, Dover, 1972 (9th printing)

8. A. Papoulis : *Probability, Random Variables, and Stochastic Processes*, McGraw-Hill, 1984 (2nd ed.)

9. D.C. Schleher : "Radar detection in Weibull clutter", IEEE Transactions on Aerospace and Electronic Systems, vol. 12, no. 6, November 1976, pp. 736-743

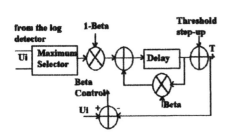

Fig. 1 - Adaptive updating clutter map

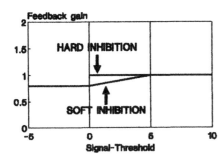

Fig.2 - Feedback control laws

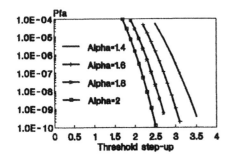

Fig. 3 - False alarm rate vs. γ (β=0.8)

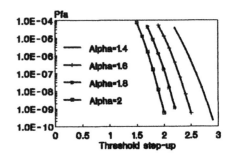

Fig. 4 - False alarm rate vs. γ (β=0.95)

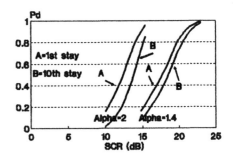

Fig. 5 - Detection Probability vs. Signal-to-clutter ratio (β=0.8)

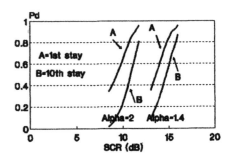

Fig. 6 - Detection Probability vs. Signal-to-clutter ratio (β=0.95)

*Time-Varying Image Processing and
Moving Object Recognition, 3* – V. Cappellini (Ed.)
© 1994 Elsevier Science B.V. All rights reserved.

Data Fusion Techniques for Remotely Sensed Images

A. Chiuderi[a] and S.Fini[b]

[a] Dipartimento di Sistemi e Informatica, Via di S. Marta 3, Firenze (Italy)

[b] Fondazione Scienza per l'Ambiente, Viale Galileo 32, Firenze (Italy)

Abstract
In this paper a new technique for multisensor data fusion for photointerpretation purposes is presented. The existing techniques are briefly reviewed, some experiments on merging of optical-infrared data with radar data are performed and the results so far obtained are compared and discussed.

1. INTRODUCTION

The process of merging multisensor data has become in the last few years more and more appealing due to the huge amount of data supplied by the different sensors observing the Earth's surface.The development of remote sensing technologies, in fact, offers the possibility to process data on the same area acquired by different sensors operating at different altitudes, with different spatial resolutions, at different times and with different technical instrumentation (wavelengths, signal to noise ratio, radiometric sensitivities...). Due to the intrinsic difficulties in remote sensing image processing and understanding, it is particulary interesting to perform the processing in an integrated way, in order to extract more information from the combination of data collected by different channels rather than from each separated channel. Remote sensing image *data fusion* [1] is thus the process by which the information collected in different channels of the electromagnetic spectrum is combined by merging the information available in each single channel into one image product.

Beside the advantages deriving from the increased amount of information on the observed scene, several other reasons make data fusion particularly important for remote sensing: data fusion can help to recover from loss of information when one sensor is damaged, the confidence on the observed features is increased when multiple separate measurements are made on the same event target and moreover extended spatial and temporal coverage is provided, since anyone sensor can detect an event where and when the others cannot.

The problem of sensor fusion has been widely studied for robotic or military purposes, but as far as remote sensing images are concerned the literature is not so extensive: the reason for this difference can be ascribed to the fact that while in the case of robotic or military applications the task is well identified (e.g the determination of a target of a given shape in the scene, or the identification of a given object to be grabbed), in the case of remotely sensed images there is not any kind of target on the observed image, but on the contrary, the purpose of the application is to produce an image of "good" quality which contains as much information as possible.

Existing literature on this subject focuses mainly on the problem of merging multispectral data (Thematic Mapper) with a high spectral resolution with panchromatic data (SPOT) with a high spatial resolution (e. g. [2,3]). The purpose of the various techniques is to preserve the spectral content of the first data set while adding the spatial resolution of the second one.

Another interesting issue in data fusion is the task of merging both data collected in different regions of the electromagnetic spectrum, such as radar data with TM data, and data of different nature, such as a digitized geological map and radar data [5]. The final product of this processing are usually images to be employed for photointerpretation.

2. COMBINING THE DATA

Merging of multispectral and multisensor image data sets oriented to photointerpretation, generates, as the final result, a false color image. The three channels used for display, must thus be as representative as possible of the data sets employed in the fusion process.

2.1. Previous approaches

The existing techniques for merging remotely sensed data can be roughly divided into 3 categories: i) "ad hoc" approaches, ii) merging by separate manipulation and iii) merging to mantain radiometric integrity [7].

Ad hoc approaches are primarily concerned with the enhancement of certain peculiarities of the obsered scene by some arithmetical operations on the data sets which are to be to merged (e.g. difference, ratio, filtering..); they are very closely related to the scene under consideration and are not of great interest as far as general techniques are concerned.

Merging by separate manipulation is definitely the most widely employed technique: the data of one data set are trasformed, by means of an invertible transformation, into a new set of data; one component of this new data set is replaced by the second data set and the inverse transformation on the set of merged data is computed. The idea underlying this kind of approach is to spread the information available from the second data set into the first one.

Finally, in the third kind of approach, a synthetic channel which is radiometrically comparable with the second set of data is generated by using only the first data set. Fusion is then performed by manipulation (as in the previous case), but the channel employed in the replacement is the synthetic one.

The most frequently employed invertible transformations are the Principal Component transformation (PC) and the Hue Intensity Saturation transformation (HIS). In the former case, the channels of the first data set are employed as input for the PC transformation, the first principal component, which accounts for the maximum variance of the transformed data set, is replaced by one channel of the second data set, and the inverse transformation is performed; three channels are then selected for display. The advantage of using the PC transformation is that more than three channels of the first data set can be involved in the transformation and thus the new, imported data, will be spread over a wider space. In the latter case, three channels are selected as the Red, Green and Blue input channels for the HIS transformation. The Intensity of the trasformed data is replaced by one component of the second data set and the inverse RGB transformation is performed. The new RGB channels are then employed for display. The advantages in the use of the HIS transformation with respect to the PC is given by the fact that the coefficients of the HIS transformation are indipendent of the data distribution. The final products of these processing techniques are images of merged data to be employed for photointerpretation; it is thus quite difficult to establish which method is the best one, the final result being subject to invididual judgement, as can be seen by a comparison between [4] and [6]: in [4] the author suggests that the HIS is more suitable for photointerpretation purposes, as shown also by [5], while in [6] the authors stress the advantages of the PC technique.

2.2. The proposed method

In this paper two new techniques for data fusion, which are particularly suitable when dealing with multisensor and multi frequency data are presented.

The data employed for these experiments were acquired during the ASI-NASA MAC-Europe 1991 airborne campaign. The study area is situated in Montespertoli, an agricultural zone in the surroundings of Florence (Italy). The data were collected by a multifrequency, multipolarization airborne SAR, the spectral bands being L, P and C bands in HH, HV and VV polarization and by an airborne Thematic Mapper Simulator, which collected information in the optical-infrared region of the electromagnetic spectrum.

Table 1
Statistics of TM data

Covariance matrix

	TM1	TM2	TM3	TM4	TM5	TM7
TM1	1689.173					
TM2	1671.734	1727.071				
TM3	1632.627	1686.478	1750.471			
TM4	-187.298	-161.354	-418.830	1721.877		
TM5	1541.152	1601.546	1614.541	-30.575	1775.009	
TM7	1530.377	1582.458	1633.737	-405.837	1667.753	1741.958

Eigenchannel	Eigenvalue	Deviation	%Variance
1	8247.3020	90.8147	79.26
2	1744.7150	41.7698	16.77
3	286.3546	16.9220	2.75
4	63.5786	7.9736	0.61
5	43.8752	6.6238	0.42
6	19.7327	4.4421	0.19

Table 2
Statistics on SAR data

Covariance matrix

	HH	HV	VV
HH	124.777 (32.13%)		
HV	121.127	164.148 (42.27%)	
VV	100.390	112.260	99.425 (25.6%)

Total Variance = 388.35

Eigenchannel	Eigenvalue	Deviation	%Variance
1	355.6550	18.8588	91.58
2	22.6239	4.7565	5.83
3	10.0715	3.1736	2.59

Tables 1 and 2 show the covariance matrices of TMS and SAR data respectively: as can be seen, the first principal component of the TMS data accounts for 79.26% of the total variance while each polarization of the L band accounts only, at the most, for 42.27%. The standard PC technique, would thus result in a general loss of information when the first principal component is replaced by anyone of the polarizations of the L band. The same observation holds in the case of the HIS transformation: a loss of information would result when the Intensity of the transformed data (the component which carries most of the spatial information of the first data set) is substituted, with one of the L band polarization.

On the other hand, see again table 2, the first principal component of the L band data accounts for 91.58% of the total variance, which means that the information concerning the three polarizations of the L band is nearly totally carried by it. The two new approaches proposed here are based on the observation of the values of the two covariance matrices.

A non-standard PC transformation is proposed for this data set: the principal components of both the TM data and the SAR data are computed, and the first principal component of the TM data is substituted by the first principal component of the SAR data before performing the inverse transformation.

The same idea can also be applied to the HIS technique: the intensity of the TM data set is replaced by the first principal component of SAR data and then, as usual, the inverse transformation is performed.

2.3. The processing

The data employed in this study have different spatial resolution: the TM data have a 25 by 25 meter resolution, while the SAR data have a resolution of 6 meter in range and 12 meter in azimuth. Due to the presence of the speckle noise the higher spatial resolution of the SAR data could not be exploited without diminishing the overall quality of the image; the SAR data were thus resampled, by averaging, to a 24 by 24 meter pixel.

Successively, in all our experiments, GCP were selected in order to coregister the SAR data to the TM data by means of a 5th order polynomial leading to residuals of 0.284 and 0.330 in the x and y directions respectively; the resampling was here performed by means of cubic interpolation.

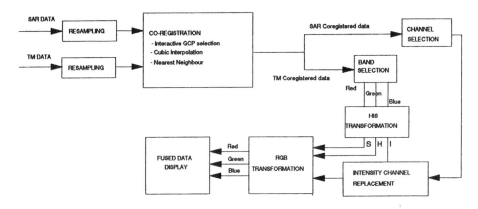

Figure 1. The HIS processing technique

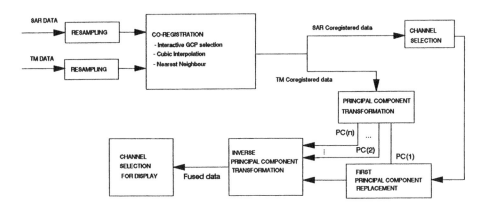

Figure 2. The PC processing technique

As a first set of reference images the standard HIS merging technique was employed: the Red Green and Blue TM channels were used as input to the HIS transformation (the intensity being the average of the three components); a histogram matching was then performed between SAR data and the intensity of the TM image; the new, stretched L band data, were then substituted with the intensity component and the inverse transformation was performed in order to obtain, as a final result, three channels R, G and B used for display. This transformation was applied to the three polarizations of the L band.

The non standard PC technique was carried out on the resampled and coregistered data: the eigenvectors of the covariance matrix were computed on the TM data set, in order to determine the orthonormal transformation matrix M; the first principal component of the TM data was replaced by the first principal component of the SAR data and the reverse transformation M^{-1}, yielding to the previous feature space, was computed on these data. Three channels of the merged data set were then selected for display.

Finally, the mixed HIS+PC was also implemented. The R, G, and B channels of the TM data were input to the HIS transformation; as in the case of the standard HIS transformation, a histogram matching was performed on the first principal component of the SAR before the replacement of the intensity component and the inverse transformation. In figures 1 and 2 the steps of the different processing techniques are reported.

The reported plates illustrate the data available and some of the obtained image products.

3. CONCLUSIONS

In this paper data fusion techniques for remote sensing image photointerpretation have been investigated. In particular two new techniques have been proposed and applied to a multisensor image data set.

The results obtained with the non standard PC and the mixed HIS+PC have been compared with the most frequently applied HIS techniques.

The photointerpretation of the images obtained by the PC and the HIS+PC transformations indicate the presence of a higher contribution of the SAR data and thus the possibility to better exploit the information arising from microwave data with respect to the image produced with the HIS transformation. This subjective opinion is in accordance with the statistical consideration on the previously reported covariance matrices .

REFERENCES

1. E. Walz, J. Llinas: Multisensor Data Fusion. Artech House, Boston London (1990)
2. V.K. Shettigara: A Generalized Component Substitution Technique for Spatial Enhancement of Multispectral Images Using a Higher Resolution Data Set. PE&RS vol. 58, N. 5, pg. 561-567 (1992)
3. G. Cliche, P. Teillet: Integration of the SPOT Panchromatic Channel into Its Multispectral Mode for Image Sharpness Enhancement. PE&RS vol. 51, N. 3, pg. 311-316 (1985)
4. M. Ehlers: Multisensor Image Fusion Techniques in Remote Sensing. ISPRS Journal of Photogrammetry and Remote Sensing, N. 46, pg. 19-30 (1991)
5. J.H. Jeff, R. Murray: IHS Transform for the Integration of Radar Imaginery with other Remotely Sensed Data. PE&RS, Vol. 56, N. 12, pg. 1631-1641 (1990)
6. P.S. Chavez, S.C. Sides, J. A. Anderson: Comparison of Three Different Methods to Merge Multiresolution and Multispectral Data: Landsat TM and SPOT Panchromatic. PE&RS, Vol. 57, N. 3, pg. 295-303 (1991)

7. C. K. Munechika, J. S. Warnik, C. Salvaggio, J.R. Schott: Resolution Enhancement of Multispectral Image Data to Improve Classifiation Accuracy. PE&RS, Vol 59, N 1, pg 67-72 (1993)

256

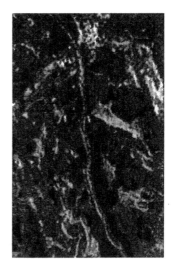

Montespertoli area - June 1991
SAR data: L band HH polarization
Pixel: 6m (range) x 12m (azimuth)

Montespertoli area - June 1991
TMS data: True color image
Pixel: 25m x 25m *

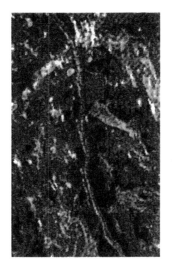

Data fusion of SAR and TMS data:
HIS transformation *

Data fusion of SAR and TMS data:
HIS+PC transformation *

* these illustrations are shown in colour on page 429

H
DIGITAL PROCESSING
OF BIOMEDICAL IMAGES

Time-Varying Image Processing and
Moving Object Recognition, 3 – V. Cappellini (Ed.)
© 1994 Elsevier Science B.V. All rights reserved.

Depth from Motion in X-Ray Imagery

G. Coppini [a], L. Castellani [b], G. Bertini [b], G. Valli [b]

[a] CNR Institute of Clinical Physiology, Pisa, Italy

[b] Dept. Electronic Engineering, University of Florence, Florence, Italy

In the paper we describe a method for estimating depth maps from a (reduced) set of radiograms acquired with small changes of the projection angle. In general, the problem is ill-posed: the formulation of adequate physical assumptions has permitted us to transform it into a well-posed one.

The method involves two phases: a) computation of optical flow as produced by the relative motion between the radiation source and the observed bodies, b) estimation of depth map from optical flow. Experiments performed both on simulated X-ray scenes and physical X-ray phantoms are described.

1. INTRODUCTION

Three-dimensional information about biological structures is of paramount importance in the understanding of medical images [1]. Several high level tasks such as diagnosis and treatment often rely upon knowledge of spatial relations among the observed entities.

Unfortunately, Conventional Projection Radiography (CPR) does not make explicitly available depth data about the scene and the physician usually utilizes (in a qualitative way) two or more views to solve position ambiguities.

Although the introduction of tomographic techniques has strongly mitigated the above problem, CR still remains one of the major imaging techniques in medicine. This is due to several factors, such as: the amount of information collected in a about a century of clinical practice, limitation of radiation dose, operational simplicity.

These facts motivate the development of CPR vision systems to assist radiologist in his

260

tasks [2]. In particular, the availability of depth information is expected to improve the accuracy and reproducibility of medical findings.

In this paper, we report on the use of relative motion between the X-Ray source and the observed scene to recover depth data. As known from computer vision studies, motion is a valuable source of 3D information, though its recovery is usually an ill-posed problem. In the following (section 2) we will introduce some general assumptions about the observed object (and the imaging system) which make the problem tractable. Recovery of depth data by optical-flow computation is described in section 3. In section 4, experimental results on computer simulations and physical objects are described.

2. PHYSICAL ASSUMPTIONS

As shown in Figure 1, let us consider two consecutive X-ray projections taken with a small displacement of the radiation source. Let such displacement measured by the angle $\delta\alpha$. In general, if the projections of a given 3D point P are known in the two views, the spatial coordinates of P can be recovered unambiguously. Thus, similarly to stereopsis, one can conceive the considered problem as a matching problem among two views. However, the peculiarities of X-Ray imaging make the direct application of stereo algorithms unpractical. In fact, in X-Ray images, gray-levels result from the integration of X-Ray attenuation along a projection line: pixel matching on the basis of their intensity would be, in general, meaningless. In fact, matching uniqueness cannot be assumed in this case. Notwithstanding, we note that significant edges are produced by the boundaries between different objects (or between objects and background). In addition, for a wide class of biological structures, we can expect that overlapping of projected boundaries occur in a limited and sparse set of points of the radiogram. Otherwise stated, we assume that pixels representing overlapping boundaries are a zero measure set (see

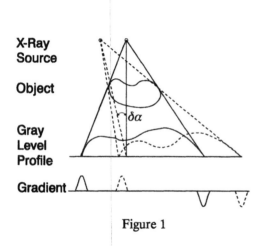

X-Ray
Source

Object

Gray
Level
Profile

Gradient

Figure 1

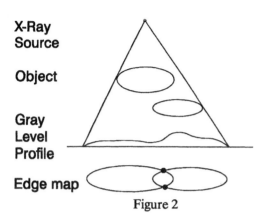

X-Ray
Source

Object

Gray
Level
Profile

Edge map

Figure 2

Figure 2). Thus, we can say that discontinuities of gray level distribution convey significant spatial information and, as depicted in Figures 1 and 2, image edges are the basic entities to be studied in order to recover depth maps.

On the ground of the above considerations, we have formulated the following physical assumptions which are valid for many organs and biological structures:

A1 - Objects have a smooth spatial distribution of X-Ray attenuation coefficient;

A2 - Object boundaries are usually regular and smooth surfaces.

These hypotheses permitted us to utilize Marr's principle of continuity of discontinuities [3].

It is worth noting that matching of primitives more complex than edges is also described in literature (see for example [4]). In our experience, this approach can be advantageous when large angle views are used. As an example, one can consider the reconstruction of vascular trees in biplane angiography: methods based on high-level knowledge are well-suited for such a task [5]. However, if $\delta\alpha$ is small enough, matching among two consecutive views can be conveniently referred back to the computation of the relative motion field produced by (uniformly) moving the radiation source with respect to the object. As a consequence, computation of optical flow at the edge-points can provide the correspondence law between consecutive frames [6]. Afterwards, straightforward computations permit the assessment of depth data for a set of 3D boundary points.

3. OPTICAL FLOW COMPUTATION

In this section we summarize the algorithmic procedure employed to perform the experiments described in Section 4. Since we are dealing with edge representations, Hildreth's theory [7] has been adopted for optical flow computation and implemented as follows.

Let $I_1 \overset{\text{def}}{=} I(\alpha)$ and $I_2 \overset{\text{def}}{=} I(\alpha + \delta\alpha)$ two consecutive views of the same scene taken with an angular increment $\delta\alpha$. Each image is filtered by a Laplacian of Gaussian mask $\nabla^2 G$ so as to produce two filtered images L_1 and L_2:

$$L_1 = \nabla^2 G \otimes I_1 \ , \ L_1 = \nabla^2 G \otimes I_1 \tag{1}$$

Afterwards, zero crossings of L_1 are detected and linked into curves using a linking algorithm based upon points connectivity (eight-connectivity has been used). In this way, from the edge map of L_1 we obtain a set of curves $\Gamma = \{\gamma_k\}$.

Assuming a unit time interval, we can scale velocities back to vector displacements. For each point of a given curve γ_k, we estimate the normal component $\tilde{u}_n$ of the displacement vector by the classical optical flow constraint [8], that is:

$$\tilde{u}_n = \frac{L_2 - L_1}{|\nabla G \otimes L_1|} \tag{2}$$

Computation of the complete displacement field $\mathbf{u}$ is achieved by minimizing the functional:

$$E(\mathbf{u}) = \int \left| \frac{\partial \mathbf{u}}{\partial s} \right|^2 ds + \beta \int \left[u_n - \tilde{u}_n \right]^2 ds \tag{3}$$

The first term on the right side of equation (3) acts as a smoothing constraint while the second term is a measure of data fitting. Parameter β is a regularization parameter which can be used to trade off between smoothness and data closeness.

The functional defined by equation (3) has been discretized as follows. Each curve has been sampled at equispaced points P_i and the displacement vectors (u_x^i, u_y^i) have been assumed as unknown quantities. The discretized functional is written into the form:

$$\Phi(\mathbf{u}^i) = \Phi_1(\mathbf{u}^i) + \beta\,\Phi_2(\mathbf{u}^i) \tag{4}$$

where:

$$\Phi_1(\mathbf{u}^i) = \sum_i \left[\left(u_x{}^i - ut_x{}^{i-1}\right)^2 + \left(u_y{}^i - u_y{}^{i-1}\right)^2 \right] \tag{5}$$

and:

$$\Phi_2(\mathbf{u}^i) = \sum_i \left(u_n{}^i - \tilde{u}_n{}^i\right)^2 \tag{6}$$

The discretized functional Φ is iteratively minimized (for each curve) by means of the conjugate gradient algorithm.

4. EXPERIMENTAL RESULTS

Experiments have been performed both on simulated radiograms and on physical X-Ray phantoms.

Simulated radiograms include scenes made up of ellipsoids with different size, spatial position and X-Ray attenuation coefficients. Figure 3a illustrates three of such scenes. In

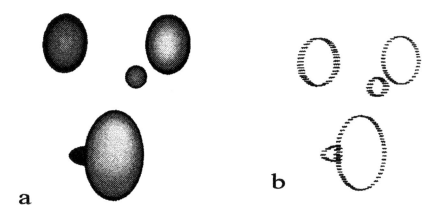

a

b

Figure 3

Figure 3b the computed optical flows for a translation displacement of the X-ray source are depicted. Finally, Figure 3c shows the surfaces of the objects as recovered from the computed 3D points. Surface recovery has been obtained utilizing a closed elastic thin surface model as described in [9].

In Figure 4a we can see two views of a physical egg-like phantom embedding a small plastic sphere. In panel 4b we see the zero crossing map and the recovered displacement map. Panel 4c shows two views of

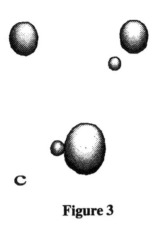

c

Figure 3

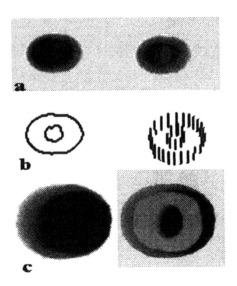

Figure 4

the recovered surfaces of the phantom using the same surface model as shown in Figure 3c.

5. CONCLUSIONS

A depth recovery method which is suitable for application in conventional radiography has been described. Spatial information is made available with a low radiation dose (only two fluoroscopic views can be used) and with a moderate computational load.

The method can be employed to assess the spatial position of lesions as well as to collect spatial data for recovering the shape of biological organs. It is worth noting that, using an entire sequence of projections (e.g. acquired by X-Ray fluoroscopy with a large angular excursion) one is able to recover a large set of points belonging to the boundaries of imaged structures. As concern the accuracy of depth maps, we have estimated that depth resolution (in the described implementation of the method) is similar to that of digital tomosynthesis [10-11].

Presently, we are working both to test the method on clinical data sets, and to integrate depth information with that produced by other processing modules.

REFERENCES

1. A. Macovsky. *Medical Imaging Systems*. Prentice Hall, New Jersey, 1983.
2. G. Coppini, M. Demi, R. Poli, G. Valli. An artificial vision system for X-Ray images of human coronary trees. IEEE Trans. on Pattern Analysis and Machine Intelligence, 15: 156-152, 1993.
3. D. Marr. *Vision*. W.H. Freeman, San Francisco, 1982
4. Y. Shirai. *Three-dimensional computer vision*. Springer Verlag, 1987.
5. G. Coppini, M. Demi, R. Mennini, G. Valli. Three-dimensional knowledge driven reconstruction of coronary trees. Medical & Biological Engineering & Computing, 29:535-542, 1991.
6. J.K. Aggarwal, N. Nandhakumar. On the computation of motion from sequences of images. Proceedings of the IEEE, 76:917-935,1988.
7. E. Hildreth. Computation underlying the measurement of visual motion. Artificial Intelligence, 23:309-354,1984.

8. B.K.P. Horn, B.G. Shunck. Determining optical flow. Artificial Intelligence, 17:185-303,1981.

9. R. Poli, G. Coppini, R. Nobili, G. Valli. LV shape recovery from echocardiographic images by means of computer vision techniques and neural networks. Proceedings Computers in Cardiology, 117-120, IEEE Comp. Soc. Ed., 1991.

10. D.G. Grant, Tomosynthesis: a three-dimensional radiographic imaging technique. IEEE Transaction on Biomedical Engineering, 19:20-28,1972.

11. G. Coppini, M. Demi, G. Valli. A class of optimized algorithms for digital tomosynthesis. Proceedings CAR-85, 247-251, Springer-Verlag, 1985.

Time-Varying Image Processing and
Moving Object Recognition, 3 – V. Cappellini (Ed.)
© 1994 Elsevier Science B.V. All rights reserved.

Motion evaluation from synthesized 3-D Echocardiograms

P. Baraldi [a], S. Cavalcanti [b], D. Del Giudice [b], C. Lamberti [b], A. Sarti [b] and F. Sgallari [c]

[a] Istituto di Fisiologia Umana, Università di Modena. Via Campi 287, 41100 Modena, Italy.

[b] Dipartimento di Elettronica, Informatica e Sistemistica, Università di Bologna. Viale Risorgimento 2, 40136 Bologna, Italy.

[c] Dipartimento di Matematica, Università di Bologna.Via Saragozza 8, 40123 Bologna, Italy.

A 2-D algorithm for optical flow (OF) computation has been generalized to evaluate the motion field of 3-D echocardiograms. This method has been tested on 3-D synthetic sequences simulating biological tissues and ultrasonic scattering. The encouraging results prompted us to also start an investigation on motion analysis of real 3-D echocardiographic volume sequences.

1. INTRODUCTION

Two dimensional (2-D) echocardiography of the heart is an effective non invasive tool in clinical cardiology. In particular, an estimation of endocardial wall motion provides a valid means to evaluate the degree of ischemia and infarction [1,2]. However, analysis of 2-D echocardiogram sequences is inadequate to study the motion of the heart, due to its rotation and translation with respect to the transducer during the cardiac cycle. Therefore our long-term aim is to provide clinicians with computer tools which will enable them to visualize, manipulate and quantify the whole 3-D heart shape, size and motion.

In this paper we generalize a 2-D method [3,4] based on the computation of optical flow to determine the motion field of 3-D echocardiograms (Section 2). The method has been tested on simple test objects and on 3-D data simulating biological tissue and ultrasonic scattering. The time-varying sequences have been obtained by applying a known motion field to the tissue model followed by the necessary transformations to obtain the 3-D synthetic echo data (Section 3). Several results are presented and discussed in Section 4.

2. OPTICAL FLOW ALGORITHM FOR MOTION EVALUATION

To obtain 3-D optical flow from 4-D data sets we extended the method proposed by Uras, Girosi, Verri and Torre [3,6]. Let $E = E(x,y,z,t)$ be the brightness pattern at time t and location (x,y,z) in the 3-D space and $V = (v_x(x,y,z,t), v_y(x,y,z,t), v_z(x,y,z,t))$ the motion field, i.e. velocity vectors at time t and location (x,y,z) in a suitable system of coordinates fixed in the 3-D space. We suppose the gradient of the brightness pattern to be stationary, that is :

$$\frac{d\nabla E}{dt} = 0 \tag{1}$$

Equation (1) can be rewritten as :

$$\mathbf{H U} = -\nabla \frac{\partial E}{\partial t} \tag{2}$$

where $\mathbf{H}$ is the Hessian matrix, with respect to the spatial coordinates, of E. Written explicitly equation (2) becomes :

$$u_x \frac{\partial^2 E}{\partial x^2} + u_y \frac{\partial^2 E}{\partial x \partial y} + u_z \frac{\partial^2 E}{\partial x \partial z} = -\frac{\partial^2 E}{\partial x \partial t}$$

$$u_x \frac{\partial^2 E}{\partial x \partial y} + u_y \frac{\partial^2 E}{\partial y^2} + u_z \frac{\partial^2 E}{\partial y \partial z} = -\frac{\partial^2 E}{\partial y \partial t}$$

$$u_x \frac{\partial^2 E}{\partial x \partial z} + u_y \frac{\partial^2 E}{\partial y \partial z} + u_z \frac{\partial^2 E}{\partial z^2} = -\frac{\partial^2 E}{\partial z \partial t}$$

$U(u_x, u_y, u_z)$ has been used instead of $\mathbf{V}$ to stress the difference between the solution of equation (2) and the true motion field $\mathbf{V}$:

$$U = V + H^{-1} (M^T \nabla E - \nabla dE/dt) \tag{3}$$

where $\mathbf{M^T}$ is the transpose of the Jacobian matrix of $\mathbf{V}$. The three components of $\mathbf{U}$ can be recovered from equation (2), using well known difference formulas, whenever the determinant of $\mathbf{H}$ (det $\mathbf{H}$) is different from zero. It is clear from equations (2) and (3) that when det $\mathbf{H}$ is "large" and the condition number of $\mathbf{H}$ is close to one, a good recovery of the 3-D motion field $\mathbf{U}$ is obtained.

3. MODELLING OF 3-D ECHO DATA AND SEQUENCE GENERATION

To generate 3-D data similar to real ultrasound scans, we adopted a model of the acoustic response of biological tissue and its interaction with an imaging system, as suggested by Bamber et al. [5].

The tissue may be represented by a continuous distribution of elementary scattering sources modelled by its compressibility function $\beta(x,y,z)$. The statistical properties of the spatial variations in compressibility are described by an autocorrelation function

$$N(R) = e^{-\frac{R^2}{\sigma^2}}$$

where σ, known as the correlation length, is a measure of the characteristic size of the inhomogeneities and R^2 represents $x^2+y^2+z^2$. Here it is assumed that the system forming the volume has a linear, separable, and space invariant point spread function (PSF) H(x,y,z). The radio frequency ultrasonic signal may be obtained by convolving the PSF with the tissue impulse response T(x,y,z) :

$$I(x,y,z) = H(x,y,z) * T(x,y,z) \tag{4}$$

where

$$H(x,y,z) = \cos(2\pi f z)e^{-\left(\frac{x^2}{a^2}+\frac{y^2}{b^2}+\frac{z^2}{c^2}\right)} \tag{5}$$

$$T(x,y,z) = \nabla^2\beta(x,y,z) \tag{6}$$

∇^2 represents the Laplacian operator; f is the central frequency of the acoustic field originated by the echo transducer; a,b,c are real parameters related to the acoustic field dimension. The Fourier Transform of the unknown $\beta(x,y,z)$ has modulus equal to the Fourier Transform of N(R) and random phase. The inverse transform of these quantities gives the desired compressibility β. The 3-D echo data E(x,y,z) is at last obtained enveloping the radio frequency signal along the propagation direction by applying a Hilbert transform.

Following this method 3-D echo data shaped by a simple geometry, such as a torus, has been generated. The time-varying sequence was obtained by applying a known motion field to each compressibility function followed by the transformations above described to obtain the final 3-D echo data (Figure 1).

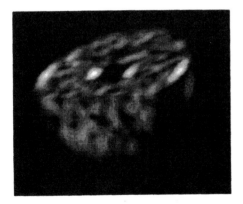

Figure 1. 3-D Echo data generated by
numerical simulation.

4. RESULTS AND DISCUSSION

This section presents the results obtained by applying the algorithm to the simulated sequences. As a first step, a simple test object sequence of an expanding torus has been generated. The pattern of luminance is radially shaped with a sinusoidal function as shown in Figure 2. The entire sequence has been generated so that each 3-D frame is obtained by applying to the previous one the velocity field shown in Figure 4 which has a radial component equal to 1 pixel/frame and a null tangential components.

Figure 2. Central section of the torus test object.

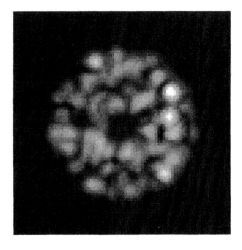

Figure 3. Central section of the 3-D Echo data generated by numerical simulation.

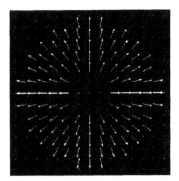

Figure 4. Velocity field imposed for sequence generation.

Figure 5. Modulus of the imposed velocity field.

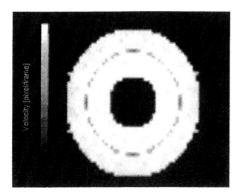

Figure 6. Modulus of the computed velocity field referred to the torus test object.

Figure 7. Modulus of the computed velocity field referred to the central section of the 3-D Echo data generated by numerical simulation.

Figure 6 shows the modulus of the computed optical flow on the central slice of the torus. Comparison between the motion field (Figure 5) and the computed field (Figure 6) shows that a good estimate has been obtained in all the points internal to the torus, whereas no values are present in the background portion, where the aperture problem arises.

Subsequently, the algorithm was tested on a sequence of simulated 3-D echo data that were generated following the procedure explained in Section 3. In particular, the first 3-D frame has been generated by convolving a simulated torus of a biological tissue with a PSF typical of real echocardiographic systems; in particular the central frequency of the ultrasonic burst was f = 3.5MHz, the axial dimension was c = 0.44 mm and the lateral dimensions where a = b = 1.32 mm. The corresponding dimensions of this torus would be an external diameter of 30 mm and an internal diameter of 10 mm. Figure 3 shows the section corresponding to the central plane x,y orthogonal to the ultrasound propagation direction z.

Due to scanty experimental data on the dimension of tissue inhomogeneities, the choice of the correlation length of the compressibility function is a non trivial point. However, following the indications of [5], σ = 4 mm has been assumed, that allows to obtain echo images very similar to the real ones. Finally, in order to generate the entire sequence, the same velocity field as in the previous case has been applied (Figure 4). Pre-processing by a Gaussian filter with S.D. = 1.5 was performed before optical flow computation.

The computed optical flow is shown in Figure 7. The percent average velocity error between the optical flow estimation and the applied velocity field is of the order of 25%; moreover, the aperture problem is present in several points internal to the tissue, and also, obviously, in the background.

Since the algorithm allows to associate to each optical flow value a measure of its reliability, it is possible to consider only the most reliable ones. In this case the resulting field appears very sparse and it is necessary to choose a compromise between estimation accuracy and field density.

Due to the encouraging results obtained by processing simulated 3-D data, experimental work is in progress on motion analysis on real 3-D echocardiographic data. Such sequences have been obtained by means of a new echocardiographic transducer that allows acquisition of 50 standard 2-D images at 3.6 degree increment of rotation around its central axis from any acoustic window and for each frame of the cardiac cycle [7]. The algorithm presented here will be one of several considered in the attempt to reach a compromise between reliability of the results, robustness of the method and required computation power.

REFERENCES

1. H. Feigenbaum, "Ecocardiography", Lea and Febiger, 3rd ed., Philadelphia, 1981.
2. D.J. Skorton, S.M. Collins, R.E. Kerber, "Digital image processing and analysis in echocardiography". In Cardiac imaging and image processing. S.M. Collins, D.J. Skorton Editors. McGraw-Hill, New York, pp. 171-205, 1986.
3. S. Uras, F. Girosi, A. Verri and V. Torre, "A computational approach to motion perception", Biol.Cybern., vol.60, pp.79-87,1988.
4. P. Baraldi, A. Guidazzoli, C. Lamberti, A. Prandini, F. Sgallari, "Computational Approaches for Evaluation of Ventricular Wall Motion from 2-D Echocardiograms". VI Mediterranean Conference on Medical and Biological Engineering. MEDICON 92. Editors : M. Bracale and F. Denoth, Vol. 2, pp. 711-714, 1992.
5. J.C. Bamber and R.J. Dickinson, "Ultrasonic B-Scanning: a computer simulation", Phys.Med.Biol., vol.25, pp.463-478, 1980.
6. P. Baraldi, E. De Micheli, G. Radonich and V. Torre, "The recovery of motion and depth from optical flow", in Image Understanding and Machine Vision, Tropical Meet. Optic.Soc.of America, Tech.Digest Series, 14, pp.54-57, 1989.
7. R. Pini, M. Costi, G.A. Mensha, L. Masotti, K.L. Novins, D.P. Greenberg, B. Greppi, M. Cerofolini and R.B. Devereux, "Computed tomography of the heart by ultrasound", Computers in Cardiology, IEEE Computer Society, pp.17-20, 1991.

I
MOTION ESTIMATION

Time-Varying Image Processing and
Moving Object Recognition, 3 – V. Cappellini (Ed.)
275

Estimating Articulated Motion by Decomposition

Richard J. Qian and Thomas S. Huang

Beckman Institute for Advanced Science and Technology
University of Illinois at Urbana-Champaign
405 North Mathews Avenue
Urbana, IL 61801
U.S.A.

This paper presents three new algorithms for estimating articulated motion by decomposition assuming perspective projection. The numerical properties of the proposed algorithms on topics such as uniqueness of real solution are discussed. And numerical experimental results are also given.

1. INTRODUCTION

An articulated object is a collection of rigid or non rigid bodies which are connected by joints. Biological vision systems have amazing power to understand the visual motion of articulated objects. Studying this type of visual motion by means of computer vision may help us to gain insight into the nature of the way biological visual systems work. Although there are many published papers studying motion analysis of purely rigid or even now purely non rigid objects [1], we found that only a few ones dealt with articulated objects which are neither purely rigid nor purely non rigid objects. Webb and Aggarwal presented an algorithm for the recovery of 3D structures from fixed-axis motion in 1981 [2]. Orthographic projection is assumed in their paper. The result is that the 3D structure of two points which execute fixed-axis rotation can be uniquely determined with four consecutive image frames. Hoffman and Flinchbaugh presented their algorithm for the recovery of 3D structures from biological motion in 1982 [3]. Because they aimed at computing the 3D structure and motion of animal limbs, a "planarity assumption" is introduced in their paper, i.e., the parts around a joint of an articulated object always move inside a single plane.

Orthographic projection is also assumed. Their results are 1) given three distinct orthographic projections of the two end points of a rigid rod which is constrained to rotate in a plane, the structure and motion compatible with the three views are uniquely determined; 2) given two distinct orthographic projections of the three endpoints of two rigid rods linked in a hinge joint to form a pairwise-rigid structure which is constrained to move in one plane, the structure and motion compatible with the two views are uniquely determined.

Understanding the visual motion of articulated objects has also broad applications. One application may be to apply the developed algorithms for 3D structure recovery from the motion of articulated objects to motion analysis of human ambulatory patterns. Studying the visual motion of human ambulatory patterns is of great importance for both vision research and applications. For example, it can be used in model-based image coding and transmission such that we may only need to transmit a few estimated motion parameters instead of sending a whole image sequence if certain human body models are available at both sender and receiver sides [4]. Another important application of understanding the visual motion of articulated objects is in robotic vision where most robot arms are ideal articulated objects.

In Section 2, we will present three new algorithms for estimating articulated motion by decomposition assuming perspective projection. In Section 3, we will discuss the numerical properties of the proposed algorithms on topics such as uniqueness of real solution based upon our numerical experimental results which will be given in Appendix. In Section 4, we will give our conclusions.

2. MOTION ANALYSIS ALGORITHMS FOR ARTICULATED OBJECTS

The articulated objects discussed in both Hoffman and Webb's papers are pairwise-rigid, i.e., each part of the articulated object is rigid; moreover each joint has at most one degree of rotational freedom. In this paper, we also emphasize this type of articulated objects. However, we will study the cases assuming perspective projection, because in practice camera projection is always perspective and, in some cases, using orthographic projection as an approximation cannot achieve the desired accuracy.

Systematically, we can write equations which describe the whole articulated object no matter how many joints it has [5]. Due to the high order of the nonlinear equations, however, it is very difficult to solve such equations and analyze uniqueness of real solution. Alternatively, here we propose a new way to solve the whole system by decomposition. Only one or two joints are considered at a time. We will first develop an algorithm for the recovery of 3D structures from fixed-axis motion. Then we will discuss the more restricted cases, i.e., those of planar motion. For the sake of simplicity, we will use identical characters to denote points and their associated coordinate vectors in 3D space. For example, if there is a point $\mathbf{p}$ associated with a 3D coordinate vector of (x, y, z), then we will still use the character $\mathbf{p}$ to denote that vector, i.e., we let $\mathbf{p} = (x, y, z)^T$. Further, we will use $\overline{\mathbf{p}_0 \mathbf{p}_1}$ to denote the vector from point $\mathbf{p}_0$ to $\mathbf{p}_1$, and use $\|\overline{\mathbf{p}_0 \mathbf{p}_1}\|$ to denote its length. Assuming perspective projection and unity focus length, we have image coordinates and their corresponding 3D coordinates related as follows:

$$X = \frac{x}{z} \tag{2-1}$$

$$Y = \frac{y}{z} \tag{2-2}$$

where (X, Y) are image coordinates, (x, y, z) are the corresponding 3D coordinates, and z is the depth. Then for any point p, we have

$$\mathbf{p} = (X, Y, 1)^T \cdot z \tag{2-3}$$

2.1. Algorithm for Three-dimensional Structures from Fixed-Axis Motion

Algorithm 2.1 : Two points in four frames

Suppose we are given an image sequence of one part of a moving articulated object; see Figure 2.1.

The only allowed motion is that rod $\overline{\mathbf{p}_0 \mathbf{p}_1}$ can rotate around an unknown but fixed axis through joint $\mathbf{p}_0$ The 3D coordinates (or equivalently the depth only) of the joint $\mathbf{p}_0$ are given. The goal is to recover the 3D coordinates of points $\mathbf{p}_1$ in 3D space from the given image coordinates.

We attack this problem by solving a set of nonlinear equations. Using four consecutive frames, we can first write three degree-2 equations based on the rigidity constraint

$$\left\|\overline{P_0 P_1}\right\| = \left\|\overline{P_0' P_1'}\right\| = \left\|\overline{P_0'' P_1''}\right\| = \left\|\overline{P_0''' P_1'''}\right\| \tag{2-4}$$

Because the rod rotates around a fixed axis through the joint, namely, the end point p_1 remains in a single plane during the rotation. Then we have another degree-3 equation

$$\overline{P_1'' P_1'''} \cdot \overline{P_1 P_1'} \times \overline{P_1' P_1''} = 0 \tag{2-5}$$

Note that we have four unknowns and have set up four nonlinear equations. From Bezout's theorem, the number of possible solutions is no larger than $2^3 \cdot 3 = 24$.

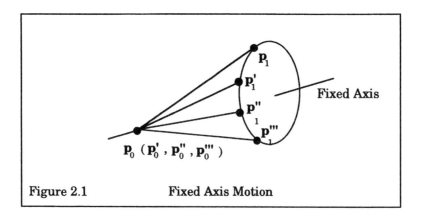

Figure 2.1   Fixed Axis Motion

2.2. Algorithms for Three-dimensional Structures from Planar Motion

Algorithm 2.2.A : Two points in three frames

Here again we are given an image sequence of one part of a moving articulated object, see Figure 2.2.A.

But rod $\overline{p_0 p_1}$ now is restricted to remain in a plane during its rotation. The 3D coordinates (or equivalently the depth only) of the joint p_0 are still given. The goal is to recover the 3D coordinates of points $\mathbf{p}_1$ in 3D space from the given image coordinates.

Using only three consecutive frames this time, we can write two degree-2 equations based on the rigidity constraint

$$\left\|\overline{\mathbf{p}_0\mathbf{p}_1}\right\| = \left\|\overline{\mathbf{p}_0'\mathbf{p}_1'}\right\| = \left\|\overline{\mathbf{p}_0''\mathbf{p}_1''}\right\| \tag{2-6}$$

Then based on the planarity assumption, we can write another degree-3 equation

$$\overline{\mathbf{p}_0\mathbf{p}_1} \cdot \overline{\mathbf{p}_0'\mathbf{p}_1'} \times \overline{\mathbf{p}_0''\mathbf{p}_1''} = 0 \tag{2-7}$$

In this case, we have three unknowns in total. Three nonlinear equations are available. From Bezout's theorem, the number of solutions is no larger than $2^2 \cdot 3 = 12$.

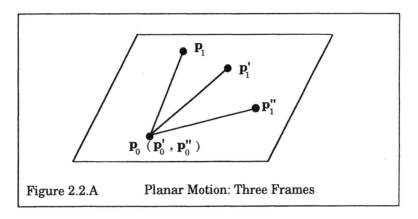

Figure 2.2.A   Planar Motion: Three Frames

Algorithm 2.2.B : Three points in two frames

Fewer frames are needed for 3D structure recovery, if we consider two rods at a time, and if the motion is planar, i.e., the two rods remain in a single plane during their rotation. See Figure 2.2.B.

With only two consecutive frames, our first two degree-2 equations still come from the rigidity constraints

$$\left\|\overline{\mathbf{p}_0\mathbf{p}_1}\right\| = \left\|\overline{\mathbf{p}_0'\mathbf{p}_1'}\right\| \tag{2-8}$$

$$\left\| \overline{\mathbf{P}_1\mathbf{P}_2} \right\| = \left\| \overline{\mathbf{P}_1{}'\mathbf{P}_2{}'} \right\| \qquad (2\text{-}9)$$

Then based on the planarity assumption, we have two degree-3 equations

$$\overline{\mathbf{P}_1{}'\mathbf{P}_2{}'} \cdot \overline{\mathbf{P}_0\mathbf{P}_1} \times \overline{\mathbf{P}_1\mathbf{P}_2} = 0 \qquad (2\text{-}10)$$

$$\overline{\mathbf{P}_1{}'\mathbf{P}_0{}'} \cdot \overline{\mathbf{P}_0\mathbf{P}_1} \times \overline{\mathbf{P}_1\mathbf{P}_2} = 0 \qquad (2\text{-}11)$$

There are four unknowns and four nonlinear equations available. From Bezout's theorem, the number of solutions is no larger than $2^2 \cdot 3^2 = 36$.

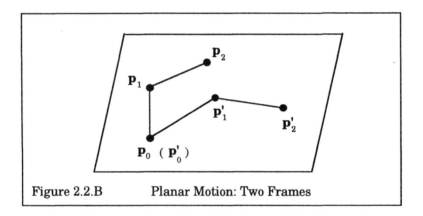

Figure 2.2.B   Planar Motion: Two Frames

3. DISCUSSION ON NUMERICAL PROPERTIES

Applying Equations (2-1)-(2-3) to Equations (2-8)-(2-11) can generate the following expanded equations in terms of z coordinates only:

$$A_1 z_1^2 + A_2 z_1{}'^2 + A_3 z_1 + A_4 z_1{}' = 0 \qquad (3\text{-}1)$$

$$B_1 z_1^2 + B_2 z_2^2 + B_3 z_1{}'^2 + B_4 z_2{}'^2 + B_5 z_1 z_2 + B_6 z_1{}' z_2{}' = 0 \qquad (3\text{-}2)$$

$$C_1 z_1 z_2 z_1{}' + C_2 z_1 z_2 + C_3 z_1 z_1{}' + C_4 z_2 z_1{}' + C_5 z_1 + C_6 z_2 = 0 \qquad (3\text{-}3)$$

$$D_1 z_1 z_2 z_1' + D_2 z_1 z_2 z_2' + D_3 z_1 z_1' + D_4 z_1 z_2' + D_5 z_2 z_1' + D_6 z_2 z_2' = 0 \tag{3-4}$$

where

$$A_1 = X_1^2 + Y_1^2 + 1 \tag{3-5}$$

$$A_2 = -(X_1'^2 + Y_1'^2 + 1) \tag{3-6}$$

$$A_3 = -2(X_0 X_1 + Y_0 Y_1 + 1) \tag{3-7}$$

$$A_4 = 2(X_0' X_1' + Y_0' Y_1' + 1) \tag{3-8}$$

$$B_1 = X_1^2 + Y_1^2 + 1 \tag{3-9}$$

$$B_2 = X_2^2 + Y_2^2 + 1 \tag{3-10}$$

$$B_3 = -(X_1'^2 + Y_1'^2 + 1) \tag{3-11}$$

$$B_4 = -(X_2'^2 + Y_2'^2 + 1) \tag{3-12}$$

$$B_5 = -2(X_1 X_2 + Y_1 Y_2 + 1) \tag{3-13}$$

$$B_6 = 2(X_1' X_2' + Y_1' Y_2' + 1) \tag{3-14}$$

$$C_1 = X_1'(Y_2 - Y_1) - Y_1'(X_2 - X_1) + (X_2 Y_1 - X_1 Y_2) \tag{3-15}$$

$$C_2 = -(X_0'(Y_2 - Y_1) - Y_0'(X_2 - X_1) + (X_2 Y_1 - X_1 Y_2)) z_0' \tag{3-16}$$

$$C_3 = (X_1'(Y_1 - Y_0) - Y_1'(X_1 - X_0) + (X_1 Y_0 - X_0 Y_1)) z_0 \tag{3-17}$$

$$C_4 = (X_1'(Y_0 - Y_2) - Y_1'(X_0 - X_2) + (X_0 Y_2 - X_2 Y_0)) z_0 \tag{3-18}$$

$$C_5 = -(X_0{}'(Y_1 - Y_0) - Y_0{}'(X_1 - X_0) + (X_1 Y_0 - X_0 Y_1))z_0 z_1 \tag{3-19}$$

$$C_6 = -(X_0{}'(Y_0 - Y_2) - Y_0{}'(X_0 - X_2) + (X_0 Y_2 - X_2 Y_0))z_0 z_0{}' \tag{3-20}$$

$$D_1 = X_2{}'(Y_2 - Y_1) - Y_2{}'(X_2 - X_1) + (X_2 Y_1 - X_1 Y_2) \tag{3-21}$$

$$D_2 = -(X_1{}'(Y_2 - Y_1) - Y_1{}'(X_2 - X_1) + (X_2 Y_1 - X_1 Y_2)) \tag{3-22}$$

$$D_3 = -(X_1{}'(Y_1 - Y_0) - Y_1{}'(X_1 - X_0) + (X_1 Y_0 - X_0 Y_1))z_0 \tag{3-23}$$

$$D_4 = (X_2{}'(Y_1 - Y_0) - Y_2{}'(X_1 - X_0) + (X_1 Y_0 - X_0 Y_1))z_0 \tag{3-24}$$

$$D_5 = -(X_1{}'(Y_0 - Y_2) - Y_1{}'(X_0 - X_2) + (X_0 Y_2 - X_2 Y_0))z_0 \tag{3-25}$$

$$D_6 = (X_2{}'(Y_0 - Y_2) - Y_2{}'(X_0 - X_2) + (X_0 Y_2 - X_2 Y_0))z_0 \tag{3-26}$$

Because the equations (3-1)-(3-4) are nonlinear, it is of great interest to study their numerical properties. Among them, the property of uniqueness of real solution may be the most important. By real solution, we mean the one which has a physically proper interpretation. For example, in a certain selected 3D reference coordinate system, the z coordinates (or the depths) of the observed objects must have positive values. In such a case, we know that the solutions which have negative or complex z coordinates cannot be real solutions because they do not have physically proper interpretation. To analyze uniqueness of real solution of the equations in the previous section where perspective projection is assumed is much more difficult than the corresponding cases where orthographic projection is assumed. At this point, we have only accomplished a number of numerical experiments.

In our numerical experiments, we first generate the synthesized image coordinates of a selected articulated object model in a number of consecutive frames. Doing that, we initially specify the 3D locations of the joints of the articulated object and their associated 3D motion in the space. Using perspective projection, we project those 3D locations of the joints onto the image plane at different time instants to obtain the image coordinates in different frames. Then with the synthesized image coordinates only, we solve the nonlinear equations in the previous sections for z coordinates (consequently x and y coordinates) by using the continuation method.

Up to now, two observations have been established from our numerical experimental results: 1) the real solution is always unique up to a reflectance; 2) the number of the obtained solutions from our continuation method based program is less than the number of possible solutions suggested by Bezout's theorem. In Appendix we reprint the results of one example of our numerical experiments. It is related to Algorithm 2.2 B in Section 2.2.. In this example, we can see that the number of the solutions is 22. Solution #22 is the real solution and Solution #6 is the twisted reflectance of the real solution.

Another numerical property of interest is the speed of convergence of the employed algorithm for solving nonlinear equations. As mentioned above, our current program is based on the continuation method. The speed of convergence is acceptable. For example, it takes about five minutes on the average on a SUN-3 workstation for the algorithm to converge for the equations in the previous sections.

4. CONCLUSIONS

Most of the previously published algorithms for motion analysis of articulated objects assume that camera projection is orthographic. In practice, however, perspective projection is more likely the actual case. In this paper, we have developed three algorithms assuming perspective projection for the recovery of 3D structures from the motion of several types of joints of articulated objects, viz., joints which allow only fixed-axis rotation and joints which allow only planar rotation. The algorithms have been tested in terms of uniqueness of real solution and speed of convergence with synthesized data. The numerical experimental results show that our algorithms give a unique real solution up to a reflectance and the speed of convergence of the algorithms is acceptable.

Studying the uniqueness of the real solution is of great importance for nonlinear algorithms. If a nonlinear algorithm cannot guarantee to always give a unique real solution, then it may be difficult to be used in practice. In that case, some additional procedures may have to be employed to select appropriate solutions. For example, repeatedly apply the nonlinear algorithm to a long image sequence and hope to obtain a unique solution which is the only one consistently surviving during the whole sequence. In the cases where noises presents, nonlinear Kalman filtering technique may also be useful. At this point, we have only tested our algorithms in a number of numerical experiments. Analytically studying uniqueness of real solution of the nonlinear equations in Section 2 is much more difficult than the corresponding cases in which orthographic projection is assumed. In fact, it has been considered as one of our research issues. In addition to uniqueness of real solution, another interesting property of our nonlinear algorithms is the maximal number of possible solutions. It is noted that the number of solutions obtained from our continuation method based program is always less than the number suggested by Bezout's theorem. In fact, this would indicate the possibility to find a tighter upper bound instead of the one from Bezout's theorem. But we have not made any significant progress on that until now.

LIST OF REFERENCES

[1]   T. S. Huang, "Motion analysis," *Encyclopedia of Artificial Intelligence*, S. Shapiro, Eds. John Wiley and Sons, 1987, pp. 620-632.

[2]   J. A. Webb and J. K. Aggarwal, "Structure from motion of rigid and jointed objects," *Proc. 1981 IJCAI*, pp. 686-691.

[3]   D. D. Hoffman and B. E. Flinchbaugh, "Interpretation of biological motion," *Biological Cybernetics*, vol. 42, pp. 195-204, 1982.

[4]   T. Kimoto and Y. Yasuda, "Hierarchical representation of the motion of a walker and motion reconstruction for model-based image coding," *Optical Engineering*, vol. 30, no. 7, pp. 888-903, 1982.

[5]   R. Featherstone, "Robot dynamics algorithms," *Robotics: Vision, Manipulation and Sensors*, T. Kanade, Eds. Kluwer Academic Publishers, 1987.

[6]   R. J. Qian and T. S. Huang, "Motion analysis of articulated objects," *Proc. 1992 IU Workshop*, San Diego, CA, pp. 549-553.

[7]   .R. J. Qian and T. S. Huang, "Motion analysis of human ambulatory patterns," *Proc. 1992 ICPR*, Hague, Netherlands.

APPENDIX

Initial Joint 3D Locations and Motion

Joint	x	y	z	Rotation Angle
0	2.50000000	1.00000000	5.00000000	17.00000000

1	6.00000000	7.50000000	15.00000000	-25.00000000
2	9.50000000	4.00000000	10.00000000	0.00000000

Calculated 3D Coordinates in Space

	Time Instant 1			Time Instant 2		
Joint	x	y	z	x	y	z
0	2.50000000	1.00000000	5.00000000	2.50000000	1.00000000	5.00000000
1	6.00000000	7.50000000	15.00000000	2.36295955	7.89891221	15.33857970
2	9.50000000	4.00000000	10.00000000	4.98037075	4.19053767	9.96297584

Synthesized Image Coordinates

	Frame 1		Frame 2	
Joint	X	Y	X	Y
0	0.50000000	0.20000000	0.50000000	0.20000000
1	0.40000000	0.50000000	0.15405335	0.51497025
2	0.95000000	0.40000000	0.49988787	0.42061104

Solutions from Continuation Method

		Time Instant 1	Time Instant 2
Solution	Joint	z	z
#1	1	0.00000000 + j-0.00000000	0.00000000 + j-0.00000000
	2	0.00000000 + j0.00000000	0.00000000 + j0.00000000
#2	1	-37.82431015 + j0.00000000	48.93419161 + j0.00000000
	2	-13.21087252 + j0.00000000	63.46146739 + j0.00000000
#3	1	0.00000000 + j-0.00000000	0.00000000 + j-0.00000000
	2	0.00000000 + j-0.00000000	0.00000000 + j-0.00000000
#4	1	-0.00000000 + j0.00000000	-0.00000000 + j0.00000000
	2	0.00000000 + j0.00000000	0.00000000 + j0.00000000
#5	1	8.31595320 + j0.00000000	1.00949518 + j0.00000000
	2	-0.57873157 + j-0.00000000	-7.99545756 + j0.00000000
#6	1	2.94895498 + j0.00000000	3.72658246 + j0.00000000
	2	2.17700403 + j0.00000000	3.06776774 + j0.00000000
#7	1	4.59578974 + j-0.64550244	4.57151068 + j-1.65816072
	2	3.71753100 + j0.92281800	4.61583271 + j-0.21911914
#8	1	-89.24434089 + j0.00000000	102.72921903 + j0.00000000
	2	-120.46201805 + j0.00000000	28.49514441 + j0.00000000
#9	1	-3.06982948 + j0.00000000	-3.31074355 + j0.00000000
	2	-3.88225276 + j0.00000000	-4.95443757 + j0.00000000

#10	1	-0.00000000 + j0.00000000	-0.00000000 + j0.00000000
	2	0.00000000 + j-0.00000000	-0.00000000 + j0.00000000
#11	1	-0.00000000 + j0.00000000	9.15506398 + j-0.00000000
	2	-0.00000000 + j0.00000000	8.30045137 + j-2.61122008
#12	1	7.02067872 + j0.00000000	2.56165473 + j0.00000000
	2	-2.65838792 + j0.00000000	12.30885040 + j0.00000000
#13	1	-17.39706413 + j0.00000000	-18.38997838 + j0.00000000
	2	-170.68259434 + j0.00000000	172.55530424 + j0.00000000
#14	1	4.59578974 + j0.64550244	4.57151068 + j1.65816072
	2	3.71753100 + j-0.92281800	4.61583271 + j0.21911914
#15	1	-54.67283086 + j0.00000000	66.56362240 + j0.00000000
	2	-24.80372133 + j0.00000000	38.45442134 + j0.00000000
#16	1	0.00000000 + j-0.00000000	0.00000000 + j-0.00000000
	2	-0.00000000 + j0.00000000	0.00000000 + j-0.00000000
#17	1	17.26243919 + j0.00000000	-8.56893372 + j0.00000000
	2	2.15270598 + j0.00000000	6.82227183 + j0.00000000
#18	1	-0.00000000 + j0.00000000	-0.00000000 + j0.00000000
	2	-0.00000000 + j-0.00000000	-0.00000000 + j-0.00000000
#19	1	-0.00000000 + j-0.00000000	9.15506398 + j0.00000000
	2	-0.00000000 + j-0.00000000	8.30045137 + j2.61122008
#20	1	-0.00000000 + j0.00000000	-0.00000000 + j0.00000000
	2	-0.00000000 + j0.00000000	-0.00000000 + j0.00000000
#21	1	0.00000000 + j-0.00000000	0.00000000 + j-0.00000000
	2	-0.00000000 + j-0.00000000	-0.00000000 + j-0.00000000
#22	1	15.00000000 + j0.00000000	15.33857970 + j0.00000000
	2	10.00000000 + j0.00000000	9.96297584 + j0.00000000

(Number of solutions = 22 ; z coordinate of Joint 0 being set as 5.0)

Acknowledgement

This work was supported by the National Science Foundation Grants IRI-89-08255 and IRI-89-02728.

Time-Varying Image Processing and
Moving Object Recognition, 3 – V. Cappellini (Ed.)
© 1994 Elsevier Science B.V. All rights reserved.

Pose from Motion: A Direct Method *

Masanobu Yamamoto
Department of Information Engineering, Niigata University
8050 Ikarashi 2-nocho, Niigata-city, 950-21, Japan

Abstract

This paper concerns a problem estimating pose and 3D motion parameters of an object from an image sequence under assumptions of monocular and perspective view as well as given 3D model of the object. We derive nonlinear equations which can simultaneously and directly estimate pose and 3D motion parameters of the object. The Newton method iteratively gives the solution to the system of nonlinear equations. Our method requires neither model to image matching nor image to image matching, while the Lowe's method requires model to image matching.

1 Introduction

Temporal-spatial gradient scheme enables us to recover 3D structure and motion of an object from an image sequence without correspondence between images. However, when the set of unknowns includes the depth information, it becomes difficult to obtain a unique solution since the number of unknowns usually exceeds the number of equations. The various constrains must be introduced to cope with this problem. When the surface of the object is planar [7][5] or quadric [6], when the depth is known [8][9][1], or when the motion is pure rotation or pure translation [1], the structure and/or motion can be recovered.

This paper, succeeded to the case of known depth in the works of Yamamoto[8][9] and Horn[1], concerns the following problem: The object has any surface structure, i.e. the scene is not limited to specific structures such as a planer or quadric world. The surface structure is known by a 3D model. However, the 3D model is deviated from the correct pose. Therefore, the model cannot give a correct depth information. This paper describes a direct method to correct the pose displacement of the model, and simultaneously estimate the 3D motion parameters from the image sequence, in the case of small pose deviation.

The method is applied to a problem of tracking object based on 3D model. Our method requires neither model to image matching nor image to image matching, while the Lowe's method [3][4] requires model to image matching.

The next Section describes image flow model represented by 3D motion parameters and 3D pose correction/displacement parameters. The 3rd Section gives a direct method to estimate these parameters. The Section 4 is devoted to an application of our method to object tracking problem.

---

*This work was supported in part by the SANKYO SEIKI MFG. CO., LTD., and in part by the Domestic Cooperative Research Center at Niigata University.

2 Image Flow Model

Let a coordinate system of the scene be a Cartesian (x, y, z) one. Now, the displacement vector $\delta = (\delta x, \delta y, \delta z)^T$ of the point $\mathbf{p} = (x, y, z)^T$ on a rigid object can be presented as eq.(1) with the angular velocity vector $\mathbf{R} = (R_x, R_y, R_z)^T$ around the center of the object, $\mathbf{p}_c = (x_c, y_c, z_c)^T$, and the translation vector $\mathbf{T} = (T_x, T_y, T_z)^T$.

$$\delta = \mathbf{R} \times (\mathbf{p} - \mathbf{p}_c) + \mathbf{T} \tag{1}$$

We can observe the motion of object of which projection lies on a image plane, $z = 1$, with the origin of the coordinate as the projection center. An object point $\mathbf{p}$ is projected on the image point (X, Y).

$$\begin{cases} X = x/z \\ Y = y/z \end{cases} \tag{2}$$

By the total derivative of eq.(2), the image flow vector $\mathbf{\Delta} = (\Delta X, \Delta Y)^T$ at the image point (X, Y) is related with the object displacement vector δ.

$$\mathbf{\Delta} = \frac{1}{z} A \delta \tag{3}$$

where

$$A = \begin{pmatrix} 1 & 0 & -X \\ 0 & 1 & -Y \end{pmatrix}$$

Substituting eq.(1) into eq.(3), the image flow vector can be represented by the 3D motion parameters.

$$\mathbf{\Delta} = \frac{1}{z} A \mathbf{T} + (B - \frac{1}{z} B_c) \mathbf{R} \tag{4}$$

where

$$B = \begin{pmatrix} -XY & 1+X^2 & -Y \\ -(1+Y^2) & XY & X \end{pmatrix}$$

$$B_c = \begin{pmatrix} -y_c X & z_c + x_c X & -y_c \\ -(z_c + y_c Y) & x_c Y & x_c \end{pmatrix}$$

The depth z in eq.(4) is given by the 3D model. However, when the 3D model deviates from the correct pose, the depth information is false. The 3D model displacement must be corrected. Pose displacements of the 3D model consist of orientation and position displacements. The orientation displacement of the 3D model can be corrected by the rotation around the axis passing through the center of the object (x_c, y_c, z_c). The position displacement of the 3D model can be corrected by the translation. Supposing the displacements to be very small, the pose corrections can be represented by the vectors $\mathbf{Q} = (Q_x, Q_y, Q_z)^T$ and $\mathbf{S} = (S_x, S_y, S_z)^T$,which denote rotation and translation, respectively.

We suppose that a depth z map written as

$$z = h(x, y) \tag{5}$$

which is obtained from the incorrect model. When the corrections $\mathbf{S}, \mathbf{Q}$ are given, the correct depth map is obtained by the equation (12) in the companion paper[11].

$$z + Q_y(Xz - x_c) - Q_x(Yz - y_c) - S_z =$$
$$h(Xz + Q_z(Yz - y_c) - Q_y(z - z_c) - S_x, Yz - Q_z(Xz - x_c) + Q_x(z - z_c) - S_y).(6)$$

Solving the eq.(6) with respect to z gives a new depth map,

$$z = G(X, Y, \mathbf{S}, \mathbf{Q}) \tag{7}$$

Furthermore, substituting X, Y of eq.(2) into eq.(7), and solving eq.(7) with respect to z, let the solution be

$$z = g(x, y, \mathbf{S}, \mathbf{Q}). \tag{8}$$

If the pose displacement are not corrected,

$$g(x, y, \mathbf{0}, \mathbf{0}) = h(x, y).$$

The depth z from the 3D model can be represented by the pose correction parameters $\mathbf{S}, \mathbf{Q}$. The image flow of eq.(4) can be represented by 3D motion parameters as well as pose correcting parameters.

3   Correcting Pose Displacements of 3D Model

Given a sequence of images and 3D information of the object from incorrect 3D model, let's present a direct method to estimate 3D motion parameters $\mathbf{T}, \mathbf{R}$ and pose corrections $\mathbf{S}, \mathbf{Q}$.

Let the intensity of the image at point (X, Y) at time t be $E(X, Y, t)$. The intensity is assumed remain constant so that

$$E(X + \Delta X, Y + \Delta Y, t + \Delta t) - E(X, Y, t) = 0. \tag{9}$$

The ΔX and ΔY of the eq.(9), as discussed in the previous Section, includes $\mathbf{T}, \mathbf{S}$ and $\mathbf{S}, \mathbf{Q}$. Therefore, the constraint equations (9) derived from image points are a system of nonlinear equations with $\mathbf{T}, \mathbf{S}$ and $\mathbf{S}, \mathbf{Q}$ as unknowns. Solving the system of the nonlinear equations, we can get 3D motion parameters as well as pose corrections of the 3D model. However, it is difficult to find the closed-form solution to satisfy the system of equations because of the nonlinearity. Minimizing the L_2 norm measure of the left hand side of eq.(9),

$$\sum_{X,Y} (E(X + \Delta X, Y + \Delta Y, t + \Delta t) - E(X, Y, t))^2, \tag{10}$$

we can get a least-square solution. Starting from an appropriate initial guess, the Newton method can give an optimal solution.

The initial guess is set as zero since displacements and movements are regarded as very small.

The iteration is expressed by

$$\mathbf{T}^{(n+1)} = \mathbf{T}^{(n)} + \Delta \mathbf{T} \tag{11}$$
$$\mathbf{R}^{(n+1)} = \mathbf{R}^{(n)} + \Delta \mathbf{R} \tag{12}$$
$$\mathbf{S}^{(n+1)} = \mathbf{S}^{(n)} + \Delta \mathbf{S} \tag{13}$$
$$\mathbf{Q}^{(n+1)} = \mathbf{Q}^{(n)} + \Delta \mathbf{Q} \tag{14}$$

The set of $\Delta\mathbf{T}$, $\Delta\mathbf{R}$ and small corrections to $\mathbf{S}, \mathbf{Q}$

$$\Delta\mathbf{S} = (\Delta S_x, \Delta S_y, \Delta S_z)^T$$
$$\Delta\mathbf{Q} = (\Delta Q_x, \Delta Q_y, \Delta Q_z)^T$$

is obtained from solving the system of linear equations

$$\mathbf{J}(\mathbf{T}^{(n)}, \mathbf{R}^{(n)}, \mathbf{S}^{(n)}, \mathbf{Q}^{(n)}) \begin{pmatrix} \Delta\mathbf{T} \\ \Delta\mathbf{R} \\ \Delta\mathbf{S} \\ \Delta\mathbf{Q} \end{pmatrix} = \mathbf{e}^{(n)} \tag{15}$$

where $\mathbf{J}(\mathbf{T}, \mathbf{R}, \mathbf{S}, \mathbf{Q})$ is a Jacobian matrix

$$\mathbf{J}(\mathbf{T}, \mathbf{R}, \mathbf{S}, \mathbf{Q}) = \begin{pmatrix} \vdots \\ \mathbf{J}_{X,Y}(\mathbf{T}, \mathbf{R}, \mathbf{S}, \mathbf{Q}) \\ \vdots \end{pmatrix}$$

$$\mathbf{J}_{X,Y}(\mathbf{T}, \mathbf{R}, \mathbf{S}, \mathbf{Q}) = \frac{\partial E(X + \Delta X, Y + \Delta Y, t + \Delta t)}{\partial(\mathbf{T}, \mathbf{R}, \mathbf{S}, \mathbf{Q})}.$$

$$\Delta^{(n)} = (\Delta X^{(n)}, \Delta Y^{(n)})^T$$
$$= \frac{1}{z^{(n)}} A\mathbf{T}^{(n)} + (B - \frac{1}{z^{(n)}} B_c)\mathbf{R}^{(n)}$$
$$z^{(n)} = G(X, Y, \mathbf{S}^{(n)}, \mathbf{Q}^{(n)})$$

Components of the Jacobian matrix, $\mathbf{J}_{X,Y}(\mathbf{T}, \mathbf{R}, \mathbf{S}, \mathbf{Q})$, include a partial derivative of z with respect to $\mathbf{S}, \mathbf{Q}$. The partial derivatives are derived from partially differentiating the eq.(6) with respect to $\mathbf{S}, \mathbf{Q}$. Therefore, the linear equation of the system (15) is described in detail

$$\frac{1}{z^{(n)}} \mathbf{a} \cdot \Delta\mathbf{T} + (\mathbf{b} - \frac{1}{z^{(n)}} \mathbf{b}_c) \cdot \Delta\mathbf{R} +$$
$$\frac{\mathbf{a} \cdot \mathbf{T}^{(n)} - \mathbf{b}_c \cdot \mathbf{R}^{(n)}}{(z^{(n)})^2} (\mathbf{c} \cdot \Delta\mathbf{S} - (\mathbf{d} - \mathbf{d}_c) \cdot \Delta\mathbf{Q}) = e_{X,Y}^{(n)} \tag{16}$$

where

$$\mathbf{a} = \begin{pmatrix} E_X \\ E_Y \\ -XE_X - YE_Y \end{pmatrix}$$

$$\mathbf{b} = \begin{pmatrix} -Y(XE_X + YE_Y) - E_Y \\ X(XE_X + YE_Y) + E_X \\ XE_Y - YE_X \end{pmatrix}$$

$$\mathbf{b}_c = \begin{pmatrix} -y_c(XE_X + YE_Y) - z_cE_Y \\ x_c(XE_X + YE_Y) + z_cE_X \\ x_cE_Y - y_cE_X \end{pmatrix}$$

$$\mathbf{c} = \frac{1}{k}\begin{pmatrix} h_x \\ h_y \\ -1 \end{pmatrix}$$

$$\mathbf{d} = \frac{z^{(n)}}{k}\begin{pmatrix} Y + h_y \\ -X - h_x \\ Yh_x - Xh_y \end{pmatrix}$$

$$\mathbf{d}_c = \frac{1}{k}\begin{pmatrix} y_c + z_c h_y \\ -x_c - z_c h_x \\ y_c h_x - x_c h_y \end{pmatrix}$$

$$k = 1 - Xh_x - Yh_y - (Y + h_y)Q_x^{(n)} + (X + h_x)Q_y^n + (Xh_y - Yh_x)Q_z^{(n)}$$

The surface gradient (h_x, h_y) on the 3D model is calculated at

$$x = Xz^{(n)} + Q_z^{(n)}(Yz^{(n)} - y_c) - Q_y^{(n)}(z^{(n)} - z_c) - S_x^{(n)}$$
$$y = Yz^{(n)} - Q_z^{(n)}(Xz^{(n)} - x_c) + Q_x^{(n)}(z^{(n)} - z_c) - S_y^{(n)}$$

Calibrating the pose of the 3D model by $\mathbf{\Delta S}, \mathbf{\Delta Q}$ at every iteration, the $\mathbf{c}, \mathbf{d}, \mathbf{d}_c, k$ among the coefficients of eq.(16) may be simplified. The $n + 1$th iteration is calculated based on the revised 3D model after the nth iteration. The Jacobian matrix can be updated by setting $\mathbf{S}^{(n)} = \mathbf{0}, \mathbf{Q}^{(n)} = \mathbf{0}$ since the 3D model has been revised. The surface gradient (h_x, h_y) on the incorrect 3D model can be replaced by a surface gradient (g_x, g_y) on the revised 3D model. The $\mathbf{c}, \mathbf{d}, \mathbf{d}_c, k$ may be simplified as

$$\mathbf{c} = \frac{1}{k}\begin{pmatrix} g_x \\ g_y \\ -1 \end{pmatrix}$$

$$\mathbf{d} = \frac{z^{(n)}}{k}\begin{pmatrix} Y + g_y \\ -X - g_x \\ Yg_x - Xg_y \end{pmatrix}$$

$$\mathbf{d}_c = \frac{1}{k}\begin{pmatrix} y_c + z_c g_y \\ -x_c - z_c g_x \\ y_c g_x - x_c g_y \end{pmatrix}$$

$$k = 1 - Xg_x - Yg_y.$$

The system of linear equations (15) has a unique solution if and only if the coefficient matrix $\mathbf{J}(\mathbf{T}, \mathbf{R}, \mathbf{S}, \mathbf{Q})$ is non-singular.

$$[\mathbf{J}^T\mathbf{J}]^{-1}\mathbf{J}^T\mathbf{e}^{(n)}$$

The iterative procedure of eqs.(11)~(14) stops when the amount of (10) does not decrease. Then, the 3D model is posed in the correct orientation and position, the 3D motion parameters are obtained.

It is, in general, difficult to obtain the surface gradient (g_x, g_y) by an analytical approach since eqs.(6, 7) are nonlinear. In practice we use, however, a geometrical modeler [2], namely SOLVER, which can give

(1) depth maps, $g(x, y, \mathbf{S}, \mathbf{Q})$, $G(X, Y, \mathbf{S}, \mathbf{Q})$,

(2) projection of the 3D model, and

(3) orientation map (g_x, g_y) on the model surface.

Figure 1: (Left) The 3D model is superimposed on the object image at starting frame. (Right) The 3D model at 100th frame without correcting pose displacement.

4 Experiments

An experimental result is described, where an object can be tracked by estimating 3D motion parameters and correcting the pose displacements of the 3D model. The target for tracking in this experiment is a soccer ball which is hanging from the ceiling using a string. The pendulum, the ball with a string, swings in the plane perpendicular to the camera axis. The ball itself is rotating around the string. The total number of frames in the image sequence is 100. The first frame of the image sequence is depicted in the left picture of Fig.1. In this case, the orientation displacement of the model can not be uniquely corrected since the ball is a sphere [10]. Furthermore, the estimation of z component of the position displacement falls in ill-condition. Therefore, we wish to correct only x and y components of the position displacement, and estimate 3D motion parameters.

First, let us show the object tracking without correcting pose displacement of the 3D model. At the beginning of tracking, the 3D model of the ball is manually overlapped on the image of the object. The manual adjustment involves the pose displacement of the model. The right picture of Fig.1 shows the 3D model and the object image at the 100th frame in tracking. The left picture of Fig.2 shows the 3D model at every 10 frames overlapped on the first frame. It is clear to increase the displacement between the object and the 3D model.

Let us show the effect correcting the pose displacement. As soon as the tracking starts, the displacement is corrected. The right picture of Fig.2 shows the 3D model at every 10 frames overlapped on the first frame. The left and right pictures of Fig.3 show tracking results at the 64th and 100th frames, respectively. These seem to be good fitting between the model and object.

How much displacement of the position could be corrected ? We made an experiment of correcting the pose displacement and estimating motion parameters from the 1st and 2nd frames. The left picture of Fig.4 shows the 3D model before correcting pose displacement. The model is the ball's radius away from the exact position. The right picture of Fig.4 shows the 3D model at every iteration in the Newton method up to 7 iterations. Making

Figure 2: (Left) The 3D model at every 10 frames without correcting pose displacement. (Right) The 3D model at every 10 frames by correcting pose displacement.

Figure 3: Tracking results by correcting pose displacement are shown as the 3D model at 64th frame (left) and 100th frame (right).

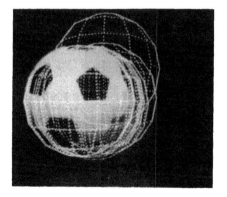

Figure 4: (Left) Initial position of the 3D model. (Right) Correcting the pose of the 3D model at every iteration.

the experiments under various positional displacements, it becomes clear to be able to perfectly correct the displacement within a half of radius of the ball.

5 Concluding

We presents a method which can directly correct the pose displacement of the model and estimate 3D motion parameters assuming that 3D shape of the object is given as a 3D model. Using this model-based method, it is possible to track the object without matching model to image and correspondence between successive images.

References

[1] B.K.P.Horn and E.J.Weldon Jr.. Direct methods for recovering motion. *International Journal of Computer Vision*, vol.2, pp.51-76, 1988.

[2] K.Koshikawa and Y.Shirai, A 3-D modeler for vision research, *ICAR*, pp.185-190, 1985.

[3] D.G.Lowe, Integrated treatment of matching and measurement errors for robust model-based motion tracking, Proc. of *3rd ICCV*, Osaka, pp.436-440, 1990

[4] D.G.Lowe, Fitting parameterized three-dimensional models to images, *IEEE*,vol.PAMI-13,no.5,pp.450, 1991

[5] S.Negahdaripour and B.K.P.Horn, Direct passive navigation, *IEEE*,vol.PAMI-9, no.1, pp.168-176, 1987.

[6] S.Negahdaripour, Direct methods for structure from motion, Ph.D Thesis , Mechanical Engineering, MIT, 1986

[7] R.Y.Tsai and T.S.Huang, Estimating three-dimensional motion parameters of a rigid planar patch. *IEEE*, vol.ASSP-29, no.6, pp.1147-1152, 1981.

[8] M.Yamamoto, Direct estimation of three-dimensional motion parameters from image sequence and depth, *IECE Japan*, vol.J68-D, no.4, pp.562-569, 1985.

[9] M.Yamamoto, A general aperture problem for direct estimation of 3-D motion parameters. *IEEE*, vol.PAMI-11, no.5, pp.528-536, 1989.

[10] M.Yamamoto, P.Boulanger, J.-A.Beraldin, M.Rioux and J.Domey, Direct estimation of deformable motion parameters from range image sequence, Proc. 3rd ICCV, Osaka, pp.460-464, 1990.

[11] M.Yamamoto and K.Ikeda, Stereoscopic correcting pose of 3D model of object, this issue, 1993.

Time-Varying Image Processing and
Moving Object Recognition, 3 – V. Cappellini (Ed.)
© 1994 Elsevier Science B.V. All rights reserved.

Tracking based on Hierarchical Multiple Motion Estimation and Robust Regression[1]

Serge Ayer, Philippe Schroeter and Josef Bigün

Signal Processing Laboratory, Swiss Federal Institute of Technology, EPFL-Ecublens, CH-1015 Lausanne, Switzerland

Abstract

This paper presents a method for detecting multiple moving objects in an image sequence. The method consists of a motion parameter estimation technique followed by a segmentation. The motion estimation problem is formulated as a problem of iterative parameter estimation, which is solved by a robust linear regression procedure and an outlier detection technique. Using the estimated motion parameters, the object boundaries are then obtained by a segmentation algorithm using multidimensional clustering and boundary refinement. The result is a robust segmentation of the different moving objects in the scene.

1. Introduction

The problem of determining motion in an image sequence has been widely addressed and several schemes have been proposed. One type of motion estimator is the *fully-parametric* estimator which describes the motion within an image region in terms of a parametric model [1]. In that case, the problem is reformulated to one of parameter estimation and, after linearization of the model, standard linear regression can be used to estimate the parameters.

The parametric methods have the advantage of using the motion model to constrain the flow field computation. In contrast to this, the computation of optical flow using general non-parametric methods involves a combination of a local brightness constraint with a global smoothness constraint. The resulting optical flow is in turn interpreted by using appropriate motion models. The benefit of the parametric approach is that the interpretation is direct and that the resulting flow fields cannot be inconsistent with the model.

However, when the scene contains multiple moving objects or transparency phenomena, one must use an estimation method that can recover both the model parameters and the segmentation of the moving objects at the same time. To achieve that goal two popular paradigms have been used : *line process* (discontinuities detection) and *outlier detection* [2, 3, 4]. Here we propose a method based on *robust regression* and *outlier detection* [5, 6]. In contrast to the method used in [4], the regression methods used here

---

[1]This work has been supported by Thomson-CSF, Rennes, France

have the highest possible breakdown point and are highly resistant to outliers. In the proposed method, a line process is thus not needed during the estimation process and the moving objects boundaries are finally obtained by a segmentation algorithm using multidimensional clustering and boundary refinement [7].

Experimental results indicate that the proposed method is robust in the presence of different moving objects and converges rapidly when detecting the object. The method is also general enough to deal with scenes with moving or static camera in a stationary or non-stationary environment.

2. Model-Based Robust Motion Estimation

2.1. Overview of the Method

The basic components involved in the implementation of the method are (see Figure 1):

- the *pyramid construction*: in our implementation, we use the Laplacian image pyramid as input to the algorithm. Consequently, below $I(t)$ denotes the Laplacian of the image, as it is obtained by the pyramid construction.

- the *motion estimation* and *outlier detection*: the motion estimation is here formulated as a parameter estimation problem. Assuming intensity constancy, i.e. the brightness of a small surface patch is not changed by motion, the problem of model-based motion estimation is posed as the minimization of the magnitude of

$$I(\mathbf{x}, t) - I(\mathbf{x} - \mathbf{u}(\mathbf{x}, \mathbf{p}), t - 1) \tag{1}$$

over an image region, where $\mathbf{p}$ denotes the model parameters, $\mathbf{u}(\mathbf{x}, \mathbf{p})$ the flow field in that region, and $I(\mathbf{x} - \mathbf{u}(\mathbf{x}, \mathbf{p}), t - 1)$ the image at time $t - 1$ warped towards t. In our method, the minimization is performed in an iterative way using a robust linear regression procedure and an outlier detection technique.

- the *image warping*: image warping is achieved by using the current values of the model parameters to compute a flow field, and then using this flow field to warp $I(t - 1)$ towards $I(t)$. Currently we use bilinear or bicubic interpolation.

- the *coarse-to-fine refinement*: this component is straightforward in case of parametric models, where parameter values are simply transmitted to the next finer level of the pyramid.

2.2. Linearization of the Parametric Model

From equation (1) we see that the problem of estimating the parameter vector $\mathbf{p}$ from the measurement I is non-linear. Here we linearize the problem around a previous estimate $\hat{\mathbf{p}}$ of the parameter vector $\mathbf{p}$. The first order Taylor expansion of equation (1) leads to

$$\Delta I(\mathbf{u}(\mathbf{x}, \mathbf{p}))|_{\mathbf{p}=\hat{\mathbf{p}}} = [(\nabla I(\mathbf{x} - \mathbf{u}(\mathbf{x}, \mathbf{p}), t - 1))^T \frac{\delta \mathbf{u}}{\delta \mathbf{p}}]|_{\mathbf{p}=\hat{\mathbf{p}}} (\mathbf{p} - \hat{\mathbf{p}}) + \mathbf{n}, \tag{2}$$

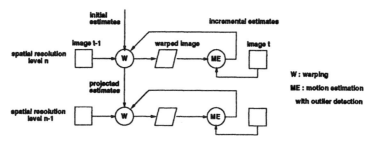

Figure 1: General framework of the robust motion estimator

where
$$\Delta I(\mathbf{u}(\mathbf{x}, \mathbf{p})) = I(\mathbf{x}, t) - I(\mathbf{x} - \mathbf{u}(\mathbf{x}, \mathbf{p}), t - 1).$$

In equation (2), $\nabla I(\mathbf{x} - \mathbf{u}(\mathbf{x}, \mathbf{p}), t - 1))$ denotes the spatial gradient vector $[\frac{\delta I}{\delta x}, \frac{\delta I}{\delta y}]^T$ of the warped image, $\frac{\delta \mathbf{u}}{\delta \mathbf{p}}$ the $2 \times n$ Jacobian matrix of the vector field $\mathbf{u}$, and $\mathbf{n}$ the measurement noise.

For the particular cases of translational and affine motion fields, the Jacobian matrices are
$$\frac{\delta \mathbf{u}}{\delta \mathbf{p}} = I$$

and
$$\frac{\delta \mathbf{u}}{\delta \mathbf{p}} = \begin{bmatrix} 1 & x & y & 0 & 0 & 0 \\ 0 & 0 & 0 & 1 & x & y \end{bmatrix}$$

After Taylor expansion, the motion estimation problem is thus reformulated to one of linear parameter estimation, and standard linear regression methods can be used to estimate the parameters.

2.3. Linear regression

The most popular regression estimator (*least sum of squares* or LS estimator) dates back to Gauss and Legendre and corresponds to

$$\min_{\hat{\Theta}} \sum_{i=1}^{n} r_i^2, \tag{3}$$

where r_i is the residual for the ith observation.

The LS estimator is known to be optimal for normal noise distribution. However, more recently attention has been given to the fact that real data usually do not satisfy the classical assumptions and that the standard LS analysis is very sensitive to minor deviations from the Gaussian noise model and to the presence of outliers in the data. In order to reduce the impact of these negative influences, robust methods that are much less affected by outliers have been developed [5]. In computer vision the problem of regression analysis is an important statistical tool and recently the interest for robust estimators have increased [6, 8, 9].

Below we discuss the application of two robust estimators to the problem of motion estimation, namely the *least median of squares* (LMedS) and the *least-trimmed squares* (LTS) given by

$$\min_{\hat{\Theta}} \text{median}_i (r_i^2) \tag{4}$$

and by

$$\min_{\hat{\Theta}} \sum_{i=1}^{h} (r_i^2)_{i:n} \tag{5}$$

where $(r^2)_{1:n} \leq \ldots \leq (r^2)_{n:n}$ are the ordered squared residuals [5].

2.4. The LMedS and LTS Estimators

Several concepts are usually employed to evaluate a regression method : the breakdown point, the relative efficiency, and the computational complexity [5, 6].

The *breakdown point* is defined as the smallest fraction of contamination that can cause the estimator to take on values outside an arbitrary range. In the case of the least squares estimator, a single outlier is sufficient to carry the estimation over all bounds, and in that case the (asymptotic) breakdown point is 0%. The LMedS and LTS estimators have been shown to have the highest possible (asymptotic) breakdown point (50%). These estimators are thus highly resistant to outliers.

The *relative efficiency* of a regression method is defined as the ratio between the lowest achievable variance for the estimated parameters (the Cramer-Rao bound) and the actual variance provided by the given method. The efficiency thus depends on the underlying noise distribution, which is usually not known. The LMedS method has been shown to have poor efficiency properties. In order to compensate for this deficiency and to improve the crude solutions of both LMedS and LTS methods, Rousseeuw and Leroy [5] proposed combining the LMedS or LTS method with a weighted least-squares procedure. The method thus combines the resistance of the robust method in the presence of outliers with the efficiency of least-squares methods once the outliers have been identified.

The last important concept for the evaluation of a regression method is the *computational complexity*. The amounts of computation required for the exact LMedS and LTS methods are $O(n^{p+1})$ and $O(n^{p+1} \log n)$. These complexities have to be reduced to practical values by means of *Monte Carlo* type speed-up technique and *parallelism*.

2.5. The Outlier Detection Technique

For the purpose of reweighted least squares procedure, one has to define a weight function, which assigns a weight to each observation in equation (2). This weight is a function of the standardized LMedS or LTS residuals r_i/σ^*, where σ^* is the robust standard deviation estimate of the regression.

Several types of weight functions may be defined for the reweighted least squares procedure. The function we utilized has binary weights and is of the form

$$w_i = \begin{cases} 1 & if \ |r_i/\sigma^*| \leq c_1 \\ 0 & otherwise. \end{cases} \tag{6}$$

Other types of weight function are less radical in the rejection scheme and are defined by means of linear or hyperbolic transitions.

2.6. The Algorithm

The different steps of the motion estimation at a given resolution level may be summarized as follows:

1. Estimate the motion parameters using LMedS or LTS regression, followed by an identification of the outliers.

2. Estimate the motion parameters using weighted LS regression, where the weights are function of the robust residuals obtained from step 1.

3. Update the parameters. Go to step 1 or stop at this level if the parameter changes are two small or after a certain number of iterations .

4. If pyramid level $\neq 0$: project down the parameters to the next pyramid level and go to step 1.
   If pyramid level $= 0$: segment (see Section 3.) the local motion error image in order to find subregions of the image in which the motion parameters are valid. Go to step 5.

5. Go to step 1 in order to compute the motion in regions where it is judged necessary according to step 4. Stop if the size of these regions is below a given threshold.

3. Segmentation of moving objects

Our goal is to find approximate boundaries of moving objects. This problem can be seen as a segmentation problem, which consists of detecting the pixels that correspond locally to the globally detected motions. In our case, a segmentation feature is first computed by means of a prediction error measure (see [2]), and the boundaries between different moving objects are obtained by a segmentation algorithm described in [7]. Note that the segmentation procedure is general enough to allow the use of other features.

Assume that at time t we have a global estimate of motion parameters. The intensity difference between the warped frame $I(t-1)$ and the frame $I(t)$ will be close to zero in the area of the object corresponding to the detected motion. In the remaining parts, the prediction error should be much different from zero except in regions with constant gray level. Thus, the problem of detecting the moving object consists of finding the parts in the image with small prediction error. For this purpose, the segmentation feature is obtained by computing the error measure at each location $\mathbf{x}$.

The multi-dimensional segmentation algorithm is applied for estimating the boundaries between the moving objects corresponding to the estimated motions (small prediction errors) and the remaining parts. This algorithm is embedded in a multiresolution framework using quad-trees [10]. Multiple resolutions are very useful in reducing the noise because at lower resolutions the class-prototypes (or properties) are better defined while higher resolutions are needed to obtain accurate borders. A pyramid of the motion features is built up to a predefined level in which each coarser level is obtained by smoothing the

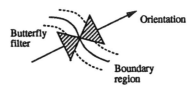

Figure 2: Illustration of the oriented butterfly filters.

preceding finer level. At the coarsest level, the amount of noise in the feature space has decreased significantly and a partition of the feature space into c classes can be obtained by applying a multidimensional clustering algorithm (see for example [11]). For most of these algorithms, the number of classes c has to be known *a priori* which is not a problem in our particular situation (c is simply set to 2: small or large prediction errors). Isolated labels are reassigned to a class nearby using an 8-connected neighborhood assumption that guarantees the spatial connectivity.

Finally, the class-boundaries are improved by gradually projecting down the class-labels and by refining the boundaries. The children of the boundary nodes (obtained at the coarsest level) define a boundary region at the next higher resolution. The non-boundary nodes at the children level are given the same labels and properties as their parents. The class-uncertainty within the boundary region is high and is reduced by applying orientation-adaptive butterfly-like filters (see figure 2). Butterfly-like shape reduces the influence of feature-vectors along the boundary. After smoothing, each boundary pixel is reassigned to its closest class.

4. Results

Various experiments on different real image sequences have been performed to test the robustness and accuracy of the tracking algorithm. The different sequences include scenes with a moving or static camera in a moving or static environment. We show here the results obtained on the mobile and calendar image sequence.

The mobile and calendar scene includes a camera motion and several moving objects. Figure 3a) shows one image from the the original image pair. Applying the algorithm described in Section 2.6. on the entire image leads to the segmentation results presented in Figure 3b). In this figure, white pixels belong to regions where the motion has been well estimated (background) and black pixels to other regions. By applying the algorithm on the connected black regions, we obtain the segmentation results shown in Figure 3c). In this case, the motion features obtained after the first two estimations were used for the segmentation procedure.

5. Conclusion and Future Work

We have presented a motion estimation method based on robust regression and outlier detection. This method used together with a multidimensional segmentation technique

Figure 3: Tracking algorithm results on the mobile and calendar sequence (from top to bottom and from left to right): a) Original image b) Segmentation results after the first motion estimation c) Segmentation results after the second motion estimation
d) Segmentation results superimposed on the original image after the second motion estimation

allows a robust segmentation of the different objects moving in the scene.

The approach has the advantage of being robust in the presence of multiple moving objects. The method also converges rapidly when detecting the object boundaries, and the use of only two frames in the sequence already allows a consistent segmentation. In order to further improve the accuracy of the object boundaries, the following approaches should be considered.

The segmentation method uses here only the prediction error as segmentation feature. The combination with other features like intensity values or responses of spatio-temporal filters would allow a more accurate definition of the boundaries.

Although the method is able to segment the motion information in just two frames, temporal integration is needed in order to improve the segmentation over time. As the technique can be applied on data of arbitrary dimension, one could simply consider data from more than just two frames. Another approach would be to consider iterative temporal integration, like a Kalman filter based approach.

A different approach for the recovering of object boundaries using the same motion estimation algorithm, is also possible. In this case, the segmentation step between each successive motion estimation is avoided, and the motion estimations are applied iteratively in a multigrid fashion on points detected as outliers. The segmentation of moving objects is finally obtained by a hierarchical classification process in a multidimensional space, in which each dimension corresponds to a prediction error image computed with the different detected motions. A description of this method can be found in [12].

6. REFERENCES

[1] J.R. Bergen, P. Anandan, K.J. Hanna, and J. Hingorani. Hierarchical model-based motion estimation. In *Second European Conference on Computer Vision*, pages 237–252, Santa Margherita Ligure, Italy, May 1992.

[2] M. Irani, B. Rousso, and S. Peleg. Detecting and tracking multiple moving objects using temporal integration. In *Second European Conference on Computer Vision*, pages 282–287, Santa Margherita Ligure, Italy, May 1992.

[3] M.J. Black. Combining intensity and motion for incremental segmentation and tracking over long image sequences. In *Second European Conference on Computer Vision*, pages 485–493, Santa Margherita Ligure, Italy, May 1992.

[4] T. Darrell and A. Pentland. Robust estimation of a multi-layered motion representation. In *IEEE Workshop on Visual Motion*, pages 173–178, Nassau Inn, Princeton, NJ, October 1991.

[5] P.J. Rousseeuw and A.M. Leroy. *Robust Regression and Outlier Detection*. John Wiley and Sons, New York, 1987.

[6] P. Meer, D. Mintz, A. Rosenfeld, and D.Y. Kim. Robust regression methods for computer vision : A review. *International Journal of Computer Vision*, 6:59–70, 1991.

[7] P. Schroeter and J. Bigün. Image segmentation by multidimensional clustering and boundary refinement with oriented filters. In *Gretsi Fourteenth symposium*, Juan les Pins, France, Septembre 1993.

[8] X. Zhuang, T. Wang, and P. Zhang. A highly robust estimator through partially likelihood function modeling and its application in computer vision. *IEEE Transactions on Pattern Analysis and Machine Intelligence*, 14:19–35, January 1992.

[9] W.B. Thompson, P. Lechleider, and E.R. Stuck. Detecting moving objects using the rigidity constraint. *IEEE Transactions on Pattern Analysis and Machine Intelligence*, 15:162–166, February 1993.

[10] W.I. Grosky and R. Jain. Optimal quadtrees for image segments. *IEEE Transactions on Pattern Analysis and Machine Intelligence*, 5:77–83, 1983.

[11] L. Kaufman and P.J. Rousseeuw. *Finding Groups in Data*. John Wiley and Sons, Inc., 1990.

[12] S. Ayer and P. Schroeter. Hierarchical robust motion estimation for segmentation of moving objects. In *Eigth IEEE Workshop on Image and Multidimensional Signal Processing*, Cannes, France, September 1993.

Time-Varying Image Processing and
Moving Object Recognition, 3 – V. Cappellini (Ed.)
© 1994 Elsevier Science B.V. All rights reserved.

Characterizing Disturbances in Dynamic Scenes using an integrated motion analysis approach

V.REBUFFEL and J.M.LETANG

LETI (CEA- Technologies Avancées), DSYS, 85X
38041 Grenoble France

Abstract

When dealing with real world image sequences, we have to take into account numerous perturbations due to sensor or environment. We have developed a motion detection method aiming to be robust to these perturbations, taking advantage of long temporal integration and statistical regularization. We propose here to use this method, extended by adding tuning parameters, to analyse the effects of the various perturbations above-mentioned, and consequently to eliminate them in a final motion detection decision, or at least to characterize them. We stress on the temporal behaviour of each phenomenon. Examples will be presented all along the discussion.

1. INTRODUCTION

Observation and control of vulnerable regions imply the detection of any moving object in any contextual conditions. Typical applications concern military and civil subjects : supervision of battlefield, traffic control of aerial zone and maritime horizon, surveillance of prohibited areas.
We consider the case of monocular static sensors either visible or infra-red, and take into account real outdoor scenes. Thus we are concerned with numerous and various disturbances, mainly due to sensor (data acquisition noise, oscillations of camera position) and to changing environment (modification of lighting, motion of natural structures as clouds). Furthermore, the moving objects we have to detect are frequently altered, with regard to their shape or apparent motion, particularly because of occlusions.
Motion analysis in an image sequence can be considered as a segmentation problem. Motion detection comes to a binary static/moving labelling, thus velocity field estimation is a vectorial segmentation. For the moment we focus on motion detection problem, because the characterization of perturbations should be done as closest as possible to the pixel level. Nevertheless extension to motion estimation will be discussed when necessary.
Moving areas in a image sequence are usually determined from the temporal variation of the intensity distribution, i.e. from variations of this distribution

between two successive frames. For temporal change detection, the simpler process to perform is the difference of two successive frames, but it is so sensitive to noise that never used alone. Hsu, Nagel and Rekers [1] have made use of statistical tests : they model linearly the local intensity function and compute a likelihood test between two successive frames in the neighbourhood of each pixel. To make this test more robust to changing illumination, some propositions have been done. In particular, Skifstad and Jain use a quadrature Picture function based on derivative of surface models [2].

To deduce motion from temporal change, additive tools are generally used. A robust and elegant technique is to consider a regularisation approach for the motion segmentation process. An energy is formulated depending on local context, and motion detection is achieved by minimising this Energy. Lalande [3] applied such an approach to get a map of binary labels static/moving using two observations : the difference between two images, and the result of a statistical test as Nagel's one.

A motion detection scheme should process a sequence and not a single image. In other terms, one can be tempted to use the result of motion detection at (t) to analyse motion at (t+1), and there are several ways to perform it, more or less easier depending on the motion detection method used, with effect on robustness but also complexity.

An important point to stress on is the velocity of each phenomenon, relatively to the acquisition frequency. As a matter of fact, methods based on spatio-temporal filters are generally sensitive to a given velocity of moving objets, and consequently are not robust to variations of the object to detect. Notice for instance Gabors' transforms applied to motion detection by Heeger [4]. Remind that the considered speed is the apparent one, which is included into the optical plane, and that this speed component can be very small for an approaching object. On the other hand, one cannot expect any method to consider any speed, for instance to deal at the same time with a fast plane and very low clouds. To be robust, motion detection methods should take into account a quite large band of speed but this band has to be adjusted depending on the context.

2. METHODOLOGICAL FRAMEWORK

This part recalls the main mathematical backgrounds of our motion detection method. To assume robustness to noise and sensitivity to low apparent motion, methods based on two successive frames are not sufficient, and a longer temporal depth is required. So we use a temporal multiscale decomposition applied on each pixel of the original sequence. More precisely, we consider the image sequence at a set of monodimensional time varying intensity signals [5]. Each signal is decomposed onto different scales using the wavelet decomposition. We choose a simple wavelet family with finite support equivalent to discrete filters of length 2. Notice that this leads to a temporal multiresolution decomposition, but not to a spatial one, as sometimes used for motion tracking. Finally at a given time (t) and a given pixel (P) we get a temporal decomposition vector.

Temporal change detection is then performed using a spatial likelihood test. Following the idea of Jain for robustness to illumination variation [2], we choose a ratio test based on the comparison of the derivative of a local linear model, expressed in terms of intensity differences with no constant term. We applied this test at each temporal scale. Therefore the decomposition gives a set of binary change maps, describing the image sequence in terms of dynamic components.

Motion detection has then to be performed from these various observations. An appropriate way to combine information while introducing local contextual knowledge is to consider the problem as a statistical labelling according to a global bayesian formulation [6]. We design an energy to minimise :

E (configuration of labels) =
$$E_{spatial} \text{ (labels) } + E_{adequation} \text{ (labels, observations) } + E_{prediction}\text{(labels)}$$

where *label* is a discrete ternary label field *(static, moving, parasitical motion)* and *observations* come from the temporal decomposition. The label *parasitical motion* has been introduced to improve robustness as illustrated later. For more details on the formulation of Energy see Letang [7].

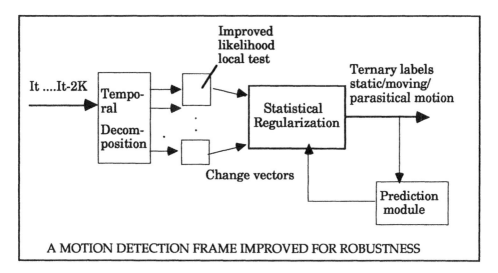

A MOTION DETECTION FRAME IMPROVED FOR ROBUSTNESS

An obvious disadvantage of most motion detection methods lies in the lack of temporal memory and prediction step. To correct it, one can try to use result at (t-1) to analyse motion at (t). An easy way in regularization techniques is to use result at (t-1) as an initial state for the optimisation process at (t), but this method preserves the past too much independently from the observations. We prefer to add a third term to the energy model, relating objects predicted from result at (t-1) to observations at (t). The prediction step is done using a temporal recursive filter relying on Kalman equations. The statistical optimisation is performed using a determinist algorithm (ICM).

The main advantages of this statistical method are the propagation of constraints it enables, the spatial smoothing that is applied conditionally to the labels, and the ability to combine various observations and properties.

3. DISTURBANCES DUE TO SENSOR

The usual acquisition noise can be considered as a random variable in space and time. Consequently it is mostly removed by the regularization process. Noise may be not restricted spatially, for instance noise line occurring in the numerisation process, in case of line array sensors. Such a perturbation happens at only one instant (no temporal depth), so it has a particular behaviour in the temporal decomposition (it only appears at the first levels). This allows us to characterize it. The sequence represented in figure 1 shows the result of the detection method for an aerial scene with an approaching plane. The noise line is correctly labelled in *parasitical motion* (grey) and the plane in *moving* (black).

 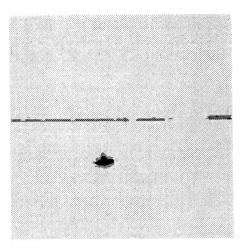

Figure 1 : initial image (left) and final motion detection map (right).

Another common disturbance comes from the uncertainty of the position of the sensor, which is not often perfectly static. The result is an apparent but irrelevant motion of the intensity edges of the images. As long as the sensor motion is oscillatory with a small amplitude, it corresponds to the case of a null temporal mean motion for the edges. Such a type of motion appears in the temporal decomposition at the first levels only. It is clear that apparent motion disappears at a particular temporal scale. By the regularization process, the corresponding pixels are labelled as *parasitical motion* - proving the efficiency of this third label. In the sequence of figure 2, a maritime scene acquired with an infrared line sensor, the horizon line has an oscillatory motion of 4 pixels high. The small images correspond to levels 1, 2 and 4 of the decomposition. Some parts of the horizon line are labelled in *parasitical motion* (grey), while a plane in the sky is correctly labelled in *moving* (black).

The speed of such oscillations is generally quite fast, which assumes the response in the first levels only. If it is not the case, the method will conclude in labels *motion*. The only way to identify a sensor motion is then to use the small spatial motion amplitude and the concordance of motion label pixels with intensity edges. If sensor motion becomes too large, the computation of the optical flow is required for compensation. Nevertheless, as long as the motion amplitude is small, the temporal decomposition and regularisation process are sufficient to identify it.

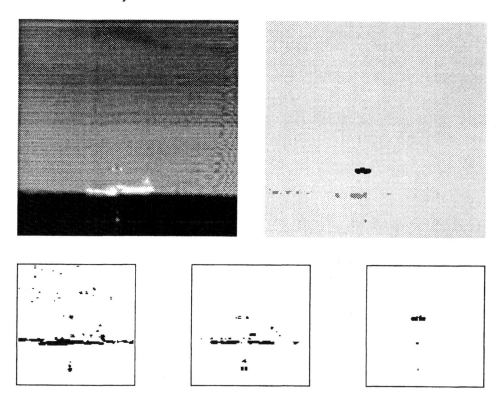

Figure 2 : initial image (up left), final map (up right), levels 1,2 and 4 (down).

The remaining defaults due to sensor are some perturbations whose effects are not restricted neither in space nor in time. It is the case for instance of power trouble. The result is a variation of illumination occurring globally (in space) and often ondulatory. It is not different from a natural change of illumination i.e. due to lighting, at least at the considered instant. It is analysed in the next part (§4). The difference lies in the persistence of the perturbation in case of a sensor cause in contrast to a quite short troubled period for a lighting cause.

4. ENVIRONMENTAL PERTURBATIONS

Well-known in image processing, the main problem in outdoor scene analysis is due to changing illumination. The observed effect is a temporal change of the whole image, or part of the image, depending locally of the nature of reflecting surface. This change is detected by all the observations based on difference (of images, or mean of images). As a result, the estimation of optical flow becomes impossible as far as it relies on the relation between change detection and spatial gradient. The solution we adopt is to ignore the constant term in the likelihood test, in other words we compare two images by their local slopes and not their absolute level. Notice that the underlying model of illumination change is an additive noise, conditionally to the local surface. The figure 3 represents a country side scene during an illumination change, and two vehicles near the horizon line. Two levels of decomposition (levels 2 and 4) are given for each two forms of the test. The test with constant term is so noisy that unusable while the improved test allows to detect correctly the vehicles (labelled in *motion* or black).

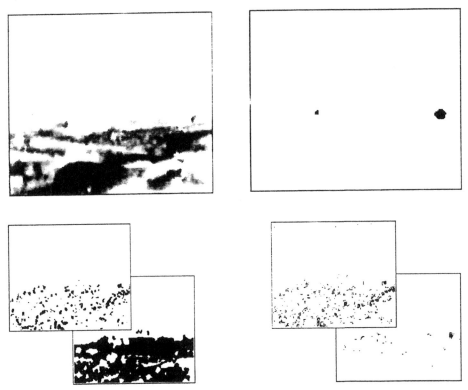

Figure 3 : initial image (up left), levels 2 and 4 using the complete test (down left) and the improved test (down right), final map with label *motion* only (up right)

Another disturbance that may occur in a natural environment is the motion of structures as clouds, smokes or tree leaves. As far as motion truly exists, our objective is not to eliminate it but to characterize it. If the perturbation corresponds to a null temporal mean motion, we can classify it by the fact that it disappears at a particular temporal scale. It is the case of oscillatory natural structures. However, the most frequent case concerns deformable and ill-defined objects such as smoke. In fact such an object looks like a set of variable shapes with irregular motion that disappear after some moments. Thanks to the method we adopt, it can be identified as a set of objects with irregular trajectory, and surrounding by *parasitical motion* labels. The figure 4 shows an outdoor scene with smoke coming from a chimney and a moving vehicle. In the three-labels resulting map, smoke is composed of such "objects" while the moving vehicle has a regular motion. In case of clouds, if the motion is regular and the cloud not deformable at the considered temporal scale, there is no way to distinguish it from truly moving object, except by the fact that the motion is slow, the concerned pixels are linked over the border of the clouds and lie in the sky (contextual global knowledge).

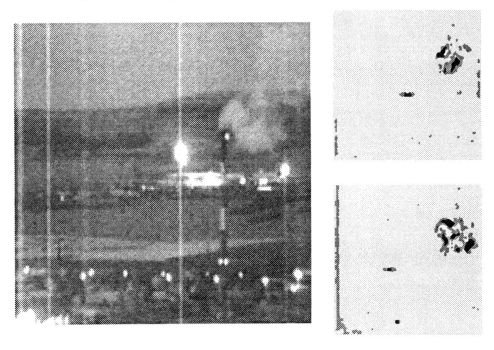

Figure 4 : initial image (left), two final motion detection maps (right)

5. VARIATIONS OF THE MOVING OBJECT SHAPE

We consider moving objects whose apparent size can be confined to a few pixels, and whose signature may change due to various reasons : variation of illumination, partial occlusion (vehicles hidden by trees). Notice for instance that the intensity signature of an approaching object varies, which is not taken into account by usual methods, and disables the determination of optical flow.

Robustness to small changing shapes can be partly assumed by conditional spatial smoothing. The sensitivity to low motion (for instance velocity along the optical axis, corresponding to an approaching object) is achieved by the long temporal depth, i.e. the number of decomposition levels. A computation shows that our method is able to detect a moving planar object from the velocity of 0.5 pixel/image in case of one level, and of $0.5/2^{K-1}$pixel/image in case of K levels.

When the moving object is so disturbed that no observation is available during several images (occlusion), the use of a prediction module, based for instance on Kalman filtering, becomes necessary to supply information. The profit of this module is proved in the country side sequence presented in figure 3, where the left vehicle is hidden by trees during some instants but kept found when using the prediction module.

6. CONCLUSION

Image sequences of real outdoor scenes acquired by common sensors are corrupted by various perturbations, due to the sensor, to the changing environment or to the moving object itself. We have shown that it is generally possible to characterize them using their temporal behaviour. By eliminating these perturbations or discriminating the corresponding effects, our method has been proved robust. Notice the importance to characterize these disturbances as early as possible in the image analysis process, that is to detect them at the step of motion detection, all the more because the corresponding contextual conditions do not allow the computation of optical flow.

We have implemented our motion detection method for ternary labelling. Computations are not very time-consuming, and real-time implementation may be considered. For the temporal decomposition, hardware could be developed because calculations are arithmetically simple and remain spatially local. For the regularization process, one may take benefit of a determinist algorithm with a few scanning, a limited number of possible configurations due to ternary labelling, and the use of an instability stack to focus on the pixels to change.

REFERENCES

1. Hsu Y.Z., Nagel H.H., Rekers G.,"New likelyhood test methods for change detection in image sequences", CVGIP, vol.26, 1984, pp.73-106.
2. Skifstad K., Jain R., "Illumination independent change detection for real world image sequences", CVGIP, vol.46, 1989, pp.387-399.
3. Bouthemy P., Lalande P, "Detection and tracking of moving objects based on a statistical regularization method in space and time", proc.ECCV Antibes, pp.307-311 , April 90.
4. Heeger D., "Optical flow using spatio-temporal filters", IJCV, vol.1, pp.279-302,1988.
5. Létang J.M., Rebuffel V., Bouthemy P., "Motion detection based on a temporal multiscale approach", proc.ICPR The Hague, pp.65-68, September 92.
6. Dubes R.C., Jain A.K., "Random field models in image analysis", Journal of applied statistics, vol.16, n°2:131-164, 1989.
7. Létang J.M., Rebuffel V., Bouthemy P., "Motion detection robust to perturbations : a statistical regularization and temporal integration framework", proc. ICCV Berlin, pp.65-68, Mai 93.

*Time-Varying Image Processing and
Moving Object Recognition, 3* – V. Cappellini (Ed.)
© 1994 Elsevier Science B.V. All rights reserved.

Noisy phenomenon in motion estimation

A. QUINQUIS [1]* and C. COLLET *

(*) ECOLE NAVALE Groupe de Traitement du Signal
[1] ENSIETA - Laboratoire du Signal et de l'Image
29240 BREST NAVAL - FRANCE

Motion estimation requires high level operators because kinematics cannot easily be extracted from a scene. Our study deals with infrared image sequences (given by a surveillance system onboard a ship) using which, we have tried to evaluate the displacement of clouds in the sky. To achieve this, we have used two principal methods (differential and integral) that allow us to estimate the velocity. This paper is concerned with noise sources in the estimation process, for both sets of methods. We will decribe their origins and compare the sensitivity between integral and differential methods. This paper lays the path for research into the combination of these two methods, in order to obtain a robust and accurate kinematic field.

1. Introduction

There are two main methods that allow us to estimate motion in infrared outdoor scenes. The first one is based on the study of the local properties of the intensity function. The variations in the illumination function with time and space are computed, linked and interpreted in order to evaluate the motion parameters. The main hypothesis is that all illumination variations in time are due to displacement phenomena. These kind of methods always refer to the optical flow constraint equation (1)(2)(3). They are usually called differential methods or gradient based methods. On the other hand, we find feature based methods that can be called integral methods. They are based on a matching process, between pixels previously selected at time t and pixels selected in the new picture, at time t+ΔT (where ΔT represents the temporal sampling rate of the surveillance system). They are called global methods, feature based methods or integral methods (4). The principle of the estimation differs between these methods. This paper presents an overview of the different sources of noise that disturb, and sometimes prevent, a correct estimation of motion. We will show that the same noisy phenomena appear in different forms in the estimation.

2. Global Methods

There are two principle techniques that allow pattern recognition and thus permit us to estimate the displacement between frames. In the general case, the results are obtained by searching for the largest coefficient in an array. The position of this value permit us to deduce the displacement of the pattern. The crosscorrelation method gives good results, but requires a long processing time and does not discriminate sufficiently between the cloud and the noisy background. Some improvement can be brought, using binarized images (6), however the problem remains the same. The pattern for which we are searching in the next frame, is not sufficiently defined. The Generalized Hough Transform (GHT) consists of an adapted filter (7) and thus maximizes the signal to noise ratio (SNR). This robust and low processing time algorithm will be presented and improved by means of a polar constraint, added during the matching process (5). We propose to observe the different sources of noise (which perturb or prevent motion parameter extraction) present in the accumulator array.

2.1. Ambient Noise

The ambient noise comes from selected pixels at time t that disappear at time t+Δt, due to propagation phenomena. It also can be caused by large changes in sun intensity (or direction) caused by a sudden break in the cloud cover. Ambient noise is not easy to quantify and is difficult to eliminate.

2.2. Self Noise of the GHT

The noise, due to the GHT transform principle, is generated by incorrect incrementations of the cells which compose the accumulator array (5). If N pixels define the witness pattern in the thresholding image at time t, then the pattern recognition at time t+ΔT will be computed using the N pixels of greatest gradient modulus. In the perfect case, (translatory motion of a rigid object, with no noise in the image), the GHT will incorrectly increment N-1 cells for each pixel previously kept. Thus, for all the selecting pixels in the witness pattern, N pixels cluster in the array, whereas N(N-1) cells are incorrectly incremented. These N(N-1) incrementations generate noise in the accumulator array, called GHT self noise.

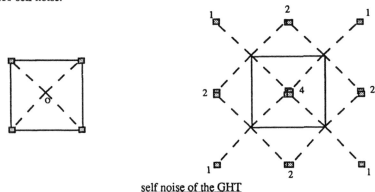

self noise of the GHT

This noise can be quantified as follows :

$$SNR_{cluster\ noise} = \left[10 \log \left(\frac{N^2}{[N(N-1)]^2} \right) = -20 \log (N-1) \right] \ dB$$

In order to minimize the preceding expression, we can choose N to be small; however the confidence associated to the motion estimation will obviously be slight. If the witness pattern is only composed of a few number of pixels, then the likelihood of the correct shape recognition decreases.

2.3. Noise due to cloud deformation

Clouds are not rigid because of their physical origin. If their displacement is composed of a weak expansion, rotational movement or stretching movement; then, some pels previously selected will be lost in the next frame. We cannot segregate the exact origin of these phenomena but we can define a criterion that measures their effect in the recognition process. All of these phenomena contribute to a reduction in the number N of pels that contribute to the emergence of a peak in the accumulator matrix A(i,j). We express these problems by means of a global criterion ξ.

$$\xi = \left[20 \log \left(\frac{\max \{A(i,j)\}}{N} \right) \right] \ dB$$

2.4. Aperture Problem

This phenomenon limits the local knowledge to the scalar product of the real velocity and the direction of the gradient. It is the main obstacle to a local estimation of movement. In the case of the global methods, if the shape does not have any gradient in a direction, then the aperture problem is again apparent in the accumulator array. The number of pels that accumulate in this direction is more important than in the perpendicular direction. The more the pattern is oriented in one direction, the more important the incertainty, in the direction perpendicular to this orientation, becomes important. This phenomenon can be very cumbersome in the case of big clouds which have an horizontal translatory motion with few horizontal gradients. The orientation of the cluster array obtained describes this phenomenon (8).

3. Improvement of the SNR

In practice, the cluster array contains the sum of the different sources of noise, with usefull information. The GHT's self noise is greater than other noise sources that corupt the cluster matrix. The noise due to deformity or stretching of the clouds is not the most important, due to the high temporal sampling rate. We select 20% of the greatest gradients in a window of 60x30 pixels that is positionned where motion has been detected (5).

3.1. Without Polar Constraint

If we take an orientated cloud (gradients are orientated in the y direction) we obtain a cluster array shown in Figure 1. For another cloud which does not present a privileged orientation, the aperture phenomenon disappears in the obtained matrix (Figure 2).

3.2. With Polar Constraint

An important improvement can be brought, by adding a polar constraint during the matching process. Six classes of gradient direction and six classes of gradient modulus have been chosen. A cell in the accumulator array will be incremented if, and only if, the pels selected at time t and t+Δt belong to the same class. In fact, we more accurately define the pattern for which we are searching and the steepness around the peak increases. Thus, as we increase the segregation parameters, the noisy cells around the peak disappear.

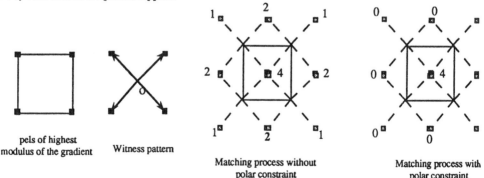

pels of highest modulus of the gradient

Witness pattern

Matching process without polar constraint

Matching process with polar constraint

removal of the GHT's self noise by means of a polar constraint

Interesting results are shown in Figure 3.

314

4. Differential Methods

Differential methods are based on the constraint equation, that links the spatio-temporal gradient $(\hat{f}_x, \hat{f}_y, \hat{f}_t)$ of the intensity function $f(x,y,t)$ to the kinematic parameters $(\dot{x}, \dot{y})$ for all the pixels (1) :

$$\hat{f}_x \dot{x} + \hat{f}_y \dot{y} + \hat{f}_t = 0 \qquad \text{(optical flow constraint equation)}$$

This equation is undetermined because there are two unknowns and only one equation. The aperture phenomenon solves this problem. As we wrote before, only the kinematics in the direction of the gradient can be locally estimated.

Thus we obtain the following relations (8) :

$$\dot{x}_i = \frac{- \hat{f}_{x_i} \hat{f}_{t_i}}{\hat{f}_{x_i}^2 + \hat{f}_{y_i}^2}$$

$$\dot{y}_i = \frac{- \hat{f}_{y_i} \hat{f}_{t_i}}{\hat{f}_{x_i}^2 + \hat{f}_{y_i}^2}$$

If we consider the intensity function : $\hat{f}(x,y,t) = f(x,y,t) + b(x,y,t)$ where $b(x,y,t)$ is a random variable and $f(x,y,t)$ is the real intensity function.
$\hat{f}(x,y,t)$ is the estimated function, given by the surveillance system.

4.1. Sensitivity of the Optical Flow Constraint Equation
Let the optical flow constraint equation be :

$$\hat{a} \dot{x} + \hat{b} \dot{y} + \hat{c} = \varepsilon$$

where

$$\hat{a} = \frac{\partial \hat{f}(x,y,t)}{\partial x} ; \; \hat{b} = \frac{\partial \hat{f}(x,y,t)}{\partial y} ; \; \hat{c} = \frac{\partial \hat{f}(x,y,t)}{\partial t}$$

and $\dot{x} = \frac{\partial x}{\partial t} ; \; \dot{y} = \frac{\partial y}{\partial t}$ the kinematics in x and y direction
We may writte

$$\hat{a} \dot{x} + \hat{b} \dot{y} + \hat{c} = (a + \Delta a) \dot{x} + (b + \Delta b)\dot{y} + (c + \Delta c) = \varepsilon$$

and the final error can be expressed as

$$\Delta a \dot{x} + \Delta b \dot{y} + \Delta c = \varepsilon$$

If we suppose that the incertitude of the gradient is as for a gaussian random variable $N(0,\sigma)$, then it is easy to show that the standart deviation of the constraint equation error is :

$$\sigma_e = |\dot{x}| \sigma_a + |\dot{y}| \sigma_b + \sigma_c$$

The incertitude ε increases linearly with velocity $(\dot{x}, \dot{y})$. As the velocity increases, the robustness of the method decreases.

4.2. Robustness of Differential Methods

The sensitivity to noise of the differential method, can be computed if we consider the possible variation of the gradients around their estimated values.

The solutions of the system are (cf IV) :

$$\left|\hat{x}\right| = \frac{ac}{a^2 + b^2} \qquad\qquad \left|\hat{y}\right| = \frac{bc}{a^2 + b^2}$$

which are equivalent to

$$\left(\begin{array}{l} \log\left|\hat{x}\right| = \log a + \log c - \log (a^2 + b^2) \\ \log\left|\hat{y}\right| = \log b + \log c - \log (a^2 + b^2) \end{array}\right)$$

The method of calculus of the varaitions allows us to obtain the following equations :

$$\left(\begin{array}{l} \dfrac{\Delta\hat{x}}{\hat{x}} = \dfrac{\Delta a}{a} + \dfrac{\Delta c}{c} - 2a\,\dfrac{\Delta a}{a^2 + b^2} - 2b\,\dfrac{\Delta b}{a^2 + b^2} \\[3mm] \dfrac{\Delta\hat{y}}{\hat{y}} = \dfrac{\Delta b}{b} + \dfrac{\Delta c}{c} - 2a\,\dfrac{\Delta a}{a^2 + b^2} - 2b\,\dfrac{\Delta b}{a^2 + b^2} \end{array}\right)$$

or

$$\left(\begin{array}{l} \dfrac{\Delta\hat{x}}{\hat{x}} = \left[\dfrac{\Delta a}{a}\left(\dfrac{b^2 - a^2}{b^2 + a^2}\right) + \dfrac{\Delta b}{b}\left(\dfrac{-2\,b^2}{b^2 + a^2}\right) + \dfrac{\Delta c}{c}\right] \\[3mm] \dfrac{\Delta\hat{y}}{\hat{y}} = \left[\dfrac{\Delta a}{a}\left(\dfrac{-2\,a^2}{b^2 + a^2}\right) + \dfrac{\Delta b}{b}\left(\dfrac{a^2 - b^2}{b^2 + a^2}\right) + \dfrac{\Delta c}{c}\right] \end{array}\right)$$

We are looking for the gradient values that give the lowest error in the kinematic parameters. We suppose that the error variation is zero, according to a and b :

$$\frac{\partial^2\left[\dfrac{\Delta\hat{x}}{\hat{x}}\right]}{\partial a \partial b} = \frac{\partial^2\left[\dfrac{\Delta\hat{y}}{\hat{y}}\right]}{\partial a \partial b} = 0 \quad \text{if} \quad (8ab^3 - 8a^3b) = 0$$

This result indicates that the error will be minimal if the spatial gradients are of the same in modulus.

316

If a = ± b, the lowest error is obtained, for the coordinate space arbitrary chosen :

$$\begin{pmatrix} \dfrac{\Delta\hat{x}}{\hat{x}} = \left[\dfrac{\Delta b}{b} + \dfrac{\Delta c}{c} \right] \\ \dfrac{\Delta\hat{y}}{\hat{y}} = \left[\dfrac{\Delta a}{a} + \dfrac{\Delta c}{c} \right] \end{pmatrix}$$

There is also a direct relation between the velocity incertitude and the gradient incertitude. The larger the incertitude of the optical flow values, the greater the velocity magnitude incertitude. We select the pixels for their high gradient modulus, because of noise and the aperture phenomenon, thus it will be very difficult to obtain a sufficient robustness for differential estimators. Nevertheless these methods are very interesting because they allow to estimate all kinds of motion and supply a dense kinematic field. In general, if results on real pictures are very bad, it means that the spatial and temporal sampling rate is too small. The constraint equation is always valid if the spatio temporal sampling limits tend to zero. In our case, we have to verify that the spatio-temporal sampling rate is sufficient in order to verify the constraint equation for all pixels.

5. Conclusion

In a conclusion, the differential methods are very sensitive to noise because the kinematic evaluation is entirely based on the differential properties of the intensity function. If the sequence of pictures is noisy, which is the case for infrared images, these estimations are limited to small diplacements because the Taylor expansion has to be verified for all the pixels. The noise in the image also prevents a high resolution for a small displacement. In addition, the kinematic parameters are limited between two closed values of displacement. Many authors have proposed to smooth the velocity field (9)(10)(11). This path however, is not always valid (like in occlusion zones) and morover, the smoothing process tends to smooth the velocity discontinuities between patterns, thus these discontinuities have to be detected. On the other hand, the integral methods do not take the kinematic field discontinuities into account and they give a global and average estimate for all pels. They allow us to estimate large displacements as long as the pattern remains within the window. This is why the opposite approaches of these methods have to be combined in order to obtain their advantages without their disadvantages.

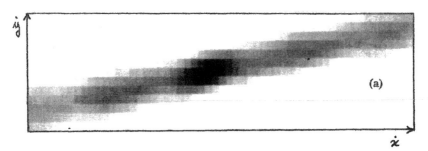

Figure 1 : accumulator obtained by GHT (a) for an oriented cloud (b)

References

1   PRAZDNY K. : On the information in optical flows
    Computer vision, graphics, and image processing 22, pp 239-259 (1983)
2   SUBBARAO M. : Interpretation of image flow : a spatio-temporal approach
    IEEE transactions on pattern analysis and machine intelligence Vol.11 N°3 (1989)
3   SINGH A. : Optical Flow Computation
    IEEE Computer Society Press (1991)
4   VEGA-RIVEROS J.F. - JABBOUR K. : Review of motion analysis techniques
    IEEE proceedings, Vol 136, Pt1, N°6 (1989)
5   COLLET C. - QUINQUIS A. - BOUCHER J.M. : Cloudy sky velocity map based on matched filter
    IAPR workshop on Machine Vision Application - Tokyo (1992)
6   BALLARD D.H. : Generalizing the Hough Transform to detect arbitrary shapes
    Pattern Recognition Vol 13, N°2, pp 111 (1981)
7   SKLANSKY J. : On the Hough technique for curve detection
    IEEE transactions on computers Vol 27, N°10, pp 923 (1978)
8   COLLET C. : Estimation de la cinématique des fonds nuageux en imagerie infrarouge
    Thèse de doctorat présentée à l'Université de Toulon et du Var (1992)
9   NAGEL H.H. : An investigation of smothness constraint for the estimation of displacement vector
    fields from image sequence
    IEEE transactions on pattern analysis and machine intelligence Vol.PAMI 8 (1986)
10 BOUTHEMY P. : Estimation of Edge Motion Based on Local Modeling
    SPIE Vol 595 Computer vision for robots (1985)
11 NAGEL H.H. : Displacement vectors derived from second order intensity variations in image
    sequences
    Computer Vision,Graphics & Image Processing 21, 85-117 (1983)

Figure 2 : accumulator obtained by GHT (a) for an non-oriented cloud (b)

318

Figure 3.a : accumulator obtained by GHT without polar constraint

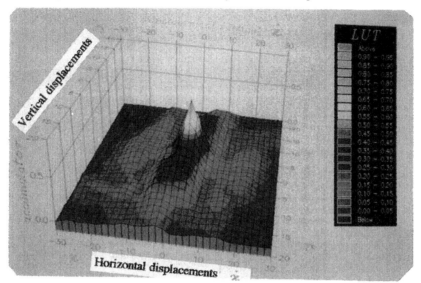

Figure 3.b : accumulator obtained by GHT with polar constraint

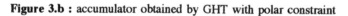

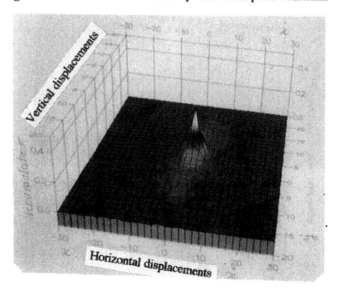

Time-Varying Image Processing and
Moving Object Recognition, 3 – V. Cappellini (Ed.)
© 1994 Elsevier Science B.V. All rights reserved.

Motion estimation of curves: experimental results using regularization techniques

Johan De Vriendt[1]

Laboratory for Communication Engineering, University of Ghent, Sint-Pietersnieuwstraat 41, B-9000 Gent, Belgium

Abstract

In this paper we compare the use of first and second order stabilizer functionals for the estimation of the motion of curves for real images. We discuss some implementation aspects which influences the performance. The algorithm is tested on a number of image sequences with known motion (quantitatively) and with motions which are only qualitatively known. The accuracy with which the motion of curves can be estimated is about 0.2 pixels for images undergoing relatively large motions (translations, expansions and rotations).

1 INTRODUCTION

Many algorithms that estimate the motion of curves are based on the motion constraint equation. This equation results from a Taylor's series expansion of the image intensity function E(x,y,t) up to first order. The motion constraint equation is given by:

$$\nabla E^T \mathbf{v} + E_t = 0 \tag{1}$$

and only gives one component of each flow vector. Consequently, additional constraints are necessary. D'Haeyer [2] proposed to minimize

$$\int \left| \mathbf{t}^T \frac{d\mathbf{v}}{ds} \right|^2 + \alpha \left| \mathbf{n}^T \frac{d\mathbf{v}}{ds} \right|^2 ds \tag{2}$$

constrained by the motion constraint equation

$$\mathbf{n}^T \mathbf{v} = v^\perp = -\frac{E_t}{|\nabla E|} \tag{3}$$

and the tangential boundary conditions in the endpoints s_1 and s_2

$$\mathbf{t}^T \mathbf{v}(s_1) = v_1^\top \quad ; \mathbf{t}^T \mathbf{v}(s_2) = v_2^\top \tag{4}$$

---

[1]The author is supported by the Belgian National Fund for Scientific Research(NFWO)

We note that t and n are the tangential and normal unit vectors of the curve. This functional is a generalization of Hildreth's functional [1] ($\alpha=1$). With this functional, all translations of open curves (with no knowledge of the tangential information in the endpoints) are exactly retrieved, and also translations and expansions of closed curves or open curves with known tangential information. Rotations are exactly retrieved if $\alpha = 0$ or if the curve is a straight line.

Recently, De Vriendt and D'Haeyer [3] proposed a second order stabilizer functional that enlarged the class of physical motion fields that are found exactly. This functional is given by

$$\int \left| t^T \frac{dv}{ds} \right|^2 + \alpha \left| \frac{d}{ds} \left(n^T \frac{dv}{ds} \right) \right|^2 ds \tag{5}$$

With this functional, all translations of open curves (with no knowledge of the tangential information in the endpoints) are exactly retrieved, and also translations, expansions and rotations (and combinations of these motion fields) of closed curves or open curves with known tangential information.

They also showed that the second order functional

$$\int \left| \frac{d^2 v}{ds^2} \right|^2 + \alpha \left| \frac{dv}{ds} \right|^2 ds \tag{6}$$

(suggested by Gong and Brady [5] as an extension of Hildreth's functional) is not satisfactory because the class of physical motion fields that can be retrieved without error becomes smaller.

Up till now we have considered that the measurements $v^\perp(s)$ and v_1^T, v_2^T are known exactly. Of course, this is not a realistic assumption as these measurements will be corrupted due to noise, sampling and quantization. Therefore, the functional will be modified by introducing two new terms: one that takes into account how good our solution should satisfy the motion constraint equation, and a second term that takes into account the accuracy of the tangential information in the endpoints. Furthermore, the image data is discrete (sampling, quantization) and therefore, the integral will be replaced by a finite sum. Furthermore, the derivatives will be replaced by finite differences.

Taking into account all the above arguments, the following function $\phi(v_1, ..., v_N)$ has to be minimized:

$$\sum_{i=1}^{N} \left[t_i^T (v_{i+1} - v_i) \right]^2 + \alpha \sum_{i=2}^{N-1} \left[-\kappa_i t_i^T (v_{i+1} - v_i) + n_i^T (v_{i+1} - 2v_i + v_{i-1}) \right]^2$$
$$+ \beta \sum_{i=1}^{N} \left[n_i^T v_i - v_i^\perp \right]^2 + \gamma \left[(t_1^T v_1 - v_1^T)^2 + (t_N^T v_N - v_N^T)^2 \right] \tag{7}$$

In [4] we tested this functional for different motion fields, different curves and for different quality of the input data. The following qualitative theoretical results were obtained. For translations and expansions, the first order functional results in a better performance than the second order functional. However, for rotations, the second order functional performs better. If the quality of the input data decreases, the optimal value of β decreases (i.e. we shouldn't force the solution too close to the measurements). If we look at the influence of the parameter α, we notice a problem for the first order functional. For translations and expansions the optimum value of α equals 1, while for rotations the optimum value of α is much smaller than 1. However, the optimum value of α for the second order functional equals 1 for translations, expansions and rotations.

In this paper we will compare the results for the first and second order functional for real images with different motions. The performances will be evaluated on images with known motions (quantitatively) and with motions which are only qualitatively known. In section 2, we describe the implementation aspects of this problem of motion estimation. In section 3, we discuss the obtained results. Finally, in section 4, we give the conclusions.

2 IMPLEMENTATION ASPECTS

The theoretical results described in the introduction were obtained for ideal curves for which t, n and κ were known. The sample points were at a constant distance and consequently didn't lie on an orthogonal grid. In our implementation we have

$$\mathbf{n} = \frac{(E_x, E_y)}{|\nabla E|} \tag{8}$$

$$\mathbf{t} = \pm \frac{(E_y, -E_x)}{|\nabla E|} \tag{9}$$

where the sign is chosen such that

$$(\mathbf{r}_{i+1} - \mathbf{r}_i)^T \mathbf{t}_i \geq 0 \tag{10}$$

κ is obtained based on the equation $dt/ds = \kappa n$ or $\kappa = n^T dt/ds$ (see e.g. [6])

$$\kappa_i = \mathbf{n}_i^T (\mathbf{t}_{i+1} - \mathbf{t}_i) \tag{11}$$

t, n and κ are obtained from measurements on Gaussian filtered versions of the images. This is done in order to reduce the influence of noise. The accuracy of the normal components of $v^\perp$ are tested by measuring $v^\perp$ for translations of $(1/2,1/2)$, $(1,1)$ and $(2,2)$ pixels, and for several images (MOBILE, BAR, ZOOM, RENATA). The way E_t and ∇E are computed is given by

$$E_t = E(\mathbf{r}, t + \Delta t) - E(\mathbf{r}, t) \tag{12}$$

and

$$\nabla E = \nabla(E(\mathbf{r}, t)) \tag{13}$$

or

$$\nabla E = \nabla(E(\mathbf{r}, t) + E(\mathbf{r}, t + \Delta t))/2 \tag{14}$$

Table 1: rmse of the measurements of the normal component of the motion vector as a function of the number of selected points, the spread of the Gaussian smoothing filter and the correct motion vector.

	image	BAR				ZOOM			
	σ	1	2	3	4	1	2	3	4
mot. vector	sel. pix.(%)								
	100	0.243	0.151	0.344	0.449	0.530	0.275	0.342	0.410
	95	0.143	0.081	0.107	0.127	0.315	0.182	0.281	0.363
(1/2,1/2)	90	0.133	0.070	0.079	0.093	0.289	0.156	0.227	0.342
	85	0.127	0.064	0.067	0.085	0.271	0.141	0.184	0.266
	80	0.123	0.058	0.062	0.078	0.259	0.129	0.163	0.231
	100	0.415	0.404	0.465	0.539	1.054	0.419	0.414	0.485
	95	0.283	0.155	0.184	0.214	0.629	0.287	0.312	0.435
(1,1)	90	0.267	0.136	0.154	0.150	0.581	0.252	0.258	0.368
	85	0.256	0.128	0.119	0.138	0.550	0.231	0.224	0.293
	80	0.250	0.120	0.112	0.131	0.526	0.217	0.211	0.256
	100	2.458	1.362	0.889	1.171	4.596	1.360	0.814	0.817
	95	1.390	0.591	0.431	0.501	2.721	0.964	0.616	0.685
(2,2)	90	1.239	0.528	0.380	0.353	2.385	0.838	0.540	0.578
	85	1.148	0.491	0.341	0.294	2.130	0.762	0.496	0.500
	80	1.088	0.460	0.308	0.266	1.978	0.711	0.452	0.456

The spatial derivatives are computed using 1x3 or 3x3 filter masks (see e.g. [8, 9]). From the experiments we made, we can conclude that much better results are obtained by using (14) instead of (13). The size of the filter mask (for the spatial derivative) does not influence the results sensitively. In table 1, the measured rmse of $v^{\perp}$ is given for different values of the spread of the Gaussian filter. A threshold is set on $|\nabla E|$. Indeed, the lower $|\nabla E|$ is, the larger the expected error in $v^{\perp}$ is. Thresholds have been introduced such that 100, 95, 90, 85 and 80% of the pixels on the curves in the image have a $|\nabla E|$ larger than the threshold. The rmse is calculated only for the selected pixels. The root of the expected value of $v^{\perp^2}$ is respectively 1/2, 1 and 2 under the assumption that the directions of n are uniformly distributed i.e. n($\cos\theta$,$\sin\theta$) with θ uniformly distributed in the interval $[0,2\pi[$. These figures should be used as a reference for the figures in table 3.

From the results we see that the larger the motion is, the larger the spread of the smoothing filter should be. Indeed, the larger the smoothing filter the better the intensity function can be approximated by a low order polynomial in an environment of the size of the motion. From the above experiment we conclude that the ∇E gives an idea about the accuracy of $v^{\perp}$. Therefore, in our implementation we adapt the value of β to the value of $|\nabla E|$. For large $|\nabla E|$ (≥ 2.0), we take a much larger value of β than for small values of $|\nabla E|$ (< 2.0). We also wanted to use these techniques for relatively large motion fields. As these gradient based techniques only have a limited range of detectable motions, we used the results of a matching algorithm (see De Vriendt [7]) as an initial estimate. The matching algorithm gives the motion vectors in a number of feature points (on the curves) with pixel accuracy.

The initial estimate for the contour points which are no feature points are obtained by linear interpolation between the feature points. We will also use this initial estimation to obtain the tangential information in the endpoints. With the initial estimate $\hat{v}_i$ given, we can substitute v_i by $\hat{v}_i + e_i$ and solve the problem for the unknown e_i's. The minimization of (7) results in a set of linear equations of the form $A.e = r$, with A a banddiagonal matrix. Such systems can be solved in a parallel way. The solution is then found by an iterative procedure by which we try to improve the estimates iteratively. The criteria to stop the iterative procedure are:

$$\sqrt{\frac{1}{N}\sum_{i=1}^{N}|e_i|^2} < Thr_1 \tag{15}$$

$$\sqrt{\frac{1}{N}\sum_{i=1}^{N}|e_i|^2} > Thr_2 \tag{16}$$

number of iterations $> Thr_3$ \hfill (17)

In all the experiments we have set these thresholds to respectively 0.01, 2.00 and 20. The average number of necessary iterations to get a convergence was about 5. The most important reasons for divergence are: occlusion (i.e. part of the contour is no longer in image 2) and a bad initial estimate (the result of the matching process was incorrect). For the contours for which there is no convergence, we try to obtain a motion estimate for part of the curve.

3 RESULTS

In this section we discuss the results obtained with the algorithm proposed in section 2 for real images with different motions. The performances are evaluated on images with known motion (quantitatively) and with motions which are only qualitatively known.

The first set of test images consists of 5 sequences. In the first and second sequence the motion field is respectively an expansion and a rotation. In the other three sequences the motion field consists of a combination of translations, expansions, rotations and noise. In Table 2, the rmse is given as a function of α and β for the first two sequences. We note the relatively good correspondence with the theoretical results. For the expansion the first order functional performs better than the second order functional, while for the rotation the second order functional has a better performance. The optimal value of α for the second order functional equals 1 for both motion fields. For the first order functional, the optimal value of α equals 1 for the expansion and much smaller than 1 (about 0.1) for the rotation. In figures 1a and 1b, the percentage of the contour points with an error in a certain interval is given for the 5 sequences and for both functionals. From this figure we see that in case of a combination of translation, expansion and rotation both functionals result in similar performances. About 70% of the points have an error less than 0.2 pixels and about 95% an error less than 0.5 pixels. The rmse for these sequences are for the first order functional ($\alpha = 0.5, \beta = 1.0, \gamma = 0.1$) respectively 0.128, 0.175, 0.218, 0.205, 0.229 and for the second order functional ($\alpha = 1.0, \beta = 1.0, \gamma = 0.1$) respectively 0.191, 0.135, 0.225, 0.204 and 0.237.

Table 2: the rmse as a function α and β ($\gamma = 0.1$) for the first and second order functional and for an expansions (motion field 1) and a rotation (motion field 2).

mot. field	$\alpha \setminus \beta$	first order functional			second order functional		
		1.0	0.5	0.1	1.0	0.5	0.1
	1.00	0.1086	0.1306	0.1580	0.1905	0.2078	0.2614
	0.50	0.1281	0.1326	0.1672	0.2209	0.2379	0.2949
1	0.10	0.2177	0.2264	0.2335	0.3030	0.3258	0.3749
	0.05	0.2885	0.2897	0.2893	0.3783	0.3819	0.3716
	0.01	0.4418	0.4563	0.4468	0.4632	0.4980	0.4802
	1.00	0.2297	0.2023	0.2378	0.1351	0.1320	0.1475
	0.50	0.1753	0.2203	0.2079	0.1406	0.1435	0.1512
2	0.10	0.1640	0.1645	0.1679	0.1615	0.1607	0.1902
	0.05	0.1683	0.1673	0.2108	0.1704	0.1765	0.1925
	0.01	0.1995	0.1992	0.2583	0.2219	0.2264	0.2352

We also tested our algorithm on a number of other sequences (TRAIN & CALANDAR, ZOOM,...) for which the motion was only qualitatively known. Of course, the results can only be qualitatively checked, though we can state that similar performances were obtained. The small number of large errors that resulted from the matching algorithm were almost all eliminated with the proposed technique (divergence of the solution, and consequently we obtain no estimate for that (part of the) curve).

We conclude with an explanation why the second order functional performs better in a noise free environment, though worse for some motion fields in a noisy environment. To explain this we make a comparison between linear regression with a first order model and a second order model. If the data is noise free, linear regression with a second order model performs equally well for data points lying on a linear curve and better for data points lying on a quadratic curve compared to linear regression with a first order model. Though, if there is noise, linear regression with a second order model performs worse for the linear curve and better for the quadratic curve if the curvature is large enough. The algorithm proposed in this paper has a similar behavior. If the data is noise free, the second order functional performs equally well for translations ($t^T dv/ds = 0, n^T dv/ds = 0$) and expansions ($t^T dv/ds = cst_1, n^T dv/ds = 0$) and better for rotations ($t^T dv/ds = cst_2, n^T dv/ds = cst_3$) as the first order functional. Though if there is noise, it performs worse for translations and expansions and better for rotations if $|cst_3|$ is large enough compared to $|cst_2|$. So, we can conclude that the second order functional will outperform the first order functional if the rotation component is rather important compared to the translation and expansion components. Using the second order functional, we can also expect a better performance for nonlinear models (i.e. $v(x, y)$ is not a linear function of x and y).

4 CONCLUSIONS

In this paper we analyzed the performance of two regularization functionals for the estimation of curves on real images. The obtained results were in correspondence with the theoretical results obtained earlier. This means that the first order functional is better suited to retrieve translations and expansions, while the second order functional is better suited to retrieve rotations in a noisy environment. For combinations of these motion fields we obtain similar results with both functionals. The obtained rmse is about 0.2 pixels. With both techniques the solution is found by solving a set of linear equations $A.e = r$. This system of equations can be solved in a parallel way as the matrix A is a banddiagonal matrix. Though, the algorithm using the second order functional has a larger computational load.

References

[1] E.C. Hildreth, "The Measurement of Visual Motion", MIT Press, Cambridge, MA.

[2] J.P.F. D'Haeyer, "Determining Motion of Image Curves from Local Pattern Changes", Computer Vision, Graphics and Image Processing, Vol. 34, pp.166-188.

[3] J. De Vriendt and J. D'Haeyer, "Analysis of second order stabilizer functionals for the regularization of the problem of motion estimation on curves", Proceedings of the "Picture Coding Symposium '93", paper 4.2, Lausanne, Switzerland, march 1993.

[4] J. De Vriendt, "Using regularization techniques to estimate the motion of curves in a sequence of images", Proceedings of the "8th Scandinavian Conference on Image Analysis", pp. 159-166, Tromso, Norway, may 1993.

[5] S. Gong and M. Brady, "parallel computation of optical flow", in O. Faugeras, editor, Proc. 1st European Conference on Computer Vision", Springer Verlag, 1990, pp.124-133.

[6] F. Bedford and T. Dwivedi, "Vector Calculus", McGraw-Hill, New York.

[7] J.A. De Vriendt, "The matching problem: heuristic approach considering both split and merge competition", Proc. of the IEE fourth intern. conf. on image processing, Maastricht, The Netherlands, pp. 53-56, 1992.

[8] R. Haralick, "Digital step edges from zero crossings of second directional derivatives", IEEE Trans. om PAMI, Vol.6, pp 58-68, 1984.

[9] Vieville, T., and Faugeras, D., "Robust and fast computation of unbiased intensity derivatives in images", Proc. 2nd European Conference on Computer Vision, Lecture notes in Computer Science 588, pp. 203-211, 1992, Springer-Verlag: Berlin, Heidelberg, New York.

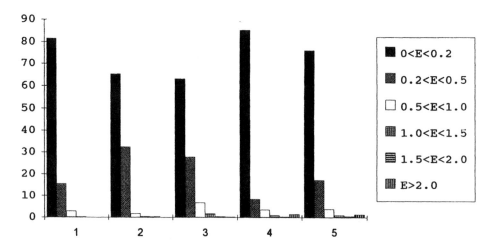

Figure 1a. Percentage of contour points with an error in a given interval for the 5 test sequences and for the first order functional ($\alpha = 0.5, \beta = 1.0, \gamma = 0.1$).

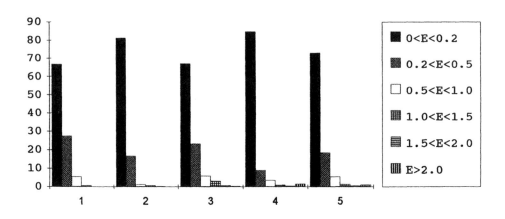

Figure 1b. Percentage of contour points with an error in a given interval for the 5 test sequences and for the second order functional ($\alpha = 1.0, \beta = 1.0, \gamma = 0.1$).

Time-Varying Image Processing and
Moving Object Recognition, 3 – V. Cappellini (Ed.)
© 1994 Elsevier Science B.V. All rights reserved.

Understanding Object Motion via Self–Organization

J. Heikkonen and P. Koikkalainen

Lappeenranta University of Technology, Department of Information Technology,
P.O. Box 20, SF–53850 Lappeenranta, Finland

Abstract
We have developed a general framework for learning and representing object motions. In this approach the self–organizing principle is utilized for building high–level representations of the characteristics of motion, without explicitly saying what these characteristics are. The obtained representation of motion is directly applicable for higher level operations, such as the determination of motion trajectories and pre-dictation of future movements. We shall also discuss how to extract features that are causes for the object motion, and about the problem of building generalized motion models to cope with situations that differ slightly from given examples. The performance of the object motion learning is demonstrated with image sequences of block–world and traffic intersection scenes where some of the objects are moving.

1. INTRODUCTION

Our objective is to use examples to understand the behavior of a moving object in a natural environment. This implies that the observed object can change the constituent parts of its movement as a function of its environment. An animal, for instance, may detect something that causes it to change its direction. In a more theoretical setting, we may say that the motion behavior is described as a sum of behaviors arising within the object itself and from the features extracted from the environment. Thus, we need to relate the affect of the features from the environment to the model that we build from the examples.

Practical uses for such an intelligent motion estimator are obvious but even the smallest of animals do much better in the task than the best of the state–of–the–art in artificial systems. Think about a fly, for example. This rather primitive animal can control and target its movements more intelligently than any modern computer hardware and software. An obvious conclusion is that the manner in which the biological nervous system is formed seems to be very suitable for these types of operations, although, the ability to learn to predict movements from examples is a reasonable assumption only when the higher animals are concerned.

To show the role of the self–organization in the context of motion understanding the usefulness of other approaches need to be examined. In a synthetic, man–made, environment it is possible to use a priori knowledge about the environment when building the model for motion behavior, as demonstrated for instance in papers [3,8]. In a real world, where the feature representation of the surroundings is much richer, we can't that easily match an individual fact to a particular behavior. Instead, a representative set of features should be matched in parallel to the inherent behaviors of the moving object. Thus, an associative neural network would be a good candidate for the task. In our case, when also the internal behavior of the object is unknown, the model for the motion behavior can be based only on the information extracted by the observer. This suggests that the object behavior is best learned via the self–organization principle, and it motivates us to use Self–Organizing Map (SOM) [5,6] to learn, to represent and to predict the object motion.

By using the SOM a certain degree of generisity can be introduced to the motion analysis. We intend to use of the same mathematical model for high–level interpretations of motion, like in [4,11] as well as for the detection of lower–level spatiotemporal behaviors. The only difference will be in the feature inputs on which the SOM is trained. In our examples we want to stress the real–world connection of the approach, and have selected examples to demonstrate image sequences of block–world and traffic intersection scenes, where some of the objects are moving. These scenes are viewed by an overhead camera in order to better facilitate the reasoning about object behaviors in space and time.

The experiments of this paper are inspired by the work of Mohnhaupt and Neumann [10,11], where *an analogical model* was used instead of SOM to represent the behavior of moving objects extracted from concrete observations. The idea of the analogical model is to use a specific data structure, a spatiotemporal buffer, which is a N–dimensional accumulator array $C(c_1, \dots, c_N)$. For example, the accumalator $C(x, y, v, d)$ represents $2 + 2$ dimensional case, where there are for each spatial coordinate pair (x, y) counter cells for the whole range of possible velocity vectors v and directions d, represented in some discretization. Observed trajectories of moving objects are stored to this buffer by accumulating the counter cells (x, y, v, d) on the corresponding trajectories. These accumulated experiences can be generalized to novel, slightly different situations, using local computing, similar to cellular automata. Some useful operations are: generalization, where nearby tragectories become one; and obstacle avoidance, where the generalized trajectory finds a path around a blocking object.

There problem of the visual buffer is the memory consumption versus resolution. Since the discretization of the accumulator array is decided by forehand, the user must represent the whole parameter space as a huge N–dimensional matrix, which takes a lot of memory or is computationally inefficient if a dynamic matrix model is used. Also, since the generalization process has some resembelance to clustering, we may compare the performance of the generalization to scalar discretization, which is from the information theoretical viewpoint much worse than in vector quantization methods. Our solution is to use the Self–Organizing Map [5,6] instead of the accumulator array. The representation differs from the visual buffer by quantizing the trajectory space instead of discretization, which results that a minimal number of units required, and no additional generalization operations are required.

2. SELF–ORGANIZING MAP

The Self–Organizing Map (SOM), introduced by Kohonen [5,6], is a method to map patterns from an input space V_I onto typically lower dimensional space V_M of the map, such that the topological relationships between the inputs are preserved. Thus, the inputs which are close to each other in V_I tend to be represented by units close to each other on the map space V_M.

The closeness between the units of the map is typically described by defining the map as a one– or two–dimensional discrete lattice of units. The training algorithm is based on iterative random sampling, where one sample, the input vector $x : x \in V_I$, is selected randomly and compared to the weight vectors of the units of the map. The best matching unit b, corresponding strongest response to given input pattern x is selected using some criterion, such as

$$\| w_b - x \| = \min_{i \in V_M} \{ \| w_i - x \| \},$$

where w_i is weight vector of any neuron i (initially all weights are set randomly to their initial positions into input space). The best matching unit b and units in its topological neighborhood $Ne(b)$ are updated according to formula:

$$w_r^{new} = w_r^{old} + \Psi_{rb}(x - w_r^{old}),$$

where $r \in Ne(b) \cup b$ and Ψ_{rb}, $0 \leq \Psi_{rb} \leq 1$, is a prespecified adjustable gain function of distance $d(b, r)$ between the units r and b. The function Ψ_{rb} may be constant for all $r \in Ne(b) \cup b$, but usually Ψ_{rb} has its maximum at b and its value decreases as distance d increases. Also $Ne(i)$ and Ψ_{rb} are usually implemented as slowly decreasing functions of the iteration counter. After a number of input vector presentations the mapping has formed i.e. the weight vectors are the centroids of clusters covering the input space and the point density function of these centers tends to follow a monotonic function of the probability density function of the input space V_I.

In our experiments we have used a sightly different algorithm, a Tree Structured SOM (TS–SOM) [7], which is computationally faster and easier to use version of the original SOM.

3. LEARNING OBJECT MOTION FROM IMAGE DATA

The task in our example is to learn the motion behavior from image sequences with as little a priori knowledge as possible. In practice we are using a method to build a high–level representation of the characteristics of movement without explicitly saying what these characteristics are. The selected approach is, in the first stage, to collect low–level dynamical features from image sequences, and in the second stage, to cluster these features by using the SOM model. Thirdly, after the learning has taken place, the SOM network can be used for predicting the changes in the constituent parts of object movements. These three stages, which are depicted in figure 1, are described in the next paragraphs.

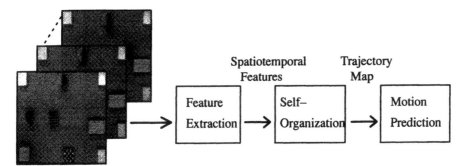

Figure 1. Block diagram of the proposed framework for object motion learning.

At the first level two or more successive frames of an image sequence are analyzed to obtain spatiotemporal features describing motions at a particular image instance. These features should include a representative set of perceptual primitives of characterizing the moving objects, for example locations, orientations, orientation changes, velocities, velocity changes, accelarations and accelaration changes. More specificly, we use a set of feature detectors such that for each spatiotemporal location where some observation criterion is matched, a vector of detector responses and spatial coordinates are memorized. It should be noted that no explicit time information is needed i.e. there is no time parameter t which relates feature vectors to certain images taken at time t, but the features themselfs may imply relative changes in time domain. It should be noted that in the litterature there are many ways of obtaining these features, and it depends on applications which method or methods are the best ones for the purpose. Clearly the selected technique should be robust and it should work well with real image sequences.

At the second stage, the SOM network is trained by using the set of memorized feature vectors. Most typically the map dimension is same or lower than the spatial dimension of our application, but this is only to make the analysis of the results easier. The usefulness of the SOM method is clearly explained by the following simple example. Assume that we have a two–dimensional view of an area where two types of trajectories are observed: a path from east to west, and another path from north to south. Let our training vectors have tree components, spatial coordinates x and y, and direction d, and let us also assume, that a very representative and equally large set of features is extracted from both paths. Since the SOM algorithm tries to cover the input set maximally, a vast majority of the units will be placed on the paths directly, and in the crossection, where there are more example vectors, also SOM produces a locally more dense population of neurons. Thus there are different units representing different directions at the same spatial location.

At the third stage the organized SOM is used for motion analysis, which in our case is the task of predicting future movements of objects. The analysis begins with an observation of the characteristics of the movement of an object for which the future behavior is estimated. The initial characteristics of the movement are extracted from the application or given, and the following time dependent behavior should correspond the behavior of a typical object in the training set.

The estimation procedure is typically reverse of the feature extractor. Let us assume that we have extracted rather simple features that represent relative changes in spatial locations i.e. our feature vector is coordinate pair (X, Y) and change of position $(\Delta X, \Delta Y)$ between two successive time steps t and $t + 1$. An obvious linear prediction at time $t + 1$ for an observation $f_t = (X_t, Y_t, \Delta X_t, \Delta Y_t)$ is then $f_{t+1}^{LINEAR} = (X_t + \Delta X_t, Y_t + \Delta Y_t, \Delta X_t, \Delta Y_t)$.

The role of the SOM is to learn and generalize the features of the training set, and thus memorize only those $(X, Y, \Delta X, \Delta Y)$ vectors that do exist in the application. In this context the SOM network can be used for correcting the linear prediction such that it corresponds the set of observations used in training phase. The corrected prediction f_{t+1}^{SOM} is then determined by the weight vector of the best matching unit b_{t+1}, corresponding the result of the linear prediction f_{t+1}^{LINEAR}:

$$\| w_{b_{t+1}} - f_{t+1}^{LINEAR} \| = \min_{i \in V_M} \| w_i - f_{t+1}^{LINEAR} \| , \qquad (3)$$

and is given by $f_{t+1}^{SOM} = (X_t + \Delta X_t, Y_t + \Delta Y_t, \Delta X_{b_{t+1}}, \Delta Y_{b_{t+1}})$. In practice the rule 3 is usually adapted to match better to the physical constraints of the application by giving different weightings for different types of information. For instance, the norm $\| w_i - f_{t+1}^{LINEAR} \|$ is propably too sensitive to differences in spatical coordinates (X, Y), compared to $(\Delta X, \Delta Y)$ type of information. The way how these modifications is done depends naturally on the type of information that is used in the application.

3.1. Experimental Results on Block–World Scenes

In this section two concrete examples are shown to demonstrate the power of our motion learning framework proposed in above section. As an application domain we have used block–world scenes viewed by overhead camera. In these scenes rigid objects undergo planar two–dimensional motions on their own trajectories.

At the lowest level of the system, successive image frames must be analyzed in order to obtain feature vectors $f_i = (X_i, Y_i, \Delta X_i, \Delta Y_i)$ capable to describe planar two–dimensional movements of objects from one image frame to the next. For this purpose, a simple an robust low–level dynamical feature detector algorithm has been developed.

Frames $F_t, t = 1..., N_F$ belonging to image sequence, are first processed with standard edge–detection technique, e.g. the Canny operator [1]. The binary edge pictures F_t' are used to remove stationary background by constructing XOR–pictures $F_t'' = F_t'$ XOR F_{t-1}', meaning that only those edge pixels are accepted which are not in both two successive images. Potential edge pixels belonging to moving objects are found by taking AND–pictures: $F_t''' = F_t'$ AND F_t''. Finally, dynamical feature vectors f_i are obtained by processing successive AND–pictures using algorithm 1 which is based on windowed mass centre calculation of edge pixels. Parameter T in the algorithm is adjustable positive integer threshold specifying what is considered as a noise point not belonging to any moving object and $(M - 1) \times (M - 1)$ (M even), is the size of the neighborhood for $(\Delta X, \Delta Y)$ calculation.

Next we will consider an example with three moving objects (figure 2). In this experience one–dimensional SOM with 64 neurons was learned according to input vectors f_i obtained from image sequence of eight 256x256 pictures (i.e. $N_F = 8$) with parameters $M = 32$ and $T = 10$ in the algorithm 1. In the figure 2 the learned short time motions are visualized by projecting every neurons of the map into image plane according to its weight vectors. A white circle is used to denote a projected location of the neuron and a short line starting from the circle is used to denote a learned direction of motion at that particular location. The results show clearly how each unit $n \in V_M$ on the map became responsible for a small subregion in trajectory space, located at (X_n, Y_n), and learned the appropriate movement $(\Delta X_n, \Delta Y_n)$ valid there.

The prediction ability of SOM based approach is demonstrated by giving starting points and starting displacements of three moving objects seen in figure 2. The one–dimensional map of 64 neurons illustrated in figure was used. We wanted to estimate object locations on the scene at time instances $t + n\Delta t$, $n = 1,..., N_F$. The prediction principles given in section 3 was employed and figure 2 shows the resulting trajectories. A white box visualize predicted object location at a particular time t and white line shows the direction of movement. Compared to real object trajectories correct results were obtained.

Algorithm 1. Extracting displacement vectors $f_i = (X_i, Y_i, \Delta X_i, \Delta Y_i)$ from images F_t'''.

For $t := 1$ to $N_F - 1$, $r := M/2$ to $R - M/2$, $c := M/2$ to $C - M/2$ do
Begin
  If $F_t'''(r,c)$ is an edge pixel then
  Begin
    Let W_t be a $(M-1) \times (M-1)$–window of image F_t''' centered at point (r,c).
    Let W_{t+1} be a $(M-1) \times (M-1)$–window of image F_{t+1}''' centered at point (r,c).
    Count the number of edge points T_t and T_{t+1} within the window W_t and W_{t+1}.
    If $(T_t \geq T$ AND $T_{t+1} \geq T)$ then
    Begin
      Calculate the mass centre (Xm_0, Ym_0) of edge points within the window W_t.
      Calculate the mass centre (Xm_1, Ym_1) of edge points within the window W_{t+1}.
      $X := Xm_0$, $Y := Ym_0$, $\Delta X := Xm_1 - Xm_0$ and $\Delta Y := Ym_1 - Ym_0$.
      Store the vector $(X, Y, \Delta X, \Delta Y)$ among the group of vectors f_i.
    End
  End
End

Finally an example of three moving blocks with overlapping trajectories is provided. The parameters in the algorithm were set to $M = 32$ and $T = 10$, and an one–dimensional SOM with 64 neurons was taught according to given principles. In figure 3 one can see every second image from the image sequence of eight 256x256 pictures ($N_F = 8$). The learned short time motions and predicted trajectories are visualized in the same manner as in figure 2. The results show that overlapped trajectories do not introduce difficulties into prediction.

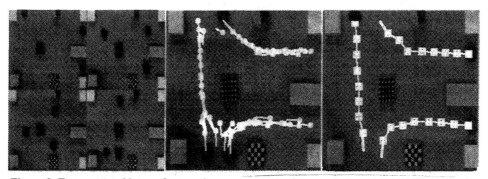

Figure 2. Every second image from an image sequence of three moving blocks (left) and projected neurons of one–dimensional SOM with 64 neurons (middle) and predicted trajectories of moving objects (right).

4. TOWARDS GENERALIZED MOTION MODELS

In section 3 we proposed a general framework for learning and predicting object trajectories from image sequences. However, a motion representation based on examples collected in a particular geometrical environment is limited to that specific domain and thus the usefulness of this kind of motion model is quite moderate. In this section we discuss the problem of generalizing the observed motions to novel and slightly different situations. For demonstration purposes we have chosen planar motions on traffic street intersections where objects typically undergo "go–straight", "turn–left" and "turn–right" movements on their own trajectories.

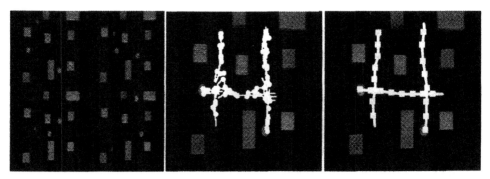

Figure 3. Every second image from an image sequence of three moving blocks (left) and projected neurons of one–dimensional SOM with 64 neurons (middle) and predicted trajectories of moving objects (right).

A natural step towards a generalized representation of motion is to remove those motion descriptions, such as spatial coordinates, which are dependent of a particular geometry (see for instance [11]) and to replace them a convenient set of invariant perceptual primitives characterizing the movements of objects. These primitives should make the learned object motions applicable to situations which may differ from the observed examples. As an example, if a "turn–right" movement is learned at a particular street intersection, the invariant primitives should make this "turn–right" representation suitable to another intersection with a different shape.

The only source of generic description for motion is the image sequence. In principle, we must extract out what circumstances in the environment i.e. image features will cause the object motion, what environment changes may happen during movement, and how the moving object should react to them while moving in its own trajectory.

There are several criteria for feature extraction technique to be useful for object motion learning. Features must be easily computed, robust, insensitive to various distortions and variations in the images, and physically meaningful. Gabor filters have been shown to possess good localization properties in both spatial and frequency domains [2] and good tolerance to various distortions in images [9]. Motivated by their good performance in applications, we use the Gabor filters as basic image feature detectors for object motion learning. The key idea of our approach is following: The basic motion representation is enriched by associating Gabor features to each units in the trajectory map. After that, both the Gabor features and trajectory map are employed to predict motions of objects.

For Gabor feature extraction each unit $n \in V_M$ on the trajectory map is projected into image plane according to their spatial coordinate weights (X_n, Y_n). Let $W_n(x, y)$ be a rectangular window in the image plane covering the neighborhood pixels of the coordinate (X_n, Y_n). The orientation of the window $W_n(x, y)$ is determined according to weights $(\Delta X_n, \Delta Y_n)$ and it is divited into an equally spaced grid of $N_x \times N_y$ nodes. If the image coordinates of the grid node (i, j) are given by (x_i, y_j) in the local window coordinate system, then the discrete Gabor features for window W_n are defined by the linear convolution

$$G_n(\theta, f) = \sum_{x, y \in W_n} I(x, y) \cdot \psi(x_i - x, y_j - y, \theta, f), \qquad (4)$$

where $I(x, y)$ is the image gray level of pixel (x, y) and $\psi(x, y, \theta, f)$ is the self–similar family of the 2–D Gabor filters. These filters are defined in spatial domain with impulse response (centered at the origin) [2]:

$$\psi(x, y, \theta, f) = \exp\left(i(f_x x + f_y y) - \frac{f^2(x^2 + y^2)}{2\sigma^2}\right), \qquad (5)$$

where $f_x = f\cos\theta$, $f_y = f\sin\theta$ and x and y are pixel coordinates in the image. Parameter f determines the central frequency of the pass band in orientation θ: $\theta = \pi k/K$, $k \in \{0,\ldots,K-1\}$ and parameter σ determines the width of the Gaussian envelope along the x– and y–directions. Here we used $\sigma = \pi$, six values of θ ($K = 6$) and two frequencies $f = \{\pi/2, \pi/4\}$. For further processing, the absolute values of the G_n are computed in each grid node (i,j) and the vectors obtained in each (i,j) by varying θ are normalized to unit length at both frequency levels separately.

Let us assume that we have extracted motion feature vector $f_t = (X_t, Y_t, \Delta X_t, \Delta Y_t)$, corresponding spatial location (X_t, Y_t) at time t and movement $(\Delta X_t, \Delta Y_t)$ between time steps t and $t + 1$. Now the task is to predict the next situation f_{t+1} in the scene at time $t + 1$. Prediction for the new spatial location (X_{t+1}, Y_{t+1}) is done as before, but the new motion pair, $(\Delta X_{t+1}, \Delta Y_{t+1})$, is is taken as a function of its environment, determined by the Gabor features as explained below.

First, Gabor features G_t are extracted from the window W_t centered at coordinate (X_t, Y_t). Next a best matching unit b_t is selected, corresponding the features G_t from the enriched map V_M:

$$\| G_{b_t} - G_t \| = \min_{i \in V_M} \| G_i - G_t \|. \tag{6}$$

(Note that only Gabor features were used to find the best matching unit b_t)

The corresponding vector $f_{b_t} = (X_{b_t}, Y_{b_t}, \Delta X_{b_t}, \Delta Y_{b_t})$ of the unit b_t is then used as a prototype movement in the Gabor space. Thus the vector f_{b_t} corresponds to the vector f_t in the concrete environment. In order to estimate the next displacement $(\Delta X_{t+1}, \Delta Y_{t+1})$ for f_{t+1}, we take the corresponding change of the direction of the movement for f_{b_t} by computing $f_{t+1}^{SOM} = (X_{b_t} + \Delta X_{b_t}, Y_{b_t} + \Delta Y_{b_t}, \Delta X_{b_{t+1}}, \Delta Y_{b_{t+1}})$ according to f_{b_t}, as explained in section 3. The change of the direction of the motion is then an angle a between displacement vectors $(\Delta X_{b_t}, \Delta Y_{b_t})$ and $(\Delta X_{b_{t+1}}, \Delta Y_{b_{t+1}})$. When this is related to the concrete movement, the estimated motion f_{t+1} in the scene can be given as $f_{t+1} = (X_t + \Delta X_t, Y_t + \Delta Y_t, \Delta X_{t+1}, \Delta Y_{t+1})$, where the displacement $(\Delta X_{t+1}, \Delta Y_{t+1})$ is determined so that the angle β between vectors $(\Delta X_t, \Delta Y_t)$ and $(\Delta X_{t+1}, \Delta Y_{t+1})$ is equal to the angle a, and the length of the movement $(\Delta X_{t+1}, \Delta Y_{t+1})$ is equal to the lenght of the vector $(\Delta X_{b_{t+1}}, \Delta Y_{b_{t+1}})$ i.e.

$$\sqrt{(\Delta X_{t+1})^2 + (\Delta Y_{t+1})^2} = \sqrt{(\Delta X_{b_{t+1}})^2 + (\Delta Y_{b_{t+1}})^2}.$$

Next, an example of the proposed motion learning approach is provided. The left picture in figure 4 shows every second picture from an "traffic intersection" image sequence of eight 256x256 pictures. In this image sequence one object performs "turn–right" movement. For low–level dynamical feature extraction, the parameters M and T in algorithm 1 were set to $M = 32$ and $T = 10$, and a SOM of 256 neurons in one–dimensional discrete lattice was taught according to principles given in section 3. Figure 4 shows the resulting map: A white circle illustrates a projected location of the neuron in the image plane and the white line starting from the circle shows the movement valid there. For illustration purposes only every fourth neuron is drawn.

The generalization ability of our motion learning approach is demonstrated in several different "traffic street" intersections. The size of the feature extraction window was 60 × 60 pixels and there were 5 × 5 grid nodes for Gabor feature extraction. In figure 4 one can see four diverse intersections and estimated "turn–right" trajectories in these domains. For every trajectory starting point (white circle) and starting direction (white tick) of motion were given. According to these results, our approach for building generalized motion models supports generalizations from a given set of a specific example paths and the learned "turn–right" behavior in one intersection can be with success used to predict the motions of objects in these example intersections with different geometrical shapes.

334

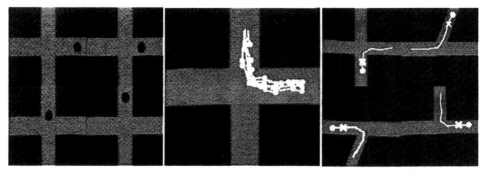

Figure 4. Every second image from an image sequence of one moving block (left) and some projected neurons of one–dimensional SOM with 256 neurons (middle) and predicted trajectories of moving objects at different intersections (right). For more information, see text.

5. CONCLUSIONS

In this paper we have described a general framework for learning and understanding object motion. We have used the Self–Organizing Map to learn and represent the motion of moving objects in image sequences. We have also discussed about the problems of predicting object trajectories and extracting features that are causes for characteristics of the motion. We did not consider some other motion issues, such as high–level motion interpretation and description of what is going in the field of the view. However, the proposed approach for object motion learning can be used to give motion information to these processes. The rather simple experiments provided in this paper show the potentials of the proposed approach and the results in this paper should be of interest to numerous areas of research including automation and robotics in industrial systems, object tracking for instance in defence systems and other image sequence analysis systems.

REFERENCES

1. Canny, J., 1986, A Computational Approach to Edge Detection, IEEE Transactions on Pattern Analysis and Machine Intelligence, Vol. PAMI–8, No. 6, 679–698.

2. Daugman, J., 1988, Complete Discrete 2–D Gabor Transforms by Neural Networks for Image Analysis and Compression, IEEE Trans. on ASSP, Vol. 36, No. 7, 1169–1179.

3. Fennama, C., Hanson, A., Riseman, E., Beveridge, J. R., Kumar, R., 1990, Model–Directed Mobile Robot Navigation, IEEE Transactions os Systems, Man, and Cybernetics, Vol. 20, No. 6, 1352–1369.

4. Howarth, R. J., Buxton, H., 1991, Analogical Representation of Space and Time, GVGIP: Image and Vision Computing, 467–478.

5. Kohonen, T., 1982, Self–organized Formation of Topologically Correct Feature Maps, Biological Cybernetics, 43, 59–69.

6. Kohonen, T.,1989, Self–Organization and Associative Memory, Springer–Verlag, Berlin, Heidelberg.

7. Koikkalainen, P., Oja, E., 1990, Self–Organizing Hierarchical Feature Maps, In proc. IJCNN–90, International Joint Conference on Neural Networks, San Diego, CA, 279–284.

8. Kosaka, A., Kak, A. C., 1992, Fast Vision–Guided Mobile Robot Navigation Using Model–Based Reasoning and Prediction of Uncertainties, CVGIP: Image Understanding, Vol. 56, No. 3, 271–329.

9. Lampinen, J., Oja, E., 1990, Distortion Tolerant Feature Extraction with Gabor Functions and Topological Coding, In proc. INNC 90, Paris, 301–304.

10. Mohnhaupt, M., Neumann, B., 1989, Some Aspects of Learning and Reorganization in an Analogical Representation, In: K. Morik, ed., Proc. International Workshop on Knowledge Representation and Organization in Machine Learning, 50–64.

11. Mohnhaupt, M., Neumann, B., 1991, Understanding Object Motion: Recognition, Learning and Spatiotemporal Reasoning, Robotics and Autonomous Systems, Vol. 8, 65–91.

Time-Varying Image Processing and
Moving Object Recognition, 3 – V. Cappellini (Ed.)
© 1994 Elsevier Science B.V. All rights reserved.

335

A Hough Transform based Hierarchical Algorithm for Motion Segmentation and Estimation

Mirosław Bober and Josef Kittler
Department of Electronic and Electrical Engineering,
University of Surrey, Guildford GU2 5XH, United Kingdom

1. INTRODUCTION

A large body of work has been concerned with solving the motion estimation and segmentation problem. Possible applications include recovery of 3D scene structure, sequence bandwidth compression and coding, object tracking and traffic monitoring.

The majority of algorithms for motion segmentation and estimation are based on optic flow equation [7]. Unfortunately, the point-wise estimates of the optical flow are rather inaccurate, and very sensitive to factors like noise or change in illumination conditions. Moreover the use of regularising techniques (smoothing), which is necessary to solve this ill-posed problem, usually distorts the optic flow on motion boundaries. Consequently, an optic flow with motion discontinuities smoothed out is very difficult to segment.

An alternative to pixel-based estimation are robust methods based on regions of pixels. For example Campani and Verri [6] proposed to compute the optic flow from an over-constrained system of linear algebraic equations based on an affine motion model. Another block-based method is a part of the framework presented by Singh [8]. The motion of the block is recovered by an extensive search within a predetermined window. Most of the region-based algorithms including the two examples quoted above suffer from at least one of the following major drawbacks. First they assume that the motion within the window is coherent, which is violated when the block covers more than one moving object. Second, they are computationally demanding.

In this paper we address the above mention drawbacks and present an approach based on the Hough Transform (HT). Due to the properties of the HT and the application of a multipass strategy the limiting assumption of a single moving object within region is no longer required. An affine motion model capable of handling rotation, translation and zoom is employed. The extra degree of freedom achieved by the use of an advanced motion model favours the motion segmentation that is coherent with the real motion present in the scene. We also address the issue of computational efficiency, which is particularly important when the search in a highly dimensional space is involved. A multiresolution gradient-based minimisation strategy that does not require an explicit computation of the support function is proposed. We also suggest the use of multiresolution in image space, in order to prevent the algorithm from being trapped in local minima. Finally, we propose a variation of the algorithm performing initial segmentation of the reference frame based on the distribution of the image intensity function. Motion parameters of each region are recovered and regions exhibiting coherent motions are then merged, forming a final segmentation map. Such strategy aids segmentation process and helps to obtain a dense estimate even within regions exhibiting small grey-level variations.

The structure of the paper is as follows. In the next section the application of the Hough Transform for motion estimation and segmentation is outlined. Details of the multiresolution minimisation strategy and multipass strategy are explained and estimate confidence measures are introduced. Experimental results are presented in the third section. We demonstrate and test the algorithm on two real-live sequences in different applications. Finally, Section 4 summarises the most important results and discusses possible extensions and directions of further research.

2. HOUGH TRANSFORM FOR MOTION ESTIMATION

Although the Hough Transform has been extensively used in solving different machine vision problems, only several examples of application for motion estimation exists. Adiv [1] employed it for the interpretation and segmentation of the precomputed optic flow and Wu [11] used an optimisation approach that segments and estimates complex motion. The second approach is appealing because it is computationally effective and can cope with multiple moving objects. Wu underlined that complex motion models can be used, however his implementations included only motion models with 2 and 3 parameters. The direct extension to more complex models proved difficult because of the convergence problems. This section begins with the formulation of the problem. Then we introduce several important modifications aiming to resolve convergence problems and improve motion segmentation. We also introduce confidence factors describing the quality of the estimate.

We formulate the problem as follows. Let us assume that after transforming any arbitrary region $\Re_0$ from the reference frame F_0 by a transformation $T_{\vec{a}}$ we obtain a region $\Re_1$ in the consecutive frame F_1. Vector $\vec{a}$ denotes the parameters of the transform T, and its dimensionality depends on the motion model. In our framework we use an affine transformation which can describe complex motions involving translation, rotation and change of scale. Pixel positions $p(x, y)$ in the frame F_0 and $p'(x', y')$ in the frame F_1 are constrained by the following equation.

$$p' = T_{\vec{a}}(p) = \quad p + (a_1 x + a_2 y + a_3, a_4 x + a_5 y + a_6) \tag{1}$$

As other motion models can be applied, a general notation $T_{\vec{a}}(p)$ is used for motion transformation. However we believe that affine model gives an optimal balance between complexity and performance for most real sequences.

The Hough Transform can be viewed as a method of segmenting the feature points into groups satisfying some parametric constraint. For the problem at hand the constraint is given by equation 1. Each pixel from the region $\Re_0$ votes for (supports) a set of motion parameters, and the estimate of the parameter values is defined by extrema in the space of votes. Under the assumption that the pixel intensity is preserved, the support given by a pixel p to a hypothetical motion vector $\vec{a}$ is defined by the kernel function $\rho(\epsilon)$:

$$h(\vec{a}, p) = \rho \left(I_0(p) - I_1(p') \right) \tag{2}$$

where I_0 and I_1 are the intensity values at the positions p and p' in the frames F_0 and F_1 respectively. Bilinear interpolation is used to approximate the grey level value $I_1(T_{\vec{a}}(p))$ at inter-pixel locations. In [11] Wu proposed to use *Quadratic* or *Absolute Value* kernel functions. All results presented in this paper are based on the latter one, however the application of *redescending* kernel functions (e.g. Hampel or Tukey kernels) gives more accurate and robust estimate [4]. Note that for the kernel used here, the support from a single pixel can take values from the interval $(0, A)$ (A denotes the number of grey-levels), with values close to zero corresponding to a strong support. The total amount of support $H(\Re, \vec{a})$ received by a motion vector $\vec{a}$ from a region $\Re$ can be expressed as:

$$H(\Re, \vec{a}) \quad = \quad \sum_{p \in \Re} h(\vec{a}, p) \tag{3}$$

The estimate of motions present within the region $\Re$ can be found by determining the positions of the minima in the Hough Space H. The rest of this section is concerned with the following issues: i) an efficient minimisation strategy that can cope with highly dimensional space, ii) recovering motion parameters of multiple objects, and iii) the definition of the estimate confidence measures.

When a search in a high dimensional space is involved, the efficiency of the search procedure is of a primary importance. The method proposed in [11] is a discrete version of the steepest descent methods, with a multiresolution in the parameter space. The procedure starts form the point $\vec{a} = \vec{0}$ in the parameter space and iteratively updates the estimate, based on the value of the partial derivatives of H.

$$\frac{\partial H}{\partial a_i} = \sum_{i \in \Re} -\frac{\partial \rho(\epsilon)}{\partial \epsilon} \frac{\partial I_1(x',y')}{\partial a_i} \tag{4}$$

$$\frac{\partial I_1(x',y')}{\partial a_i} = \frac{\partial I_1(x',y')}{\partial x'} \frac{\partial x'}{\partial a_i} + \frac{\partial I_1(x',y')}{\partial y'} \frac{\partial y'}{\partial a_i} \tag{5}$$

The parameter space has a discrete grid, defined by the current resolution in the parameter space. The algorithm records the path and when oscillations are detected (that is the path is retraced) it switches to a finer resolution in the parameter space. The search terminates when the finest resolution is reached. The minimisation on a discrete grid facilitates the detection of the minima without computation of the Hough function values after each step. On the discrete grid the path is retraced around the minimum, and this condition is easy to detect.

Since the search for the minimum is based on gradient information, the problem of the solution being trapped in a local minimum becomes very important. We found that algorithm presented by Wu may, in certain circumstances, give an incorrect estimate due to this problem. When multipass strategy is used, a biased or wrong estimate propagates and prevents the algorithm from recovering motion of remaining regions. We established two factors contributing to the creation of the insignificant minima in the support function: aliasing of high spatial frequencies and noise. Experiments proved that the more complex motion model is, the more sensitive it is to the above factors. We deal with the problem by introducing a hierarchical strategy. Figure 1 presents two contour plots of the Hough Space computed for a real sequence involving a pattern moving with translational motion (the estimated displacement is 5 pixels/frame). Numerous local minima can be observed in the left graph. In the right graph (b) we show the contour plot for the same sequence, but subsampled by 8. The axes of the parameter space are expanded 8 times to obtain the graph of the same size. Clearly all insignificant minima disappeared. However the minimum related to motion and its position is preserved. This effect can be explained on the grounds of signal processing. When large displacements are involved, the aliasing of high spatial frequencies (corresponding to fine structure of the image) produces insignificant minima

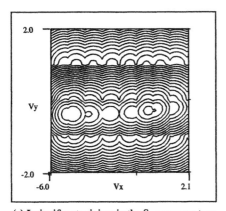

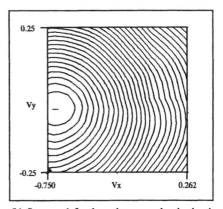

(a) Insignificant minima in the S space may trap the walk to the solution

(b) S space defined on the coarse level - local traps have been removed

Figure 1. Insignificant minima in S space are removed

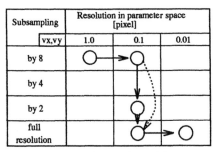 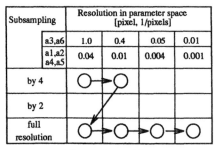

Translational motion model   Affine motion model

Figure 2. Algorithm changes resolution and subsampling following the above schedule

in the support function. In the subresolution image high spatial components are removed (or the motion range is scaled down) and minima disappear. The price to pay is the lower accuracy of the estimate. However we only need an initial starting point that can be refined at full resolution (and that is close enough to true minimum to avoid traps). At a coarse image resolution the noise is suppressed, so the second factor contributing to the creation of traps is also removed.

Now we can formulate the minimisation (search) strategy. The input sequence is filtered and subsampled in order to construct a Gaussian pyramid [5]. Then, the motion parameters are estimated at successive resolutions in image and parameter space, as defined by the schedule presented in Figure 2. As one can notice, not all resolutions have to be explored in order to obtain a final estimate. Our comparative studies have shown that between the methods analysed (the presented method, referred to as steepest descent, conjugate gradient and multiresolution exhaustive search) the steepest descent is the most efficient. A sample of execution times is presented in Table 1.

The important feature of our approach is that it not restricted to a single moving object within the region of interest. This is achieved by adopting a multipass strategy. The principle is that once the motion parameters of the first object are estimated, the pixels that supported this motion are removed. This procedure (e.g. estimation and segmentation) is applied recursively to the remaining pixels until the majority of pixels are assigned to a region. Figure 3 presents an example in which three regions, moving with different translational motions, are segmented. The minima in the Hough Space corresponding to moving regions are subsequently uncovered (Subfigures (1a)(2a)(3a)). Subfigures (2a-2c) show the pixels remaining for processing.

The final segmentation is performed when all motion parameters are estimated. For each pixel p the likelihood $L_k(p)$ that it belongs to the object with label k is computed.

$$L_k(p) = exp\{-0.5[TFD_f^k(x, y) / \sigma]^2\} \qquad (6)$$

Table 1
Comparison of computational efficiency of search methods

	2D Space	4D Space	6D Space
Steepest Descent (Multires)	2.7	4.7	68
Conjugate Gradient	10.6	$-*$	$-*$
Exhaustive Search (multires)	47.0	430	6200

Execution time in seconds for SPARC 10 workstation for 256 square image.

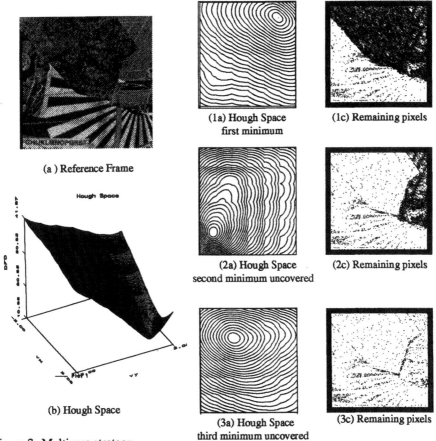

(a) Reference Frame

(b) Hough Space

(1a) Hough Space
first minimum

(1c) Remaining pixels

(2a) Hough Space
second minimum uncovered

(2c) Remaining pixels

(3a) Hough Space
third minimum uncovered

(3c) Remaining pixels

Figure 3. Multipass strategy

where TFD_f^k is a smoothed transformed frame difference computed using the motion vector $\vec{d}_k$ of the kth object. Gaussian smoothing is applied with a kernel size equal to 3 pixels. An additional region $k = 0$ absorbs any pixels with unresolved motion parameters (e.g. occluded or uncovered regions) with a constant likelihood value assigned to all pixels. The assumption is made that each pixel belongs to one of the regions, and the likelihood functions L_k are normalised so that their sum (including "unknown" region) is equal 1. Finally, each pixel is classified to the region with the highest probability.

Our framework also incorporates the estimate confidence factors. Since the complete estimate consists of motion segmentation map and a set of motion parameter vectors $\vec{a}$ there are two distinct confidence measures. The *segmentation confidence map* has the form of the image, where the pixel values reflect the probabilities of correct segmentation. The probabilities are taken from the segmentation procedure outlined above. The *motion estimate confidence factor* is defined for each distinct region, and has a form of covariance matrix the size of which depends on the motion model. The heuristics we use is that the accuracy of the estimate depends on the shape of the Hough Space around the minimum. In general, a deep minimum will give a more reliable estimate than a shallow one. The shape of the minimum is also important. For example when a translational motion model is used, a long deep valley extending along x axis gives a displacement estimate $\vec{v} = (v_x, v_y)$ with a high confidence in value v_y and low confidence in value v_x. The above heuristics can be supported by the following reasoning. In an ideal case, when noise is not present in the sequence, and quantisation and interpolation errors are omitted, the value of the Hough Space at the minimum should approach zero. The effect of the factors mentioned above on the Hough Space can be compared to an additive random noise that shifts the mismatch function upward and distorts its shape. This may result in displaced minima and biased estimates. The obvious measure of the sensitivity of the estimate is defined by the increase in value of the mismatch function H caused by a finite displacement from the minimum in one of the directions a_i. The larger the value of the increase, the less probable the displacement is. Thus, we may define the estimate covariance matrix as

$$C_M = \begin{vmatrix} c_{1,1} & \cdots & c_{1,n} \\ \cdots & \cdots & \cdots \\ c_{n,1} & \cdots & c_{n,n} \end{vmatrix} \qquad where \quad c_{i,j} = \frac{\delta H}{\delta a_i} \frac{\delta H}{\delta a_j} \qquad (7)$$

Note that in order to compute coefficients of the confidence matrix, it is not necessary explicitly to evaluate the shape of the function H. It is sufficient to take several samples in the neighbourhood of the detected minimum. Thus, this approach is in line with the fast strategy used to detect minima.

The presented approach may be implemented in a number of ways. Firstly, it can be applied to the entire image, as a global method. This approach could be used in a vision system that may use motion segmentation, in addition to motion estimate, for some high level processing. Secondly, it can be applied to regular blocks of image (e.g. 8 by 8) for example for the purpose of sequence coding. In that case the benefits over standard approaches include a fast search strategy and correct estimates even for blocks including motion boundaries. The block-based estimates can be used for motion segmentation or optic-flow estimation in a relaxation algorithm like the one presented in [3]. We also implemented and tested a variant of the presented method, introducing initial segmentation of the reference frame, based on the following strategy. Large areas with uniform grey-level or small variations form separate regions. The remaining well-structured area is arbitrary divided into a number of regions of a standard size. The algorithm is then applied to each region. Merging of the adjacent regions is performed on the basis of their motion estimates and their confidence factors. The benefit of this approach is that the number of moving objects within the analysed region is reduced. Wu [11] showed that multiple objects can result in displaced minima, and thus prevent the convergence. Pre-segmentation reduces the danger of that happening. However, the segmenting and merging process and the necessary book-keeping is computationally demanding, slowing down the algorithm substantially.

In this section we explained our approach to motion estimation and segmentation. Experimental results demonstrating its performance are provided in the next section.

3. EXPERIMENTAL RESULTS

Below we present examples of the experimental results for two applications.

Motion Segmentation Algorithm. In the sequence, two objects (a cheese box and sugar box) move on the table plane (zoom and translation). The camera is also moving, so the background is not stationary. The objective here is to segment the reference frame (Fig. 4(a)) into regions exhibiting coherent motions and estimate the motion parameters. In the preprocessing we segment the reference frame employing a relaxation based adaptive edge detection algorithm [2]. The segmentation image is displayed in Figure 4(b). All regions with the area smaller than several pixels are merged together and form so-called well-structured region. In the next stage, the Hough transform based algorithm is applied to each region in turn. The post-processing involves merging neighbouring regions with similar motion parameters and creating the final segmentation map (Fig. 4(c)). The segmentation confidence map is presented in Subfigure (d).

Tracking Algorithm. The algorithm performs well in applications involving detection of motion and tracking. It is used as a knowledge source in the Active Vision System developed

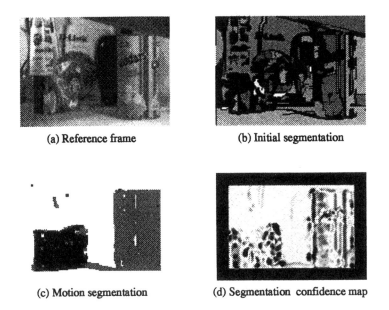

(a) Reference frame

(b) Initial segmentation

(c) Motion segmentation

(d) Segmentation confidence map

Example results for the "Lamp" sequence

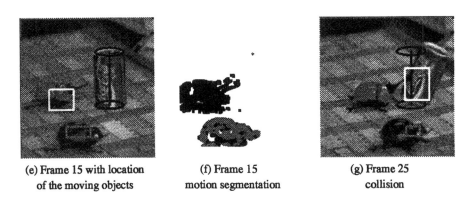

(e) Frame 15 with location
of the moving objects

(f) Frame 15
motion segmentation

(g) Frame 25
collision

Results of the tracking algorithm in the "Bugs" sequence

Figure 4. Experimental results:(a-d) Motion estimation and segmentation;(e-f) Tracking

at Surrey [1]. Figure 4(f) presents the 15 th frame from an experimental sequence called "Bugs". Three objects of interest are present in the scene, a stationary 'Coke' can and two mechanical toy-bugs. The task given to the system is to locate the 'coke' (the model of the can is present in the database) and to monitor all activities around. The black frame around 'coke' depicts the location established by the supervisor. Two light frames placed on the bugs show the estimated position of the moving objects detected by the algorithm. A three dimensional knowledge of the scene is used to reconstruct the 3D motion and decide which moving object may pose a danger to the can. The danger is depicted by white frame around the upper bug. A simplified segmentation algorithm was used, thus segmentation is not perfect 4(f). During experiments we found that the presented algorithm is very robust to change in illumination, and even to a non-rigid deformations of the target.

4. CONCLUSIONS

In the paper we presented a Hough Transform based motion estimation algorithm and demonstrated it on a range of applications. The method is very attractive since it is fast, and can cope with multiple objects and complex motions. The study of the behaviour of the support (Hough) function in the parameter space helped us to understand and overcome the convergence problems. As a result we proposed a hierarchical and multiresoluion minimisation procedure. We also defined confidence measures associated with the output. Finally, we proposed a version of the algorithm, performing initial segmentation based on grey-level distribution. The experimental results included motion segmentation and tracking applications. Further research will investigate multiresolution segmentation strategies and the possibility of a real time implementation of the algorithm on the parallel architecture "DATACUBE".

REFERENCES

1.   Adiv G. *Determining three-dimensional motion and structure from optical flow generated by several moving objects*, IEEE Trans. PAMI-7, pp. 384-401, 1985.

2.   Blake A. and Zisserman A. *Visual Reconstruction*, Cambridge, MA: MIT Pres 1987.

3.   Bober M. Kittler J. *General Motion Estimation and Segmentation. V Simposium Nacional de Reconocimiento de Formas y Analisis de Imagenes*, Valencia, 21-25 September 1992.

4.   Bober M. and Kittler J. *Estimation of Complex Multimodal Motion: An Approach based on Robust Statistics and Hough Transform*, British Machine Vision Conference 1993 ,Guildford, Sept. 1993.

5.   Burt P.J. and Adelson E. *The laplacian pyramid as a compact image code*, IEEE Trans. on Communications, 31,pp. 532-540,April 1983

6.   Campani M. Verri A. *Computing Optical Flow from an Overconstrained System of Linear Algebraic Equations*, ICCV-90, December 4-7 1990, Osaka, Japan, pp. 168-177.

7.   Horn B.K.P Schunck B. *Determining optical flow*, Artificial Intelligence , No 17 1981, pp.185-203.

8.   Singh A. *An Estimation-Theoretic Framework for Image-Flow Computation*, Proc. ICCV-90, December 4-7 1990, Osaka, Japan, pp. 168-177.

9.   Schunk B. G. *Image Flow: Fundamentals and Future Research*, Proc. IEEE Conf. on Comp. Vision and Pattern Rec., CVPR-85, June19-23 1985, San Francisco/CA pp 560-571.

10. Terzopoulos D. *Image Analysis Using Multigrid Relaxation Methods*, IEEE Trans. PAMI, vol. PAMI-8, No.2, March 1986 pp 129-139.

11. Wu F. *General Motion Estimation and Segmentation*, PhD thesis, Departament of Electronic and Electrical Engineering, University of Surrey, U.K.; November 1990.

---

[1]Partially sponsored by ESPRIT Project BRA 7108 "Vap II"

Time-Varying Image Processing and
Moving Object Recognition, 3 – V. Cappellini (Ed.)

Detecting a velocity field from sequential images under time-varying illumination

A.Nomura[a], H.Miike[a] and K.Koga[b]

[a]Department of Electrical and Electronic Engineering, Yamaguchi University, Tokiwadai 2557, 755 Ube, Japan

[b]Department of Computer Science and Systems Engineering, Yamaguchi University, Tokiwadai 2557, 755 Ube, Japan

Abstract

A generalized approach based on the gradient method is developed. Two algorithms are proposed to detect velocity fields from sequential images udner non-uniform (spatial or temporal) illumination. According to preliminary knowledges of the illumination characteristics and of the velocity fields, practical and accurate methods are tested by utilizing simulation images and real image sequences. The usefulness and reliability of the methods are confirmed.

1 Introduction

Detecting velocity vector fields from sequential images is important for computational vision, since it is useful for recovering 3-dimensional structure. In the research field of fluid physics the method measuring the velocity vector fields from sequential images of fluid flow visualized by particles is expected [2]. And it is important for meteorology to evaluate the velocity vector fields of cloud motions [3]. Several approaches detecting the velocity vector fields have been proposed. Most of all approaches are classified into correlation method [2,3] and gradient method [4,5]. These methods are based on the idea tracing a brightness pattern (point) during a small period.

On the other hand, two methods, which is different from the idea of the matching method and the gradient method, have been proposed. One of the methods is a spatio-temporal correlation method proposed by Miike et al. [6]. In the method, temporal brightness changes at a fixed area (each pixel site) are observed. The idea observing at the fixed area leads to a continuity equation of brightness. The other method [7] is based on the continuity equation. The equation can be regarded as an extended equation of the gradient method. Since both of the equations in the differential and integral forms contain a generation term of brightness, it can be expected that they are applied to the detection of the velocity fields under the non-uniform illumination. Time-varying brightness distribution infuluenced by non-uniform illumination is observed in the TV camera system which controls its gain automatically. The sunrise and sunset also cause temporally non-uniform illumination out of door. Incorrect velocity vectors are often obtained under these circumstances. Within a sphere of our knowlegde, however, there are few methods detecting velocity vectors correctly under non-uniform illumination.

In this paper, we propose two approaches detecting a 2-dimensional velocity field from sequential images under non-uniform illumination. These approaches are based on the continuity equation with the generation term of brightness and on preliminary knowledges of illumination. In traditional method considering non-uniform illumination [8] and our previous method, direct assumptions on the generation term are introduced to deal with the non-uniform illumination. In general, however, the generation term is expressed by a brightness distribution under uniform illumination, intensity distribution of the non-uniform illumination and a velocity vector (see Eq.10). Here, we propose two methods without any direct assumptions on the generation term of brightness but with the assumptions on the illumination and the velocity vector fields. One of the methods assumes spatially non-uniform illumination and a stationary velocity field. The other does temporally non-uniform illumination and local constancy of velocity vectors. Applying the methods to artificial and real image sequences, the usefulness of the methods is confirmed.

2 Theory

2.1 Basic constraint equations

Considering an analogy with the continuity equation of fluid, the following constraint equation is derived by observing a temporal change of total brightness in a local area δS,

$$\frac{\partial}{\partial t} \int_{\delta S} f ds = - \oint_{\delta C} f \boldsymbol{v} \cdot \boldsymbol{n} dc + \int_{\delta S} \phi ds, \tag{1}$$

where $f(x, y, t)$ is spatio-temporal brightness distribution of image sequence, δC is the contour surrounding δS, $\boldsymbol{v} = (u, v)$ is a velocity vector to be determined, $\boldsymbol{n}$ is a unit length normal vector pointing to outside of the δC and ϕ is rate of generation of brightness at a pixel in δS (Fig.1). The generation means increasing or decreasing brightness on an image plane under non-uniform illumination. In Eq.(1) the integral calculus along δC is transformed into integral calculus over δS by Gauss's divergence theorem. Then a differential formula of Eq.(1) is obtained as follows,

$$\frac{\partial f}{\partial t} = -f div(\boldsymbol{v}) - \boldsymbol{v} \cdot grad(f) + \phi. \tag{2}$$

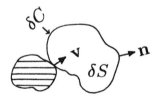

Figure 1: Definition of symbols for the constraint equation

2.2 Additional constraint equations

While Eq.(2) contains three unknown variables; $v = (u, v)$ and ϕ, we have only one constraint equation at a point in sequential images. Other two constraint equations on v and ϕ are necessary to solve the equation. Kearney et al. proposed the following additional constraint equation to obtain an instantaneous velocity field [5],

$$\frac{\partial v}{\partial x} = \frac{\partial v}{\partial y} = 0 \quad in \ a \ small \ area \ \delta S. \tag{3}$$

The method considering the above assumption is called the local optimization method (LOM). Recently the authors have proposed the following constraint equation to determine a stationary velocity field [7],

$$\frac{\partial v}{\partial t} = 0. \tag{4}$$

We call tentatively this approach as a temporal optimization method (TOM). On the parameter ϕ Cornelius et al. proposed a spatial smoothness constraint of ϕ [8]. But ϕ doesn't directly correspond to intensity of illumination as described in the following section. So we have to clear the relationship between the generation term of brightess ϕ and the intensity of the illumination and then we should make assumptions on the intensity.

2.3 Detecting velocity fields under non-uniform illumination

In this paper, the following equation is utilized for detecting a velocity field under non-uniform illumination.

$$\frac{\partial f}{\partial t} + v \cdot grad(f) = \phi \tag{5}$$

Equation (5) is obtained from Eq.(2) under the assumption $div(v) = 0$. Let's $f(x, y, t)$ be actual brightness distribution observed through a camera system under the non-uniform illumination and $g(x, y, t)$ be virtual brightness distribution under uniform illumination. A relationship between $f(x, y, t)$ and $g(x, y, t)$ is expressed with the effect of the non-uniform illumination $r(x, y, t)$, as

$$f(x, y, t) = r(x, y, t) \cdot g(x, y, t). \tag{6}$$

Substituting Eq.(6) for the $f(x, y, t)$ in Eq.(5), we get

$$\frac{\partial(rg)}{\partial t} + v \cdot grad(rg) = \phi. \tag{7}$$

Expanding the left side of Eq.(7) for rg, the following equation is reduced,

$$g\{\frac{\partial r}{\partial t} + v \cdot grad(r)\} + r\{\frac{\partial g}{\partial t} + v \cdot grad(g)\} = \phi. \tag{8}$$

Here we assume that $g(x, y, t)$ is constrained by the equation;

$$\frac{\partial g}{\partial t} + v \cdot grad(g) = 0. \tag{9}$$

That is to say, it is supposed that there is no occluding boundary and no objets movement parallel to a camera's optical axis in sequential images. Then we obtain the relationship;

$$g\{\frac{\partial r}{\partial t} + v \cdot grad(r)\} = \phi. \tag{10}$$

It is clear that ϕ depends on $g(x, y, t)$, $r(x, y, t)$ and v. If we want to determine high-accurate velocity vector fields under non-uniform illumination, we should not make any assumption on ϕ directly but make some assumptions on $r(x, y, t)$ and v separately.

As the first circumstance, we assume that the illumination is spatially non-uniform and constant with respect to time; $r = r(x, y)$. From Eq.(10) and $\partial r / \partial t = 0$, ϕ is expressed as,

$$\phi = g\{v \cdot grad(r)\} = f\frac{v \cdot grad(r)}{r}. \tag{11}$$

With the idea of TOM in Eq.(4), the fraction of the right side of Eq.(11) is also constant with respect to time. If the constraint in Eq.(4) is satisfied from the time: $t = t_0$ to the time: $t = t_0 + n$ for a digitized image sequence and a symbol a is used as an unknown constant, a velocity vector v is detected by minimizing the following error function with the least square method.

$$E = \sum_{t=t_0}^{t_0+n} (f_t + uf_x + vf_y - fa)^2. \ (a \stackrel{\text{def}}{=} \frac{v \cdot grad(r)}{r}) \tag{12}$$

As the second circumstance, let's consider the case of temporally non-uniform illumination. We assume that r is constant with respect to space; $r = r(t)$. From Eq.(10) and $grad(r) = 0$, we get

$$\phi = g\frac{\partial r}{\partial t} = f\frac{\partial r / \partial t}{r}. \tag{13}$$

With the idea of LOM in Eq.(3) the fraction of the right hand side of Eq.(13) is regarded as constant with respect to space. If a symbol b is used as an unknown constant, the following error function can be defined.

$$E = \sum_{x=x_0-l}^{x_0+l} \sum_{y=y_0-m}^{y_0+m} (f_t + uf_x + vf_y - fb)^2. \ (b \stackrel{\text{def}}{=} \frac{\partial r / \partial t}{r}) \tag{14}$$

Here, l and m represent x and y width of the local neighborhood; δS. In this paper, both of them are set to 3 pixels. Consequently the local neighborhood has an area of 7×7 pixels. A velocity vector v is determined by minimizing Eq.(14) with the least square method.

3　Simulation experiments

Artificial sequential images are created from an image shown in Fig.2 which is acquired as a static real image through a TV-camera. The static image has the resolution of 120 $\times$ 120 pixels. Brightness is quantized into 256 steps. Error of a detected velocity vector field is evaluated by

$$E = \frac{1}{I \cdot J} \sum_{i=1}^{I} \sum_{j=1}^{J} \frac{\|v_t(i,j) - v(i,j)\|}{\|v_t(i,j)\|} \times 100(\%), \tag{15}$$

where i and j represent x and y coordinates respectively, $v_t(i,j)$ is a given(true) velocity vector, $v(i,j)$ is the calculated velocity vector and I and J represent x and y width of an image. As typical examples velocity fields are detected from two kinds of artificial sequential images as follows.

The first example of the sequential images $f_1(x,y,t)$ simulating spatially non-uniform illumination is created by the equation;

$$f_1(x,y,t) = \alpha \cdot xy \cdot g_1(x,y,t), \tag{16}$$

where $\alpha = 1.416 \times 10^{-4}$ and $g_1(x,y,t)$ is obtained by rotating the static image (see Fig.2) clockwise around its center axis with constant angular velocity (1 degree/frame). The size of the sequential images is 84×84 pixel and 64 frames. The term $\alpha \cdot xy$ represents the effect of spatially non-uniform illumination. At $(x = 0, y = 0)$ the value of the term is zero and at $(x = 84, y = 84)$ the value becomes to 1.0.

The traditional TOM [7], which has the assumptions of uniform illumination and of a stationary velocity field, and the proposed TOM based on Eq.(12) are applied to detect velocity fields from the sequential images $f_1(x,y,t)$. The error of a velocity field detected by the traditional method is evaluated as 36.8(%). The error of a velocity field detected by the proposed method is evaluated as 23.2(%). These results show a superiority of the proposed method in accuracy.

The second example of the sequential images $f_2(x,y,t)$ simulating temporally non-uniform illumination is created by the equation;

$$f_2(x,y,t) = (\zeta t + \eta) \cdot g_2(x,y,t), \tag{17}$$

where $\zeta = 5.0 \times 10^{-2}$, $\eta = 0.2$ and $g_2(x,y,t)$ is obtained by translating the static image (see Fig.2) at a constant velocity vector $v = (0.7, 0.8)$ pixel/frame. The size of the sequential images is 84×84 pixels and 16 frames. The term $(\zeta t + \eta)$ increases linearly with time. At $t = 0$ the term is 0.2 and at $t = 16$ the term becomes to 1.0. The traditional LOM [5], which has the assumptions of uniform illumination and of local constancy of velocity vectors, and the proposed LOM based on Eq.(14) are applied to detect velocity fields from the sequential images $f_2(x,y,t)$. The velocity fields are averaged with respect to time. The error of a velocity field detected by the traditional method is evaluated as 90.8(%). On the other hand the error of a velocity field detected by the proposed method is evaluated as 13.7(%). According to these error evaluations, it is confirmed that the proposed method is more reliable than the traditional one.

Figure 2: A static image

348

4 Actual scene analysis

In order to confirm the usefulness of the proposed TOM, we pick up sequential images of a real flow field affected by spatially non-uniform illumination as an example. The flow field is visualized by polystyrene particles and a slit-light illumination. A dynamic scene of the visualized flow is acquired into a computer system through a TV camera with sampling frequency of 30 Hz. The size of the acquired images is 128×128 pixels and 64 frames. Brightness is quantized into 256 steps. Figures 3 (a) and (b) show the first and the 64th frames of the images. A cylinder is located at the upper center in the images. The flow

direction is from right to left in the images. The velocity field is stationary except for the lower stream of the cylinder. Spatially non-uniform illumination is observed just under the cylinder. A velocity field detected by the traditional TOM is shown in Fig.4, and a velocity field detected by the proposed TOM is shown in Fig.5. Note that the difference of the fields can be seen at under the cylinder. Fig.4 incorrect velocity vectors affected by spatially non-uniform illumination are detected under the cylinder, however in Fig.5 velocity vectors are detected correctly.

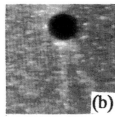

Figure 3: Real sequential images of fluid flow under spatially non-uniform illumination. (a)1st image and (b)64th image.

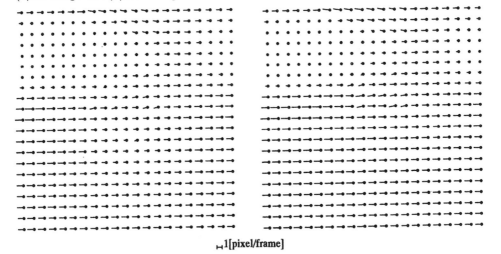

$\vdash\dashv$1[pixel/frame]

Figure 4: A velocity field detected by the traditional TOM from the real sequential images of fluid flow (see Fig.8).

Figure 5: A velocity field detected by the proposed TOM from the real sequential images of fluid flow (see Fig.8).

In order to confirm the usefulness of the proposed LOM, we took sequential images of cloud motions at the evening through a TV camera with sampling frequency of 2(Hz) as an example of temporally non-uniform illumination. The size of the images is 160 × 64 pixels and 32 frames. Brightness is quantized into 256 steps. The 1st and 32nd image are shown in Fig.6 (a) and (b), respectively. The sunset brings temporally non-uniform illumination. According to our judgment by naked eyes, the cloud seems to move toward the top of the left end with a constant velocity. Figures 7 and 8 show the velocity fields by the traditional LOM and by the proposed LOM, respectively. Both velocity fields are averaged with respect to time. The field detected by the proposed LOM is consistent with the judgment; most of the velocity vectors are uniform, and they indicate same direction from the bottom of the right side to the top of the left side in the field. While, the field detected by the traditional LOM includes several incorrect velocity vectors which is apparently different from the judgment.

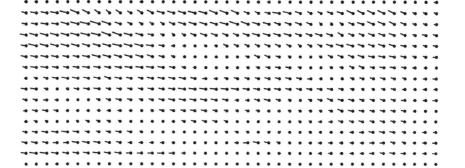

Figure 6: Real sequential images of cloud motions in the evening sky. (a)1st image and (b)32nd image.

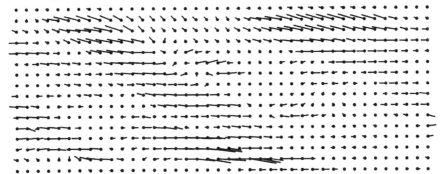

Figure 7: A velocity field detected by the traditional LOM from the real sequential images of cloud motions (see Fig.11).   ⊢⎯⎯⊣1[pixel/frame]

Figure 8: A velocity field detected by the proposed LOM from the real sequential images of cloud motions (see Fig.11).   ⊢⎯⎯⊣1[pixel/frame]

5 Conclusion

Two methods detecting velocity fields from sequential images under spatially or temporally non-uniform illumination were proposed. The methods are based on the constraint equation which is obtained by observing total brightness change in a fixed local closed area. One of the methods assumes the spatially non-uniform illumination and a stational velocity field. The other one assumes temporally non-uniform illumination and local constancy of velocity vectors. The usefulness of the proposed methods is confirmed by applying them for detection of velocity fields from artificial and real sequential images. If we have preliminary knowledges such as spatially or temporally non-uniform illumination on the images, the proposed methods act as a reliable tool for detecting velocity fields.

Acknowlegements

The sequential images of the fluid flow are offered by Dr.H.Yamada in Yamaguchi University. We thank him for offering the images and his students for the experimental help.

REFERENCES

1. K.Kanatani, Computer Vision, Graphics, and Image Processing, **35** (1986) 181.

2. S.E.Englert, Z.Sheng and R.L.Kirlin, Pattern Recognition, Vol.23, No.3/4 (1990) 237.

3. J.A.Leese, C.S.Novak and V.R.Taylor, Pattern Recognition, Vol.2 (1970) 279.

4. B.K.P.Horn and B.G.Schunck, Artificial Intelligence, 17 (1981) 185.

5. J.K.Kearney, W.B.Thompson and D.L.Boley, IEEE Trans. Pattern Anal. Mach. Intell., VOL.PMAI-9, No.2 (1987) 229.

6. H.Miike, Y.Kurihara, H.Hashimoto and K.Koga, Trans. of the Institute of Electronics, Information and Communication Engineers, VOL.E 69, No.8 (1986) 877.

7. A.Nomura, H.Miike and K.Koga, Pattern Recognition Letters, 12 (1991) 183.

8. N.Cornelius and T.Kanade, Proc. of ACM SIGGRAPH / SIGGART Interdisciplinary Workshop Motion: Representation and Perception, Toronto / Canada (April 4-6, 1983)

Time-Varying Image Processing and
Moving Object Recognition, 3 – V. Cappellini (Ed.)
351

Optical Flow Estimation by Using Classical and Extended Constraints

A. Del Bimbo[†], P. Nesi[†*], J. L. C. Sanz[!*]

[†]Dipartimento di Sistemi e Informatica, Università di Firenze, Firenze, Italy.
[*]Computer Science Department, IBM Almaden Research Center, California, USA.
[!]Computer Research and Advanced Applications Group, IBM Argentina.

1 Introduction

Motion analysis from image sequences addresses the estimation of the relative movements between the objects in the scene and the TV-camera. One of the most important approaches for motion estimation is the gradient-based. This is based on the observation of image brightness changes, and leads to the estimation of motion of image brightness features [1], [2], [3], [4], [5]. The flow field of these features is commonly referred to as "optical flow". Generally speaking, the optical flow differs from the perspective projection of the 3-D motion on the image plane [6], [7], [8]. Nevertheless, the optical flow is a good approximation of the true motion in the image sequence.

Many approaches are available in the literature for optical flow estimation. Typically, they use the so-called Optical Flow Constraint (OFC). More recently, a different constraint equation has been introduced, which also contains the divergence of the flow field of the image brightness features [9]. In the following, this equation will be referred to as Extended Optical Flow Constraint (EOFC). EOFC has been analyzed by a few authors [6], [8], [5]. There is not an agreement in the literature about which of these two constraints provides the best estimation of the velocity field. Two main approaches for the computation of optical flow can be found in the literature: *regularization-* [1] and *multiconstraint-based* [2], [3], [4], [10], approaches.

In this paper, two solutions for the estimation of optical flow are proposed, which are based on an approximation of the partial differential equation modeling the changes in the image brightness features (see Section 3). A comparison is carried out with some gradient-based selected solutions previously presented in the literature (see Section 4).

2 Basic Notions

If the image brightness of each point in the image is supposed to be stationary with respect to the time variable (i.e., $dE/dt = 0$), then the following expression holds:
$$E_x u + E_y v + E_t = 0, \tag{1}$$
where the abbreviation for partial derivatives of the image brightness has been introduced, and u, v correspond to dx/dt, dy/dt, representing the components of the local velocity vector $\mathbf{V}$ along the x and y directions, respectively. Equation (1) is usually called: "Optical

Flow Constraint" (OFC), and its solutions are referred to as "optical flow". Optical flow is thus defined as the field of image gray value pattern displacements.

According to the observation that the OFC is very similar to the Continuity Equation of Fluid Dynamics, a more general motion constraint equation, relating image brightness derivatives and components of the local velocity vector and its derivatives, was presented by Schunck in [9]:

$$E_x u + E_y v + E u_x + E v_y + E_t = 0; \qquad (2)$$

it will be referred to as Extended Optical Flow Constraint (EOFC) in the rest of this paper, where the adopted notation recalls that of OFC. From the structural point of view, the EOFC equation (2) differs from the OFC equation (1) only in the term involving the divergence of the optical flow field vector $(E \nabla \cdot \mathbf{V})$. A derivation for the EOFC was presented by Nagel in [8], by choosing the same approach as Schunck [6], [11].

In general, the estimation of the optical flow suffers from two main drawbacks. The first consists of the discontinuities in the local velocity, due to image brightness discontinuities which are originated by the presence of noise, patterns, and/or occlusions between moving objects, etc. The second problem is the so-called "problem of aperture" which also exists in the human vision. This is related to the impossibility of univocally recovering the direction of motion if the object is observed through an aperture which is smaller than the object size. In this case, the reference elements of the object under observation (such as patterns) are not enough to perceive the transversal component of the object motion, and only the component of the apparent velocity, which is parallel to ∇E, can be perceived and estimated:

$$\mathbf{V}_\perp = -\frac{E_t}{\|\nabla E\|} \frac{\nabla E}{\|\nabla E\|}, \qquad (3)$$

where $dE/dt = 0$ and $\|\nabla E\| \neq 0$ have been assumed.

3 New Techniques Based on EOFC and OFC

In this Section, two solutions for optical flow estimation based on first-order approximations of OFC and EOFC equations are presented.

3.1 EOFC-based solutions

Let $\mathbf{p} = (x, y)$ be the perspective projection on the image plane of a point P in the 3-D space. Assume that the optical flow changes following a law which is approximately linear with $\mathbf{p}$. A linear approximation of the EOFC model around the point under consideration $\mathbf{p}_o$ can be obtained by considering that the changes of every element of the EOFC equation have at most a linear dependence on $d\mathbf{p}$ in each location $(\mathbf{p}_o + d\mathbf{p})$ at time t. Therefore, the EOFC can be rewritten as:

$$E_t(\mathbf{p}_o + d\mathbf{p}, t) + \nabla E(\mathbf{p}_o + d\mathbf{p}, t) \cdot \mathbf{V}(\mathbf{p}_o + d\mathbf{p}, t) + E(\mathbf{p}_o + d\mathbf{p}, t)(u_x(\mathbf{p}_o + d\mathbf{p}, t) + v_y(\mathbf{p}_o + d\mathbf{p}, t)) = 0. \qquad (4)$$

By using the first-order approximation of every term, and considering the components up to the second-order in dx, dy, equation (4) can be written as:

$$(E_t + E_x u + E_y v + E u_x + E v_y) +$$
$$(E_{tx} + E_{xx} u + E_{yx} v + 2E_x u_x + E_y v_x + E_x v_y + E u_{xx} + E v_{yx}) dx +$$
$$(E_{ty} + E_{xy} u + E_{yy} v + E_y u_x + E_x u_y + 2E_y v_y + E u_{xy} + E v_{yy}) dy = 0. \qquad (5)$$

Numerical solutions are obtained in the discrete domain, since images are sampled on a fixed grid of points at regular time intervals. If the optical flow follows a law which is approximately linear in (x, y), a smoothed solution for optical flow estimation can be obtained by using a linear approximation of the constraint in the $N \times N$ neighborhood of each point on the grid.

A first-order solution is obtained assuming that (5) vanishes for all the $d\mathbf{p}$ components. Moreover, neglecting the second-order derivatives of the velocity field $(u_{xx}, v_{yx}, u_{xy}, v_{yy})$, the following system of equations for each pixel is obtained:

$$E_t + E_x u + E_y v + E u_x + E v_y = 0,$$
$$E_{tx} + E_{xx} u + E_{yx} v + 2E_x u_x + E_y v_x + E_x v_y = 0, \qquad (6)$$
$$E_{ty} + E_{xy} u + E_{yy} v + E_y u_x + E_x u_y + 2E_y v_y = 0.$$

These equations are used to build an over-determined system of $3(N \times N)$ equations in 6 unknowns $(u, v, u_x, v_y, u_y, v_x)$. The over-determined system is solved by using the least-squares technique and LU decomposition.

3.2 OFC-based solutions

The OFC-based solutions have been obtained by following the same approach adopted for the EOFC-based solutions. In this case, the constraint equation (5) takes the form:

$$(E_t + E_x u + E_y v) +$$
$$(E_{tx} + E_{xx} u + E_{yx} v + E_x u_x + E_y v_x)\, dx +$$
$$(E_{ty} + E_{xy} u + E_{yy} v + E_x u_y + E_y v_y)\, dy = 0. \qquad (7)$$

The first-order OFC-based solution is obtained assuming that (7) vanishes for all the $d\mathbf{p}$ components. If this condition is satisfied, three constraint equations for each pixel will be defined:

$$E_t + E_x u + E_y v = 0,$$
$$E_{tx} + E_{xx} u + E_{yx} v = 0, \qquad (8)$$
$$E_{ty} + E_{xy} u + E_{yy} v = 0,$$

where the first-order derivatives of the optical flow field (u_x, u_y, v_x, v_y) have been neglected. These equations are used to define a multipoint solution of $3(N \times N)$ equations on 2 unknowns. This solution is similar to the second solution presented by Tretiak and Pastor in [3]. However, in the present approach the three constraint equations are used in the neighborhood $(N \times N)$ of each pixel, thus defining a multipoint solution, while in [3] a system of 3 equations in 2 unknowns is solved for every pixel.

4 Experimental Comparisons

In this section, a comparison of the new solutions with respect to the OFC-solutions proposed by Horn and Schunck [1], Haralick and Lee [2], and the first solution by Tretiak and Pastor [3] is proposed. In order to highlight the robustness of each technique, solutions are obtained without any specific method to improve optical flow quality such as filtering of the image sequence (for smoothing discontinuities), thresholding of the optical flow field magnitude (for trimming discontinuities), or using a large neighborhood around the pixel under analysis.

The structure of the OFC equation and the presence of discontinuities make the problem of optical flow estimation ill-posed. Discontinuities arise from the presence of noise, too crisp patterns on the moving object surfaces, occlusions between moving objects or among the moving objects and the background, and too fast object velocities with respect to the measurement system. These difficulties can be overcome (or simply attenuated) by convolving the image with a 2-D or 3-D Gaussian smoothing operator.

As a general consideration, the approaches based on second-order derivatives of the image brightness (especially derivatives with respect to time t) are very sensitive to discontinuities [2], [3]. The solutions presented in this paper use an over-determined system of equations to reduce the effects of discontinuities, so that a final filtering step is not needed. The smoothness of the solution can be improved by augmenting the size of the $N \times N$ neighborhood area, even though at the expense of loss of resolution of object boundaries. The new algorithms are less sensitive to the discontinuities than the others used for comparison with the same neighborhood dimension, N.

4.1 Object occlusions

The synthetic sequence shown in Fig.1, where two objects with a superimposed plaid pattern move in different directions, was chosen to test the behavior of the solutions in the presence of discontinuities due to occlusions. The estimation results are shown in Fig.2. For the new solutions, the presence of some optical flow field irregularities on the object boundaries can be observed, even though object profiles are still maintained. The first-order EOFC-based solution provides a better estimation with respect to the other multiconstraint solutions. Approaches by Haralick and Lee, and Tretiak and Pastor produce large estimation errors at the occlusion profiles. The approach by Horn and Schunck produces less satisfactory results due to the undesirable effect of optical flow field propagation on the occluded objects.

 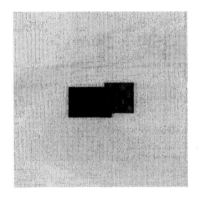

1   8

Figure 1: Sequence of images where two objects with a superimposed plaid pattern move in opposite directions, (180 and 45 degrees with respect to the X-axis, respectively), (1st, and 8th frame, 128×128 image resolution).

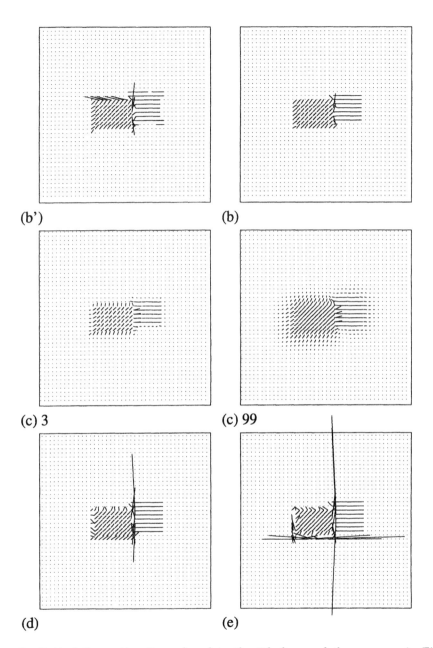

Figure 2: Optical flow estimation referred to the 5th frame of the sequence in Fig.1 obtained with: (b) First-order EOFC-based solution, $N = 3$; (b') First-order OFC-based solution, $N = 3$; (c) Horn and Schunck OFC-based solution (iterations: 3, 99), $\alpha = 0.6$, 3×3 final averaging; (d) Haralick and Lee multiconstraint OFC-based solution, 3×3 final averaging; (e) Tretiak and Pastor OFC-based direct solution, 3×3 final averaging.

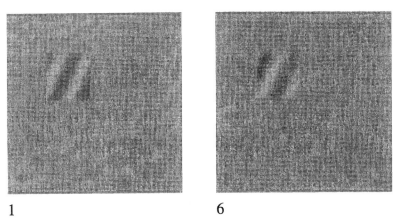

1 6

Figure 3: Sequence of images where a transversal sinusoidal pattern moves at 180 degrees through an aperture (1st and 6th frame, 128 × 128 image resolution).

4.2 The Problem of Aperture

As previously noticed, to univocally estimate optical flow several constraint equations must be defined. In our comparison, two distinct cases are considered. In the first case (Tretiak and Pastor solution) a determined system of equations is used, while in all the other cases under examination an over-determined system is employed. The solution of Tretiak and Pastor estimates velocity vectors only in those point where the Gaussian curvature is different from zero. All other techniques led to some estimation independently of the Gaussian curvature value. It should be noted that, a Gaussian curvature close to zero denotes the absence of enough references to perceive the transversal component of motion, thus leading to the problem of aperture. Therefore, the Gaussian curvature can be assumed as a measure of the ill-conditioned nature of the optical flow estimation problem. In the following, a more appropriate test sequence (very similar to the famous "barber pole" sequence) has been used (see Fig.3), in which a transversal sinusoidal pattern moves in translational motion. In each point of the aperture, the expected curvature of this pattern is equal to zero, while the second-order partial derivatives of the image brightness are different from zero. Moreover, the Gaussian curvature is equal zero, even if central numerical approximations to estimate second-order partial derivatives of the image brightness are used. Results obtained are presented in Fig.4.

It can be observed that all the multiconstraint solutions, except the first solution by Tretiak and Pastor [3], provide some estimations of the optical flow. These fields cannot be assumed as valid, since they are estimated in the presence of the problem of aperture, but they correspond to the estimation given by humans in the same conditions. The proposed solutions and the Haralick and Lee's give quite similar results. A less satisfactory behavior on the aperture boundaries can be observed with the Haralick and Lee solution. The first solution of Tretiak and Pastor estimates the optical flow only in few points in the image, corresponding to the aperture boundaries where the Gaussian curvature is different to

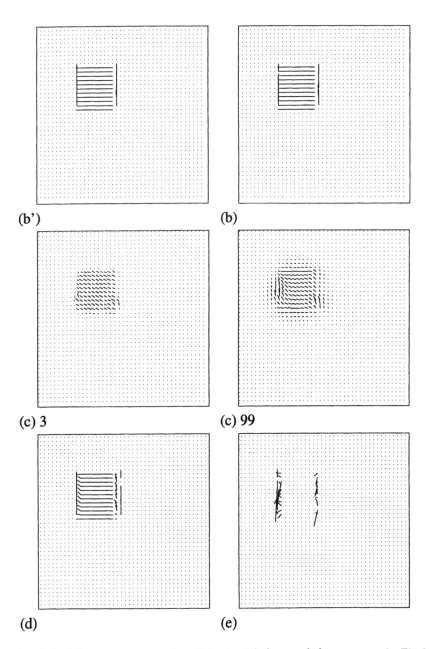

(b')   (b)

(c) 3   (c) 99

(d)   (e)

Figure 4: Optical flow estimation referred to the 5th frame of the sequence in Fig.3 (in the presence of the problem of aperture) obtained with: (b) First-order EOFC-based solution, $N = 3$; (b') First-order OFC-based solution, $N = 3$; (c) Horn and Schunck OFC-based solution (iterations: 3, 99), $\alpha = 0.6$, 3×3 final averaging; (d) Haralick and Lee multiconstraint OFC-based solution, 3×3 final averaging; (e) Tretiak and Pastor OFC-based direct solution, 3×3 final averaging.

zero due to the image brightness irregularities (see Fig.4e).

In the early iterations, the regularization-based solution gives an optical flow which is parallel to the spatial gradient ∇E, according to equation (3). After a large number of iterations the estimation leads to incorrect results with respect to the actual motion of the pattern inside the aperture (see Fig.4c).

5   Conclusions

In this paper, two new solutions for the estimation of optical flow fields have been proposed. Results achieved have been compared with selected solutions available in the literature. As was shown, the proposed solutions are better ranked with respect to the other selected solutions in almost all motions analyzed. Several differences have been observed between the solution using EOFC and that using OFC. The EOFC-based solution appears to achieve a better performance in the presence of discontinuities.

References

1. B. K. P. Horn, B. G. Schunck. Determining optical flow. Artificial Intelligence, 17, 1981.
2. R. M. Haralick, J. S. Lee. The facet approach to optical flow, (L. S. Baumann, ed.), Image Understanding Workshop (Science Applications Arlington), 1983.
3. O. Tretiak, L. Pastor. Velocity estimation from image sequences with second order differential operators, 7th IEEE ICPR 1984.
4. A. Verri, F. Girosi, V. Torre. Differential techniques for optical flow. J. Opt. Soc. Am. A, 7(5), 1990.
5. P. Nesi, A. Del Bimbo, J. L. C. Sanz. Multiconstraints-based optical flow estimation and segmentation, International Workshop on Computer Architecture for Machine Perception, Paris, 1991.
6. B. G. Schunck. Image flow: Fundamentals and future research, IEEE CVPR'85, San Francisco, Ca, USA, 1985.
7. A. Verri, T. Poggio. Motion field and optical flow: Qualitative properties, IEEE Trans. on Patt. Ana. and Mac. Intel., 11(5), 1989.
8. H.-H. Nagel. On a constraint equation for the estimation of displacement rates in image sequences, IEEE Trans. on Patt. Ana. and Mac. Intel., 11(1), 1989.
9. B. G. Schunck. The motion constraints equation for optical flow, 7th IEEE ICPR 1984.
10. M. Campani, A. Verri. Computing optical flow from an overconstrained system of linear algebraic equations, 3rd IEEE ICCV'90, Osaka, Japan, 1990.
11. B. G. Schunck. Image flow continuity equations for motion and density, Workshop Motion: Representation and Analysis, Charleston, USA, 1986.

Time-Varying Image Processing and
Moving Object Recognition, 3 – V. Cappellini (Ed.)
© 1994 Elsevier Science B.V. All rights reserved.

Enhancement of local optic flow techniques

F. Bartolini [a], V. Cappellini [b], C. Colombo [c] and A. Mecocci [d]

[a]Dipartimento di Ingegneria Elettronica, Via S. Marta 3, 50139 Firenze, Italy

[b]Dipartimento di Ingegneria Elettronica, Via S. Marta 3, 50139 Firenze, Italy

[c]ARTS Lab, Scuola Superiore S. Anna, Via G. Carducci 40, 56127 Pisa, Italy

[d]Dipartimento di Elettronica, Via Abbiategrasso 209, 27100 Pavia, Italy

In this paper an algorithm that enhances the performance of local optic flow techniques at motion boundaries is presented. Velocity vectors are computed embedding the local technique into a multiwindow superstructure. Any local technique providing an *a posteriori* measure of the reliability of the estimate could be used; least squares is chosen here due to its simplicity. Vector median filtering is used to regularize the field. Results on both synthetic and real-world sequences are shown, and a comparison with classical least squares is made. The algorithm is fast, non-iterative and non-parametric and demonstrates to have a nice behavior in terms of both noise rejection and motion boundary preservation.

1. INTRODUCTION

Optic flow fields are produced in the image plane due to the relative motion of the scene objects with respect to the camera, thereby encoding the geometry and the kinematics of the imaged scene, and also embedding the temporal correlation between successive image frames. Estimating optic flow fields is thus useful for both image compression and understanding.

The 2D piecewise continuous structure of optic flow fields reflects the nature of the 3D velocity fields of the world objects. Specifically, the points of discontinuity of the field— also known as *motion boundaries*—correspond to the points of occlusion among different moving objects, and provide a powerful cue for their spatial location and discrimination ("motion-based segmentation"). Unfortunately, the accurate evaluation of the optic flow field at motion boundaries is found to be a serious problem, in that usually in the neighborhood of these points the estimated field is either noisy or incorrectly smooth. This clearly emerges from the mathematical characterization of optic flows. The classical optic flow constraint equation [1]:

$$E_x u + E_y v + E_t = 0 \tag{1}$$

—$E(x, y, t)$ is the image brightness, (u, v) are the unknown components of the flow and subscripts stand for differentiation—is obviously not sufficient to determine (u, v) univocally; optic flow computation by this equation is thus an underconstrained problem, whose solution needs the introduction of additional constraints. This can be done by imposing optic flow to be either *globally* [1], *piecewise globally* [2–4] or or *locally* [5–8] *smooth*.

Global assumptions usually lead to an iterative solution based either on variational or bayesian estimation principles. The imposition of smoothness constraints over the whole image space introduces a correlation among the field vectors that has not any physical reason across motion boundaries; therefore variational techniques usually produce an incorrect flattening of the computed field. Although the more recent bayesian techniques explicitly include the concept of motion boundaries into the model, they nevertheless have the disadvantage of being parametric and computationally demanding.

Conversely, local techniques are known to be simple and fast. They exploit the information from a small neighborhood around the examined pixel, either by collecting constraints of neighboring points and solving the resulting overconstrained set of linear equations, or by determining at each pixel more than one constraint. Local techniques do not impose any smoothness over large patches of the image, thus offering also the advantage of eliminating undesirable flattening effects. The main drawback of such techniques is *ill-conditioning* in certain regions of the image plane—e.g. regions of approximately uniform brightness or where spatial image gradient information is nearly unidirectional— where a quite noisy flow field is evaluated, thus creating the need of a post-processing regularization step [8, 9]. It is worth noting that regularizing optic flow fields is not an easy task because field discontinuities have to be preserved; the risk is to introduce *a posteriori* that undesired flattening of motion boundaries which global techniques introduce *a priori* in the model.

In this paper a multiwindow approach to enhance the performance of local techniques in terms of motion boundary preservation is presented. The algorithm exploits the characteristics of many local techniques of providing an *a posteriori* measure of the reliability of the solution. Special attention is given to processing speed with the development of a fast computational architecture. The flow regularization is accomplished by means of a nonlinear technique, a fast approximation of a vector median filter [10]. In the following the idea behind the approach and the nonlinear regularization filter are described, the fast implementation is then discussed, and finally conclusions and directions for future research are outlined.

2. THE MULTIWINDOW APPROACH

Local techniques assume a locally continuous optic flow model. In practical implementations, the optic flow is considered either constant or linear (affine) in the small neighborhood of each pixel where information is collected. However, the continuity assumption is going to fail every time the examined neighborhood is crossed by a motion boundary, that is when information inside it is relative to different moving objects. Avoiding errors in the model would imply to suitably shape the neighborhood in order to grant that information is gathered fromy one object only; unfortunately, such a strategy could not be realized in the context of motion-based segmentation, where optic flow information is exploited to distinguish one object from the other.

A similar problem is encountered in average filtering for image processing, where it is needed to distinguish between the variations of image brightness due to noise and those due to discontinuities. Nagao and Matsuyama proposed in 1979 a solution to this problem, in which a square neighborhood around the pixel is partitioned in a small number of subwindows so that at least one of them entirely belongs to the same uniform region as the center of the window [11].

In this paper it is argued that a similar *multiwindow approach* can be used as a *computational superstructure* and applied to a suitable local technique to compute optic flow preserving motion boundaries. An $N \times N$ neighborhood—$N = 4k+1$, k positive integer— of the generic pixel is partitioned in a set of nine $(N+1)/2 \times (N+1)/2$ overlapping asymmetrical subwindows (in Figure 1 the case for $k = 2$ is shown). In each subwindow, a local technique is used to estimate optic flow, provided that it can produce an *a posteriori* measure of the reliability of the result. The (nine) different solutions obtained are then compared on the basis of the reliability measure, and the one that maximizes it is chosen as the optic flow for the center of the neighborhood. The idea here is that if the examined pixel is at the border of an object, it is likely that at least one of the subwindows is entirely inside of the object itself; that subwindow will then almost surely have the maximum reliability measure. The performance at pixels far from motion boundaries is also improved, in that a solution from a selected and more reliable subset of neighboring points is derived.

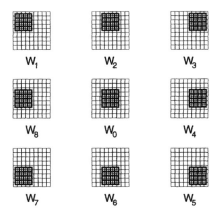

Figure 1. A 9×9 neighborhood is split into nine 5×5 subwindows.

In the actual implementation, a local technique based on a least squares computation (as described for example in [5]) is applied to the nine subwindows W_i, $i = 0, \ldots 8$. This technique assumes a locally uniform flow, and gets a solution (u_i, v_i) by minimizing the error in the set of linear constraints 1 in the whole subwindow:

$$\mathcal{E}(u,v) = \sum_{W_i} [E_x u + E_y v + E_t]^2. \tag{2}$$

As a measure of the reliability of the solution the inverse of the *residual error* $\mathcal{E}(u_i, v_i)$ is taken. Least squares are attractive for being quite noise insensitive and fast. However, if applied to the whole neighborhood, the least squares technique would produce an unsatisfactory flow field at motion boundaries. The use of the multiwindow approach can maintain the good properties of least squares and eliminate its poor edge behavior. (Further implementation details can be found in [12].)

3. VECTOR MEDIAN REGULARIZATION

As mentioned above, some gross errors result in certain image regions due to ill-conditioning. The computed field appears quite irregular also due to image noise. It is therefore needed to perform an optic flow *regularization step*.

For this purpose, nonlinear filtering is used in the place of the traditional averaging techniques, thus achieving a better edge response. Specifically, the regularized optic flow vector is computed as the *vector median* of the optic flow vectors $\mathbf{v} = u\mathbf{i} + v\mathbf{j}$ in a suitable neighborhood of each pixel. The median of a vector set $\mathcal{V} = \{\mathbf{v}_l;\ l = 1, \ldots M\}$ is defined as [13]:

$$\mathrm{VM}(\mathcal{V}) = \mathbf{v}_m \in \mathcal{V} : \quad \sum_{\mathbf{v}_l \in \mathcal{V}} \|\mathbf{v}_m - \mathbf{v}_l\|_{\mathrm{L}} = \text{minimum}, \tag{3}$$

where $\| \cdot \|_{\mathrm{L}}$ is a suitably defined 2D norm.

The characteristics of vector median filtering heavily depend on the choice of the norm, usually resulting in a trade-off between the performance in edge preservation and in noise removal. The most widely used are the L_2^2, L_1 and L_2 (Euclidean) norms. It has been found that the L_2 norm is the best for edge preservation, while the L_2^2 norm is the best for medium-tailed noise removal, vector median filtering becoming in this case quite similar to an averaging operation. Using an L_1 norm yields a performance which is intermediate between the two.

Vector median filtering is quite useful for the task of achieving motion boundary preservation and restoration, since the closure constraint in 3 allows to substitute the incorrect optic flow vectors with one of the (possibly) correct vectors in the neighborhood. It is worth noting that the good edge response characteristics of vector median filters are not achievable by a simple componentwise smoothing of the vector field—that is, by applying a scalar median filter to the field components separately.

4. FAST IMPLEMENTATION

Both the multiwindow estimation and vector median regularization phases are carried out with a fast implementation.

The idea for the fast multiwindow implementation comes from the observation that each $(N+1)/2 \times (N+1)/2$ subwindow belongs actually to nine different $N \times N$ neighborhoods; as a consequence, the results of computations into a subwindow contribute to the evaluation of optic flow at nine different points. The multiwindow processing can thus be accomplished in two steps (Figure 2 shows the local processing for the case $N = 9$). First, the local technique is applied to all the $(N + 1)/2 \times (N + 1)/2$ windows of the image. Then for each pixel the results from nine nearby windows—the subwindows of the neighborhood of the pixel—are collected and compared. As a result, the computational cost of processing $N \times N$ neighborhoods with the multiwindow approach is made only slightly bigger than that of the local technique working on $(N+1)/2 \times (N+1)/2$ windows, the difference being in the time needed for the comparison step.

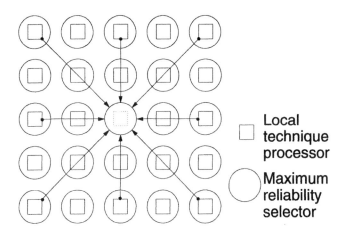

Figure 2. The fast architecture for multiwindow computation.

Computing vector medians by their definition 3 is time consuming, due both to the high number of norm evaluations and comparisons. A fast algorithm for the approximation of the L_1 norm vector median filter is proposed in [10]. It is based on the observation that a vector which minimizes the sum in 3 is the one whose components are obtained by applying a scalar median filter—for whose implementation fast algorithms exist—componentwise. Such a vector is not necessarily belonging to the original set, thereby violating the closure constraint of vector medians. However, this constraint is maintained if the set vector nearest—in the sense of the L_1 norm—to the componentwise solution is chosen, this resulting in an approximation of the exact (vector median) solution. It can be shown that the reduction in complexity is of the order of the linear dimension of the filter window.

5. RESULTS

The effectiveness of the approach was tested on several 128×128 image sequences. Results for a synthetic and a real-world case are discussed here. The regularization filter has a window size of 5.

Concerning the synthetic sequence—featuring two partially occluding moving textured squares and superimposed Gaussian noise ($\sigma^2 = 30$)— a quantitative performance evaluation and also a comparison with classical least squares was made, due to the avalaibility of the real flow field ("ground truth"). Table 1 shows the root mean square error and

Table 1
Results for the synthetic sequence

Algorithm	Root mean square error		Processing Time (s)
	Boundaries	Overall	
Classical least squares			
No regularization	0.933	0.354	12.0
Regularization	0.931	0.346	15.0
Multiwindow least squares			
No regularization	0.789	0.291	3.8
Regularization	0.718	0.246	6.8

the processing time for a neighborhood of 9×9 pixels (that is, the comparison is brought out keeping fixed the amount of information locally used by the two algorithms). Results indicate that the multiwindow approach performs better than classical least squares; specifically, a significant improvement is achieved at motion boundaries.

Figure 3. Three frames from a real-world sequence.

Figure 3 shows three consecutive frames of a real-world sequence in which two toy objects are moving; the dog is approaching the camera, while the train is "departing" in the transversal direction. Figures 4 and 5 show the results for the classical least squares ($N = 5$ and $N = 9$) and the multiwindow least squares ($N = 9$). The use of two different neighborhood dimensions for the classical least squares enables us to perform a qualitative comparison with multiwindow least squares by keeping constant in one case ($N = 5$) complexity [1] and in the other ($N = 9$) the actual amount of locally used infromation. It is evident from figures in either cases the multiwindow approach gives better results in terms of noise rejection and motion boundary preservation.

---

[1] In this case classical least squares was 15% faster due to the absence of the comparison step.

Figure 4. Results with the classical least squares algorithm. *Left*: 5 × 5 neighborhood. *Right*: 9 × 9 neighborhood.

6. CONCLUSIONS AND FUTURE RESEARCH

In this paper a multiwindow superstructure that enhances the performance of local optic flow techniques in terms of motion boundary preservation was presented together with a nonlinear regularizing filter. The multiwindow approach is non-iterative and non-parametric, and can be virtually used with any local technique providing an *a posteriori* measure of the reliability of the solution; least squares was chosen here due to its simplicity. Special attention was also given to processing speed with the development of a fast computational architecture for the multiwindow approach and the use of a fast approximation of a vector median filter. Results on both synthetic and real-world sequences together with a comparison with classical least squares carried out in terms of both complexity and amount of information demonstrated the effectiveness of the approach also for noise rejection.

It is currently under investigation the possibility of exploiting the reliability information provided by local techniques not only to achieve a better edge response but also as a means to switch among image resolution levels to reduce temporal undersampling effects by pyramidal motion processing.

REFERENCES

1. B.K.P. Horn and B.G. Schunck, Artif. Intell. 17 (1981) 185.
2. J. Konrad and E. Dubois, IEEE Trans. Patt. An. Mach. Int. 14 (1992) 910.
3. F. Heitz and P. Bouthemy, Proc. IEEE Int. Conf. Patt. Rec. (1990) 378.
4. C. Schnörr, Int. J. Comp. Vis. 8 (1992) 153.
5. K. Wohn, L.S. Davis and P. Thrift, Patt. Rec. 16 (1983) 563.
6. B.G. Schunck, IEEE Trans. Patt. An. Mach. Int. 11 (1989) 1010.

Figure 5. Result with the multiwindow least squares algorithm: 9×9 neighborhood.

7. A. Mitiche, Y.F. Wang and J.K. Aggarwal, Patt. Rec. 20 (1987) 173.
8. S. Uras, F. Girosi, A. Verri and V. Torre, Biol. Cyb. 60 (1988) 79.
9. A. Mitiche, R. Grisell and J.K. Aggarwal, IEEE Trans. Patt. An. Mach. Int. 10 (1988) 943.
10. M. Barni, V. Cappellini and A. Mecocci, Proc. EUSIPCO '92 (1992) 1485.
11. M. Nagao and T. Matsuyama, Comp. Graph. Im. Proc. 9 (1979) 394.
12. F. Bartolini, V. Cappellini, C. Colombo and A. Mecocci, Opt. Eng. 32 (1993) 1250.
13. J. Astola, P. Haavisto and Y. Neuvo, Proc. IEEE 78 (1990) 678.

Time-Varying Image Processing and
Moving Object Recognition, 3 – V. Cappellini (Ed.)
367

OPTIC-FLOW BASED TECHNIQUES IMPLEMENTATION FOR LANDSLIDE MOTION ESTIMATION

P.F.Pellegrini[a], M.Frassinetti[a], F.Leoncino[a], P.Palombo[a]
P.Ciolli[b], P.Pandolfini[b]

[a]Electronic Engineering Department - University of Florence

[b]I.R.O.E.-C.N.R., Florence

1. INTRODUCTION

The approaches to the study of motion in image sequences lie in two main points, which can be formulated in what can be called a "Feature-based" approach and an "Optic Flow-based" approach. The former is based on extracting in each image a set of features describing the objects in the scene and in establishing inter-frame correspondences between these features. The latter is based on the calculation of instantaneous velocities of brightness value in the image plane [1].

The present work describes the application of Optic Flow-based techniques to the study of environmental image sequences in which landslide phenomena are present. These methodologies have been employed because of image peculiarity, and particularly, because of the fluid flow related to landslides that, in the different frames of images, does not present clear and unambiguos features.

The aim of this work is to evaluate if these methodologies can discriminate phenomena of land movement. We have investigated - in related literature - in order to identify those algorithms of Optic Flow evaluation which are able to supply the best results for the class of images that have been recorded at Funés (an alpine locality).

2. OPTIC FLOW METHODOLOGIES

Various algorithms gradient based that have been proposed in order to evaluate the Optic Flow are related to the Optic Flow Constraint Equation (OFCE), which, for its part, relates the velocity vectors in the image plane [(u,v)], with spatial and temporal variation of the brightness function [g(x,y,t)].

$$\frac{\partial g}{\partial x}u + \frac{\partial g}{\partial y}v + \frac{\partial g}{\partial t} = 0 \tag{1}$$

Obviously, this constraint is not sufficient to determine both components of the velocity field, and therefore, to obtain a complete solution of the problem, supplementary hypotheses are required.

2.1 The local optimization method

In this method, developed by Kearney, Thompson and Boley [2], we solve the following overdetermined system that is built when evaluating the OFCE in n close points:

$$C\omega = -b \tag{2}$$

where
$$C = \begin{bmatrix} g_x^{(1)} & g_y^{(1)} \\ g_x^{(2)} & g_y^{(2)} \\ . & . \\ g_x^{(n)} & g_y^{(n)} \end{bmatrix}, \quad \omega = \begin{bmatrix} u \\ v \end{bmatrix} \quad \text{and} \quad b = \begin{bmatrix} g_t^{(1)} \\ g_t^{(2)} \\ . \\ g_t^{(n)} \end{bmatrix}$$

The solution of (2) is calculated using the pseudo-inverse formalism:

$$\hat{\omega} = C^+ b \tag{3}$$

where

$$C^+ = \left[C^T C\right]^{-1} C^T \qquad \text{and } C^T \text{ is the transpose matrix of } C.$$

2.2 The global optimization method

In this method, developed by Horn and Schunck [3], the solution is obtained by a regularization process.

In particular, one minimizes the sum of the error E_b and E_c defined here

$$E_b = g_x u + g_y v + g_t \tag{4}$$

$$E_c^2 = \left(\frac{\partial u}{\partial x}\right)^2 + \left(\frac{\partial u}{\partial y}\right)^2 + \left(\frac{\partial v}{\partial x}\right)^2 + \left(\frac{\partial v}{\partial y}\right)^2 \tag{5}$$

At last, a constant *parameter k* is introduced that takes into account the relative weight of the two errors factors E_b e E_c. The global error that has to be minimized is then:

$$E^2 = \iint (k^2 E_c^2 + E_b^2) \, dx \, dy \tag{6}$$

Following the (6), by using the calculus of variation, it is possible to obtain a system with two equations and two unknown values that is solved by using the Gauss-Seidel iterative method.

2.3 Multiconstraint Method

In this method, developed by many authors (Verri, Torre and Poggio [4]; Tretiak and Pastor; Haralick and Lee [5]), additional constraint equation are obtained by deriving the OPCE with respect to x, y and t.

$$g_{xx}u + g_{xy}v = -g_{xt} \text{,}$$
$$g_{yx}u + g_{yy}v = -g_{yt}$$

$$g_x\, u + g_y\, v = -g_t$$
$$g_{xx}u + g_{xy}v = -g_{xt}$$
$$g_{yx}u + g_{yy}v = -g_{yt}$$

$$g_x\, u + g_y\, v = -g_t$$
$$g_{xx}u + g_{xy}v = -g_{xt}$$
$$g_{yx}u + g_{yy}v = -g_{yt}$$
$$g_{tx}u + g_{ty}v = -g_{tt}$$

Verri, Girosi e Torre   **Tretiak e Pastor**   **Haralick e Lee**

3. PRACTICAL APPLICATIONS

3.1 Developed Analysis
The different kinds of analyses that have been carried out in this work can be summarised as follows:

1. *Qualitative analysis*
In this case a strategy of velocity field analysis is employed in order to extract parameters useful to identify a dangerous situation and then to generate a *warning*. The basic hypothesis is that the land movement is ruled by the gravitational influence and, according to the view angle and to the morphology of the slope, we can identify a main direction for the development of the land movement. The discrimination between static and dynamic situations is made by comparing the histograms showing the velocity directions. By observing the differences we have, with relation to the input of the main direction of motion, this discrimination becomes evident. A very basic and simple analysis can be carried out by analysing the resultant vector obtained by the sum of all velocity vectors present in the motion field.

2. *Quantitative analysis*
The algorithms for a correct evaluation of the optic flow are subject to uncertainties caused by various phenomena. A feasible way to reduce errors in image processing can be achieved by using an iterative refinement in the optic flow evaluation which supplies more accurate and useful data, for instance, for geological studies. One can also obtain a rough estimate of the error degree made upon each field vector in order to acquire suitable confidence values which supplies information for the reliability and accuracy of the computed velocities.

4. EXPERIMENTATION

Optic flow methodologies presented in section 2 have been applied to image sequences concerning a land-slide phenomenon near an alpine locality (Funés).

A procedure has been developed to perform an iterative refinement in order to obtain, in some cases, an enhancement of the precision of the estimated fields.

The image sequences used in this work have been obtained by using a portable station assembled by the technical staff of the I.R.O.E (Electromagnetic Waves Research Institute) of the National Research Council. In the centre of the images it is possible to identify a zone object of a motion caused by a landslide (see fig. 1). Because the evolution of the phenomenon is quite slow, the temporal sampling of the images is performed every three minutes.

We have carried out a series of tests, considering both pairs of images where there was no land motion and pairs where there was land motion. Images have been pre-processed, and, in particular, we have extracted the subimage shown in fig. 1.

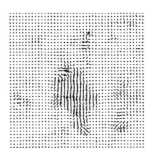

Fig. 2 Optical Flow calculated using the Global Optimization method

Fig. 1 Landslide image. The box encloses processed area.

4.1 The global optimization method

The application of this algorithm foresees the use of the *k parameter* (see par. 2.2), who's value depends on the Signal to Noise Ratio present in the images. For the images at our disposal, a high value of this parameter is required (50-100), which mean that the evaluated field can be underestimated. In any case, however, it is possible to perform a sharp distinction between images in which land motion is present and that in which there is no motion. In order to discriminate dangerous situations from static cases, one can apply a simple threshold process to the direction histogram.

We must also notice how the use of the iterative refinement yields a noticeable improvement of the obtained results.

4.2 The local optimization method

The quality of the results which follow from this algorithm depends on the conditioning of the over-determined system that has to be resolved in order to determine the velocity field. Essentially, this conditioning is related to the gradient trend. Therefore is not possible to determine a reliable value of the velocity for every point in the image. Our experimentation implies that, unlike the velocity fields from the global optimization method, these fields should not be as underestimated but are, however, less suitable for a qualitative analysis.

4.3 The multiconstraint method

With regard to the images we have recorded, one can notice that algorithms, based upon the computing of the second derivative, do not give reliable results because of serious errors that occur in computation.

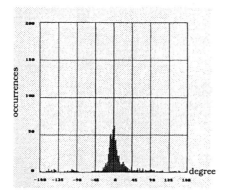

Fig. 3 Direction histogram in the case of land movement.

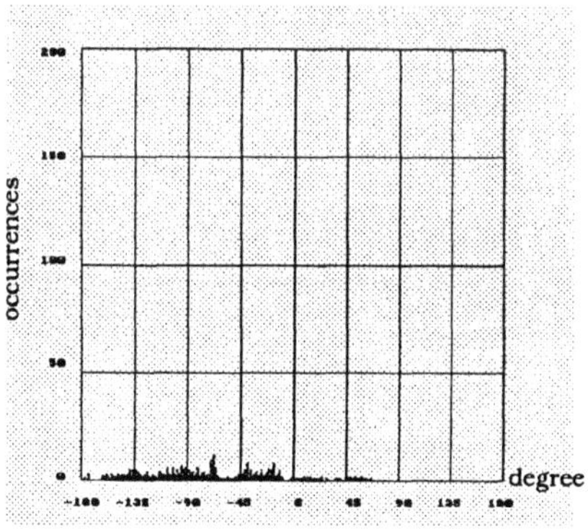

Fig. 4 Direction histogram in the case that a land motion not occur.

5. CONCLUSION

With reference to the tests that have been carried out, it can be noticed that for the algorithm of Horn and Schunck, in the case of land motion in the image, the histogram of the velocity directions is characterized by a peak in the vertical direction (see figg. 3, 4), which is the direction of the main development of the landslide. This peak can be easily detected with unsophisticated algorithms as well. In the majority of cases, it is enough to compute the direction of the resulting velocity vector in the image to generate the alarm signal.

In the local optimization algorithms, the peak is less evident (because of computation errors), but the alarm signal can still be generated by limiting the analysis to those directions that we know *a priori* as those preferential for landslide development, and a morphological analysis of the field can be can be useful in order to acquire this knowledge.

At the present time our research is oriented towards the study and exploration of methods for the reduction of errors that occur in the estimation of the velocity fields.

REFERENCES

1. J.K.Aggarwal, N. Nandhakumar, On the Computation of Motion from Sequences of Images, Proceeding of the IEEE, Vol 76, No. 8, pp. 917-935, August 1988.

2. J.K.Kearney, W.B.Thompson, D.L.Boley, Optical Flow Estimation: An Error Analysis of Gradient-Based Methods with Local Optimization, IEEE Transaction on PAMI, Vol.9, No. 2, pp. 229-244, March 1987.

3. B.K.P.Horn, B.G.Schunck, Determining Optical Flow, Artificial Intelligence, Vol. 17, pp. 185-203, 1981.

4. A.Verri, F.Girosi, V.Torre, Differential techniques for optical flow, J Opt. Soc. Am. A/Vol. 7, No. 5, May 1990.

5. Nagel, On the estimation of optical flow: Relations between different approaches and some new results, Artificial Intelligence, Vol. 33, pp. 229-324, 1987.

J
RECOGNITION AND TRACKING
OF MOVING OBJECTS

Time-Varying Image Processing and
Moving Object Recognition, 3 – V. Cappellini (Ed.)
© 1994 Elsevier Science B.V. All rights reserved.

Detecting Multiple Moving Objects by the Randomized Hough Transform

Heikki Kälviäinen

Lappeenranta University of Technology
Department of Information Technology
P.O. Box 20, SF-53851 Lappeenranta
Finland
Tel: +358-53-5743445
Telefax: +358-53-5743456
E-mail: Heikki.Kalviainen@it.lut.fi

Abstract

The algorithm to detect one moving object using the Randomized Hough Transform (RHT) has earlier been proposed by the author and his coworkers. This new non-model based method, called Motion Detection Using Randomized Hough Transform (MDRHT), was shown be to applicable to translational and rotational motion detection of one moving object. In this paper, the method is generalized to detect also multiple moving objects. The novel approach is tested with both synthetic and real-world sequences of time-varying images. The tests give promising results in 2D motion detection.

1 Introduction

Time-varying image processing has two main fields, feature-based approach and optical flow based approach [2]. In this paper, a new feature-based approach to detect multiple moving objects is presented. The approach combines the Randomized Hough Transform (RHT) [21],[20] and motion estimation techniques. Although the Standard Hough Transform (SHT) [9] has been applied to several motion detection tasks [6],[7], [8],[10],[22],[19],[5],[4],[3],[1],[18],[11],[16],[17], the Randomized Hough Transform has new properties which make it more simple to apply motion detection than the SHT. The algorithm to detect a single moving object, called the Motion Detection using Randomized Hough Transform (MDRHT) [14], is extended to explore also several moving objects in this contribution. This extended version inherits the clear merits of the basic MDRHT [14],[15],[12]. The MDRHT-like methods are suitable to the detection, recognition and tracking of moving objects. They can be applied to industrial computer vision tasks, automatic guarding systems in shops, detection and recognition of air vehicles etc.

In Section 2 the novel version for multiple moving object detection is presented. Tests with synthetic and real-world image sequences are shown in Section 3. Section 4 gives some discussion.

2   Motion Detection Algorithm for Multiple Moving Objects

The ideas of the MDRHT rely on the basis of the RHT. The RHT method samples pixels randomly from a binary edge picture to detect curves in an image. Let us consider line detection as an example. First, two edge points are sampled at random, and secondly, line parameters are solved from two line equations with these two sampled points. Finally, only one cell is accumulated in the 2D parameter space. Random sampling is repeated until an evident maximum is found in the parameter space, also called the accumulator space. The RHT algorithm is as follows:

Algorithm 1 : *The kernel of the RHT to line detection*

1. Form the set P of all edge points in a binary edge picture.

2. Pick a point pair (p_i, p_j) randomly from the set P.

3. If the points do not satisfy the predefined distance limits, go to Step 2; Otherwise continue to Step 4.

4. Solve the parameter space point (a, b) using the curve equation with the points (p_i, p_j).

5. Accumulate the cell $A(a, b)$ in the accumulator space.

6. Continue to Step 2.

To define the distance limits in Step 3 means that the points p_i and p_j are not allowed to be picked too near each other or too far away, i.e., $d_{min} \leq d(p_i, p_j) \leq d_{max}$ where $d(p_i, p_j)$ is a distance between the points p_i and p_j. If the edge picture is complex the use of distance limits is necessary. See more detailed in [21] and further developments on the RHT in [20].

In the MDRHT, these points are used to estimate motion. Only a group of single pixels is considered instead of some wider segmented areas. Hence, the approach is basically non-model-based. Usually, in the motion estimation applications of the SHT more complex structures are formed, for instance, using detected lines. Instead, in the MDRHT edge points are picked randomly from two consecutive pictures in a sequence of time-varying images. When the sampling produces enough similar, e.g., translations of two points in two pictures, the similar translation is estimated motion. After motion detection an object is segmented from the pictures, and the estimated translation is verified to be "true" motion. This procedure is repeated frame by frame to recognize moving patterns. See more detailed in [14],[15]. In this paper, not only one moving object is tracked but multiple moving groups of points are detected. The following algorithm describes the kernel of the MDRHT for motion detection of multiple moving objects.

Algorithm 2 : *The kernel of the MDRHT for multiple moving objects*

1. Form the sets B and C of edge pixels in the two pictures. Initialize the accumulator space A.

2. Pick point pairs (b_i, b_j) and (c_i, c_j) randomly from sets B and C.

3. If the point pairs correspond with their displacements calculate the x- and y-translations $dx = c_{ix} - b_{ix}$ and $dy = c_{iy} - b_{iy}$ and goto Step 4; otherwise, goto Step 2.

4. Accumulate the cell $A(dx, dy)$.

5. If a global maximum in the accumulator space is detected, motion (dx, dy) has been found and goto Step 6; otherwise, goto Step 2.

6. Form the sets B_m and C_m containing points which have motion (dx, dy) from the sets B and C.

7. If the sets B_m and C_m satisfy a motion detection criterion remove the points of the sets B_m and C_m from the sets B and C, and set $A = null$; otherwise goto Step 2.

8. If a stop criterion is satisfied then stop; otherwise goto Step 2.

In Step 3 of Algorithm 2, the correspondence of point pairs means that the displacement vector between b_i and b_j is the same as that between c_i and c_j. This matching criterion can be generalized as in [14]. The accumulation of the cell $A(dx, dy)$ in Step 4 means that its value is increased by one. In Step 5, the global maximum can be detected with a predefined threshold t. When the value of $A(dx, dy)$ is equal to t the global maximum has been found. The motion detection criterion in Step 7 verifies the moving object, in this case the group of moving points, to be real. In Step 8, the algorithm is stopped, for instance, when the predefined number of moving objects has been found, or the number of points in the sets B and C has reduced too low.

The ideas of multiple moving objects detection discussed above can be applied to rotational motion estimation like in the case of one moving object [15]. In this case the object, hence the point pair, has been assumed to rotate rigidly with a fixed rotation point and an angle. These parameters are found by sampling points from two images. See more detailed in [15]. The accumulator space is accumulated with a weight value w instead of a constant increment by one. The weight value w can be a function of the strength of point correspondence. Algorithm 3 below describes the kernel of the rotational approach:

Algorithm 3 : *The kernel of the MDRHT for multiple moving objects for translation and rotation*

1. Form the sets B and C of edge pixels in the two pictures. Initialize the accumulator space A.

2. Pick point pairs (b_i, b_j) and (c_i, c_j) randomly from sets B and C, and calculate rotational correspondence parameters of the point pairs.

3. If the point pairs correspond, calculate the rotation point (x_r, y_r) and the rotation angle θ and goto Step 4; otherwise, goto Step 2.

4. Accumulate the cell $A(x_r, y_r, \theta)$ by a weight value w.

5. If a global maximum in the accumulator space is detected, rotation (x_r, y_r, θ) has been found and goto Step 6; otherwise, goto Step 2.

6. Form the sets B_m and C_m containing points which have rotation (x_r, y_r, θ) from the sets B and C.

7. If the sets B_m and C_m satisfy a motion detection criterion remove the points of the sets B_m and C_m from the sets B and C, and set $A = null$; otherwise goto Step 2.

8. If a stop criterion is satisfied then stop; otherwise goto Step 2.

Since motion estimation of Algorithm 3 is more complex than of Algorithm 2, using gradient information [13] is hoped to help computing in new extensions in the future. Using gradient information of pixels can move the problem of defining accurate values of the rotation parameters x_r, y_r and θ to defining accurate gradient values.

3   Test Results

The tests were run on a standard SUN SPARCstation IPX. A part of a test image sequence Pic1 is shown in Fig. 1.a. The tests were performed with three different noise levels (0%, 15% and 30%). In this example in Fig. 1.a, there is about 15 % noise in the binary edge pictures of Pic1; in this case noise means points which do not belong to the moving objects. The last picture of the sequence presents the detected moving objects and their displacement vectors in two first frames.

Fig. 1.b illustrates the size of the accumulator in the tests with Pic1 without noise points. The first detected object needs naturally more size than the second and third one because there are more edge points to randomize in the pictures when detecting the first moving object. This is obvious because the points of the first detected moving object are removed from the edge pictures after segmentation of the first detected object and before calculation is continued. The computation time and accuracy of the tests with different noise levels are displayed in Figs. 1.c- 1.e. Computation time grows linearly and accuracy is high even with a low error tolerance of "exact" dx, dy-values ($dx \pm tol, dy \pm tol$). Noise seems to have stronger influence on computation times than on accuracy, in particular, with higher thresholds.

In Fig. 2.a a part of a real-world test sequence is displayed. Again, The last picture of the sequence presents the detected moving objects and their displacement vectors in two first frames. The number of hits and size in the accumulator increases when the threshold grows (Fig. 2.b). Here, the number of hits means a number of accumulations into the accumulator. The results of Fig. 2.b are similar to the ones of Fig. 1.b. The computation time and accuracy are presented in Figs. 2.c and 2.d. Computation time grows linearly, and accuracy is high even with a low error tolerance of "exact" dx, dy-values when a threshold is not very low.

4 Discussion

The method to estimate motion of multiple moving objects suggested in this paper was applied to translational and rotational motion detection. The tests with both synthetic and real-world images gave encouraging results. Despite of more complex random sampling compared to a single moving object detection, multiple moving objects could be tracked. Since motion estimation and object segmentation are done object by object using the RHT several problems are avoided. The MDRHT avoids the problem of overlapping of many objects in the XOR-picture of two consecutive frames. Besides, although objects would be partially covered by another object or a part of static background, the MDRHT usually still works. So, only partial information is also enough. There is no need of small motion (i.e., displacement and rotation) as in optical flow approaches. The determination of motion parameters is not so difficult as usually in feature-based approaches. And detected motion can really be a movement of a group of pixels, sometimes even rather separate pixels.

Unfortunately, the MDRHT has also clear demerits and limitations. Moving objects are assumed to be rigid and to have rather constant gray-level values. Moreover, inner boundaries in objects make computations more difficult. The MDRHT is a 2D motion estimation method which has been tested to translational motion estimation in this paper, and therefore, rotational concepts should also be considered more like in the case of one moving object [15]. Moreover, extensions to solve scaling problems are needed to explore. The MDRHT could be expanded to 3D motion using the RHT to determine focus of expansion. The use of more than two frames will be studied further. Gradient information of edge points in one object detection [13] can also be adopted to detect multiple moving objects.

References

[1] Adiv, G., "Recovering Motion Parameters in Scenes Containing Multiple Moving Objects", *Proc. of IEEE Computer Vision and Pattern Recognition Conf.*, Washington, USA, 1983, pp. 399-400.

[2] Aggarwal, J.K., Nandhakumar, N., "On the Computation of Motion from Sequences of Images - A Review", *Proc. of the IEEE*, vol. 76, no. 8, 1988, pp. 917-935.

[3] Ballard, D.H., Kimball, O.A., "Rigid Body from Depth and Optical Flow", *Computer Vision, Graphics, and Image Processing*, vol. 22, 1983, pp. 95-115.

[4] Cowart, A.E., Snyder, W.E., Ruedger, W.H., "The Detection of Unresolved Targets Using the Hough Transform", *Computer Vision, Graphics, and Image Processing*, vol. 21, 1983, pp. 222-238.

[5] Darmon, C.A., "A Recursive Method to Apply the Hough Transform to a Set of Moving Objects", *Proc. of IEEE Int. Conf. of Acoustics, Speech and Signal Processing*, Paris, France, May 1982, pp. 825-829.

[6] Falconer, D.G., "Target Tracking with the Hough Transform", *Proc. of 11th Asilomar Conf. on Circuits, Systems, and Computers*, Pacific Grove, USA, November 1977, pp. 249-252.

[7] Falconer, D.G., "Target Tracking with the Fourier-Hough Transform", *Proc. of 11th Asilomar Conf. on Circuits, Systems, and Computers*, Pacific Grove, USA, November 1979, pp. 471-482.

[8] Fennema, C.L., Thompson, W.B., "Velocity Determination in Scenes Containing Several Moving Objects", *Computer Graphics and Image Processing*, vol. 9, 1979, pp. 301-315.

[9] Illingworth, J., Kittler J., "A Survey of the Hough Transform," *Computer Vision, Graphics, and Image Processing*, vol. 44, 1988, pp. 87-116.

[10] Jayaramamurthy, S.N., Jain, R., "An Approach to the Segmentation of Textured Dynamic Scenes" *Computer Vision, Graphics, and Image Processing*, vol. 21, 1983, pp. 239-261.

[11] Kenner, M., Pong, T.-C., "Motion Analysis of Long Image Sequence Flow", *Pattern Recognition Letters*, vol. 11, no. 2, 1990, pp. 123-131.

[12] Kälviäinen, H., "Motion Detection by the RHT Method: Probability Mechanisms and Extensions", Research Report, no. 32, Department of Information Technology, Lappeenranta University of Technology, Lappeenranta, Finland, 1992.

[13] Kälviäinen, H., "Motion Estimation and the Randomized Hough Transform (RHT): New Methods with Gradient Information", accepted to *5th Int. Conf. CAIP'93 on Computer Analysis of Images and Patterns*, Budapest, Hungary, September 1993.

[14] Kälviäinen, H., Oja, E., Xu, L., "Motion Detection using Randomized Hough Transform", *Proc. of 7th Scandinavian Conf. on Image Analysis*, Aalborg, Denmark, August 1991, pp. 72-79.

[15] Kälviäinen, H., Oja, E., Xu, L., "Randomized Hough Transform Applied to Translational and Rotational Motion Analysis", *Proc. of 11th Int. Conf. on Pattern Recognition*, The Hague, The Netherlands, August-September 1992, pp. 672-675.

[16] Leung, M.K., Huang, T.S., "Detecting Wheels of Vehicle in Stereo Images", *Proc. of 10th Int. Conf. on Pattern Recognition*, Atlantic City, USA, June 1990, pp. 263-267.

[17] Nakajima, S., Zhou, M., Hama, H. and Yamashita, K., "Three-Dimensional Motion Analysis and Structure Recovering by Multistage Hough Transform", *Proc. of SPIE Conf. on Visual Communications and Image Processing*, vol. 1605, pt. 2, Boston, USA, November 1991, pp. 709-719.

[18] Radford, C.J., "Optical Flow Fields in Hough Transform Space", *Pattern Recognition Letters*, vol. 4, no. 4, 1986, pp. 293-303.

[19] Samy, R.A., Bozzo, C.A., "Moving Object Recognition using Motion Enhanced Hough Transform", *Proc. of the Int. Conf. on Digital Signal Processing*, Italy, September 1984, pp. 770-775.

[20] Xu, L., Oja, E., "Randomized Hough Transform (RHT): Basic Mechanisms, Algorithms, and Computational Complexities," *CVGIP: Image Understanding*, vol. 57, no. 2, 1993, pp. 131-154.

[21] Xu, L., Oja. E., Kultanen P., "A New Curve Detection Method: Randomized Hough Transform (RHT)", *Pattern Recognition Letters*, vol. 11, no. 5, 1990, pp. 331-338.

[22] Yam, S., Davis, L.S., "Image Registration using Generalized Hough Transforms", *Proc. of IEEE Computer Society on Pattern Recognition and Image Processing*, Dallas, USA, August 1981, pp. 526-533.

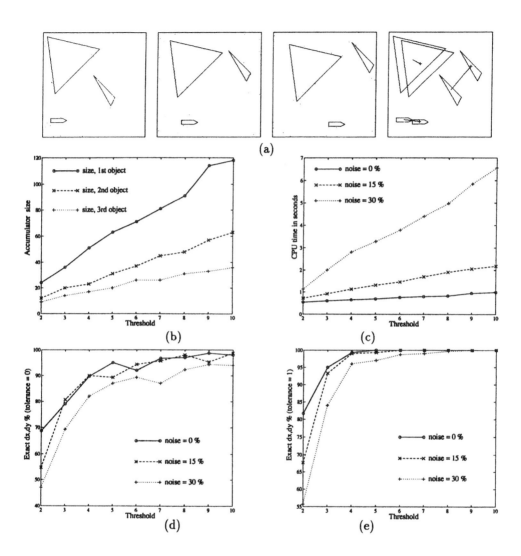

(a)

(b)

(c)

(d)

(e)

Figure 1: Test sequence Pic1: (a) The sequence Pic1 of three moving objects with 15 % noise points, (a last frame shows detected objects in first two frames); (b) Accumulator size with 0 % noise points (tests with first two frames); (c) Computation time; (d)-(e) Accuracy with tolerance 0 and 1.

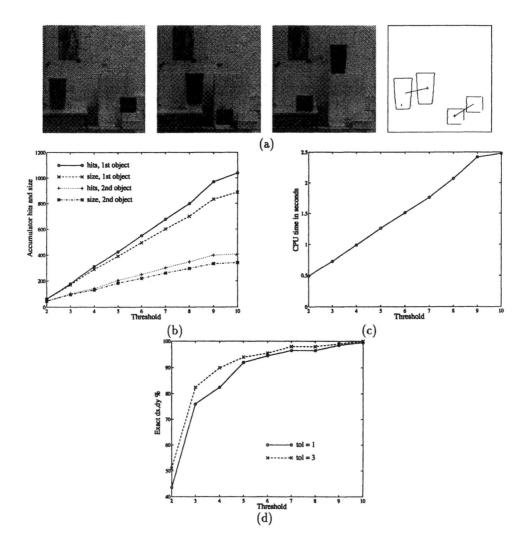

Figure 2: Test sequence Pic2: (a) The sequence Pic2 of two moving objects (a last frame shows detected objects in first two frames); (b) Accumulator hits and size (tests with first two frames); (c) Computation time; (d) Accuracy.

Time-Varying Image Processing and
Moving Object Recognition, 3 – V. Cappellini (Ed.)

Moving Object Recognition from an Image Sequence for Autonomous Vehicle Driving

G.L. Foresti, V. Murino

Department of Biophysical and Electronic Engineering
University of Genova - Via All'Opera Pia 11A, 16145 GENOVA - ITALY
Fax +39 10 3532134 Phone +39 10 3532792
E-mail forfe@hp835.dibe.unige.it

Abstract

This paper is focused on the recognition of moving objects for an autonomous vehicle driving. In particular, a method for detecting straight lines and their correspondences in successive image frames by means of a straight-line extraction and matching process is proposed. The main advantage offered by the described approach is the possibility of extracting and matching straight lines directly in the feature space, without requiring complex inverse transformations.

In particular, the Direct Hough Transform (DHT) algorithm for straight-segment is proposed. It aims to avoid the most important problems associated with the classical HT: (a) spatial information loss, (b) spurious peaks, and (c) discretization effects. Then, a Straight-Line Matching (SLM) algorithm is presented, which utilizes only four attributes to describe a segment: (a) position ρ, (b) orientation θ, (c) length l, and (d) midpoint $\mathbf{m}$.

Finally, results of some experiments performed using synthetic and real monocular image sequences are reported.

1. INTRODUCTION

In autonomous vehicle driving systems the most difficult task is the one related to machine vision. An autonomous vehicle must be able to recover the structure of the surrounding environment, the position of the road and the presence of obstacles on it. Furthermore, if other vehicles are present in the scene, it should be able to estimate with precision the position and the motion of each of them. The specific problem of moving object recognition from an image sequence is addressed in this paper. Image are used in many computer vision systems [1,2] to infer the structure and the shape of the 3D environment. Various steps, including detection of image features and matching of such features between image frames, are performed in motion analysis to find moving objects in complex scenes [3].

Motion correspondences in a sequence of image frames can be detected by using different types of features: image points (i.e., pixels), characteristic points (e.g., corner points, junction points, etc.), groups of connected pixels with similar brightness, straight lines, etc. The most used features are straight lines [2,3].

The importance of detecting and characterizing straight lines stems from the great informative content that they provide about the characteristics of regular objects (e.g., roads, cars, houses, railways, etc.). The presence of 3D man-made objects in structured environments is usually associated with the presence of groups of 2D straight segments on the image plane.

This paper describes a method for detecting straight lines and their correspondences in successive image frames by means of a straight-line extraction and matching process performed directly in the feature space. The issue of rigid objects with planar surfaces is addressed.

The problem of extracting straight lines has been faced in the literature by using two main approaches: image-space methods [3] and feature-space methods [4]. Several image-space methods have been developed to extract straight lines. Such methods include edge-tracking [3], contour-following [2], curve fitting [2] and relaxation algorithms [3]. The most widely used feature-space approach to extract straight lines is the Hough Transform (HT) method [4]. The HT is based on the mapping of each curve point into a parameter space by means of a parametric equation. Duda and Hart [4] suggested that straight lines can be parametrized by the length ρ and the orientation θ of the normal vector to the line from the image origin. This has notable advantages over the (m,c) (i.e., slope and intercept) parametrization which has a singularity for lines with large slopes (i.e., m→∞). Ballard [2,4] generalized the HT for the detection of arbitrary shapes (GHT) by decomposing complex shapes into simpler shapes.

The problem of matching straight lines between image frames has been investigated by several authors. McIntosh and Mutch [5] implemented an algorithm that relies on eight parameter values attached to each line. The most important values are compared in a single pass by means of a matching function that determines the strength of the similarity between a pair of lines. Liu and Huang [6] developed a method of two-step matching for straight-line correspondences. A matching function is utilized that characterizes the similarities between the edge lines of two images, and that is based not only on the geometrical relations between lines but also on the information derived from the intensity images.

2. SYSTEM DESCRIPTION

Images are acquired at successive time instants (e.g., t and t+1) are processed by a Canny operator to extract significative gray level discontinuities (i.e., edges) (Fig. 1). To detect straight lines an improved version of the standard Hough Transform, called Direct Hough Transform, is proposed. This algorithm which aims to avoid the most important problems associated with the classical HT: (a) spatial information loss; (b) spurious peaks (i.e., false maxima in the parameter space and, consequently, false rectilinear segments in the image space); (c) discretization effects. In order to solve the above problems, the search space has been modified to take into account the local distributions of segments both along a straight line and along different lines charachterized by comparable directions and positions.

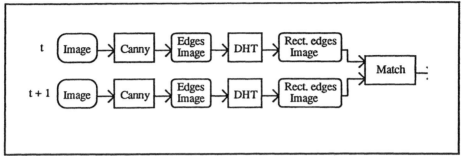

Fig. 1 General scheme of the system

To match straight segments a non-iterative algorithm which performs matching operations directly in the feature space is used. A straight segment is described by means of four attributes: (a) position ρ, (b) orientation θ, (c) length l and (d) midpoint $\mathbf{m}$. Fig. 2 shows the straight-line attributes: each i-th segment is identified univocally by an angle θ and a distance ρ from the origin (0,0) of the image reference system (X,Y), by a length l_i, and by a midpoint $\mathbf{m_i}=(x_{mi}, y_{mi})$.

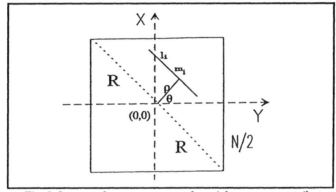

Fig. 2 Image reference system and straight-segment attributes

A straight line belonging to an object boundary will be constrained to perform small rotations and translations in a short time. 2D constraints are computed for a segment motion on the image plane: the attributes ρ, θ, l and $\mathbf{m}$ must be included into limited ranges to reduce the matching space.

3. STRAIGHT-SEGMENT DETECTION

The Direct Hough Transform is a dynamic segmentation algorithm defined as the process of mapping an edge-image space $E=E(\mathbf{x})=E(x,y)$ into a dimensional space $\{\Pi, M\}$, where $\Pi=\{\pi\}$ is a parameter space and $L=\{l(\pi)\}$ is a label space. All edge points $\mathbf{x}=(x,y)$ belonging to different straight segments on the same straight line are mapped into the same parameter space point π, but are associated with different labels. The following mapping equation is used by the algorithm:

$$g(x,y,\pi)=\rho - x{\cdot}\cos\theta - y{\cdot}\sin\theta \qquad (1)$$

The novelty of this approach lies in separating the contributions of the straight segments belonging to the same straight line by partitioning, in an incremental way, the traditional HT parameter space into a set of disjoint subspaces:

$$l(\pi)=\{l_k(\pi){:}k{\in}[1,K(\pi)]\} \qquad (2)$$

each associated with a different local label $l_k(\pi)$. $K(\pi)$ is the number of straight segments contained in π. The parameter space Π is represented by an accumulator array $C(\rho,\theta)$ with $(\Theta{-}1){\cdot}(2R{-}1)$ cells (Θ and R indicate the maximum resolutions of the parameters θ and ρ, respectively). In particular, these parameters are included in the ranges:

$$\theta \in [0,\Theta) \qquad \rho \in [-R,+R] \qquad (3)$$

where $\Theta = 180$ and $R = \dfrac{N}{2}\sqrt{2}$ (i.e., $N{\cdot}N$ is the image size).

The main steps of the DHT algorithm are: (a) voting; (b) clustering; (c) local maxima detection.

(a) Voting

The first step of the DHT algorithm is the voting, which is a mapping operation of every edge pixel from the edge image into the parameter space Π according to the equation (2). Straight lines present in an image are located in the parameter space by the intersections of many sinusoidal curves (i.e., the presence of many pixels with the same ρ and θ parameters). The main procedure is:

> For each edge pixel (x,y) **do**
>> For each pair (ρ,θ) satisfying eq. (1) **do**
>>> Add a vote to location (ρ,θ) in the parameter space
>> **end**
> **end**

From an implementation point of view, the advantages of the DHT involve the addition of two basic tasks: a) local accumulators related to different spatial areas belonging to the same π must be introduced; b) a decision criterion has to be defined to assign points to different local accumulators. We chose an adaptive clustering method to accomplish the above tasks.

(b) Clustering

The method uses to classify and cluster collinear edges into different groups belonging the same rectilinear line different reference points $P_{ref}(\pi)$ placed on the image borders [7]. A monodimensional distance $d_i{=}d(\rho,\theta,x_i)$ between the current edge point $x_i{=}(x,y)$ and the related $P_{ref}(\pi)$ is computed at each step on a 1D reference system $S(\rho,\theta)$ with origin at $P_{ref}(\pi)$ and passing through x_i. To this end, a dynamic list is maintained, which takes into account the different subclasses $l_k(\pi)$, $k=1..K(\pi)$ which can be created for each local maximum. A subclass $l_k(\pi)$ is defined as a set of edge points belonging to the straight line identified by π and to the interval $[x^k{}_{min}(\pi), x^k{}_{max}(\pi)]$ that represents the current start and end points of the k-th segment, i.e.:

$$l_k(\pi) = \{x_i : g(x_i,\pi)=0, \ x_i \in [x^k{}_{min}(\pi), x^k{}_{max}(\pi)]\} \tag{4}$$

In this way, each subclass $l_k(\pi)$ can be uniquely identified by two image points, $x^k{}_{min}(\pi)$ and $x^k{}_{max}(\pi)$, in the image reference system.

(c) Local maxima detection

Local maxima π^* in the parameter space are detected by means of the following procedure. A local maximum π^* is defined as an accumulator with the highest value with respect to neighbouring parameters and with respect to a prefixed threshold C_{th}. The set of neighbours of a straight line π is made up of all straight lines described by the parameters (ρ,θ) belonging to a mask of A·B dimensions centred in a point π. The maxima-detection phase is performed exactly in the same way as for the standard HT: a higher C_{th} value must be used to avoid spurious peaks in the parameter space. In order to handle the problem of spurious peaks, the DHT performs a merging operation between a local maximum and its adjacent accumulators. The sites in the parameter space that are characterized by a high accumulator value and that are adjacent to local maxima are recovered, and their subclasses are merged with the ones of the central maximum. As a general rule, two subclasses related to a different straight line are fused if their extremes are almost coincident.

4. STRAIGHT-SEGMENT MATCHING

The problem of straight-line matching on an image sequence is a problem with high computational complexity. In fact, to solve the matching problem for a large number of segments, it would be necessary to perform a considerable search. Consequently, the primary aim of the matching algorithm is to reduce drastically the match space by eliminating impossible matches. Motion constraints are imposed on straight-line displacements in the 2D space (e.g., image plane) in order to derive the maximum point displacements on the parameter space and to reduce the number of hypothetical matches.

We suppose that a 2D line σ can rotate, in the successive frame, around its midpoint $\mathbf{m}=(x_m,y_m)$ by $\alpha \leq \bar{\alpha}$ degrees and translate of $\mu \leq \bar{\mu}$ pixels (i.e., rigidity constraints). Then, a mask D of dimensions, $(\Delta\rho)_{max}$ x $(\Delta\theta)_{max}$, can be utilized to reduce the search space where to detect the correspondences:

$$(\Delta\theta)_{max} = |\theta'-\theta|_{max} = \bar{\alpha} \tag{5}$$

$$(\Delta\rho)_{max} = |\rho'-\rho|_{max} = \{max_\alpha [g(\theta^*,\alpha,\mathbf{m})]\} + \bar{\mu} \tag{6}$$

where α and μ are the current rotation and translation parameters, and $[g(\theta,\alpha,\mathbf{m})]$ is a real function that approximates the behaviour of the parameter ρ according to the position of the segment in the image space. The function g depends on θ, a and $\mathbf{m}$ parameters and is upper limited [8]:

$$g(\theta,\alpha,\mathbf{m}) \leq |x_m\cdot[\cos(\theta+\alpha)-\cos\theta]| + |y_m\cdot[\sin(\theta+\alpha)-\sin\theta]| + 2\sqrt{2}\,\bar{\mu} \tag{7}$$

4.1 Matching function

Let us define (Π, L) and (Π', L') as two-dimensional feature spaces at two successive time instants, t and $t'=t+T$ (e.g., $T=1/25$ sec for a video rate). The matching operation is performed in two steps by means of a hierarchical matching function: (a) directly in the parameter spaces Π and Π' to search for the corresponding maxima (i.e., straight lines in the image space); (b) in the label spaces L and L' to search for the corresponding local structures $l_i(\pi)$ and $l_j(\pi')$ (i.e., straight segments in the image space).

(a) First matching step

For each local segment $l_k(\pi)$ belonging to a maximum $\pi=(\rho, \theta)$ detected in the first parameter space Π, the dimensions $(\Delta\rho)_{max}x(\Delta\theta)_{max}$ of the search mask D are computed. Moreover, the mask D is centred in the point $\pi'=(\rho, \theta)$ of the second parameter-space frame: all maxima belonging to this area will be selected as candidates.

(b) Second matching step

After the computation of the optimal mask dimensions, a further operation is performed to reduce the number of candidate segments contained in the candidate maximum points by eliminating impossible matches. The lengths λ_t and the midpoint position m_t of the target segment and the length λ_c^v and the midpoint position m_c^v of the candidate segments are compared by means of the following rule:

if $\{\, |\lambda_t - \lambda_c^v| < \lambda^{th}, \text{ AND } d(m_t, m_c^v) < \bar{\mu} \text{ then}$

the c-th candidate segment of the v-th candidate maximum is preserved

otherwise it is eliminated

where $\lambda^{th} = \dfrac{1}{5}\lambda_t$ is an experimental value of the length threshold.

A matching function f_M is used to compare the target segment $l_t(\pi)$ with all segments contained in the candidate maxima π'_v $v=1..V$. The criterion adopted to define the function f_M is based on the minimum distance between the endpoints of the target segment and of the candidate segment, and on the ratio between their length, multiplied by the corresponding weight values D_i. The sum of all weights is such that: $\Sigma_i D_i=1$ where D_i is the weight of the i-th parameter. We have defined the function $f_M = f_M(l_t(\pi), l_c(\pi'_v))$ as:

$$f_M[l_t(\pi), l_c(\pi'_v)] = D_1[(x_{ti}-x^v_{ci})^2+(y_{ti}-y^v_{ci})^2] + D_2[(x_{tf}-x^v_{cf})^2+(y_{tf}-y^v_{cf})^2] + D_3\frac{\lambda_t}{\lambda_c^v} \quad v=1..V \quad (8)$$

where $x_{ti}=(x_{ti},y_{ti})$ and $x_{tf}=(x_{tf},y_{tf})$ are the initial and final points of the target segment, $x^v_{ci}=(x^v_{ci},y^v_{ci})$ and $x^v_{cf}=(x^v_{cf},y^v_{cf})$ are the initial and final points of the current candidate segment, $D_1=0.4$, $D_2=0.4$, and $D_3=g0.2$ are the weights of the parameters.

5. RESULTS

The algorithms described were tested on an HP9000/835 workstation. Test images were 256x256 in size and their gray levels ranged from 0 to 255. We made experiments on both synthetic and real images.

(a) Synthetic images

Synthetic images were obtained from a ray-tracing program and represented a cube translating and rotating in a 3D space. Fig. 4a to 4d show the original images. The first and second frames represent a simple translation on the x-axis, the second and third frame represent a simple rotation, and the third and fourth frame are the composition of a rotation and a translation together. Fig. 5a to 5d show the matched lines, giving that all the three cases are handled correctly.

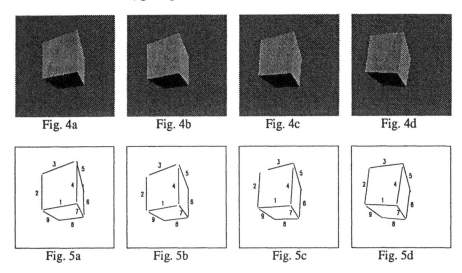

Fig. 4a	Fig. 4b	Fig. 4c	Fig. 4d
Fig. 5a	Fig. 5b	Fig. 5c	Fig. 5d

(b) Real images

To evaluate the robustness of the algorithms we chose a more complex (and noisy) pair of real images representing a moving car. Fig. 6a and 6b show the edge images (extracted by means of the Canny operator), and fig. 6c and 6d show the matched pairs of all the segments extracted by the DHT. We can see that all the main segments belonging to the road and to the car are correctly matched.

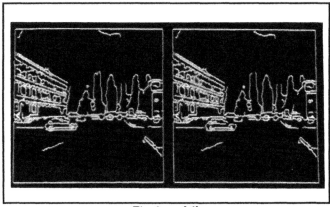

Fig. 6a and 6b

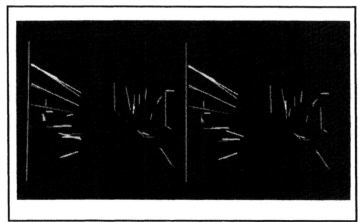

Fig. 6c and 6d

6. CONCLUSIONS

This paper is focused on the recognition of moving objects for an autonomous vehicle driving. In particular, a method for detecting straight lines and their correspondences in successive image frames by means of a straight-line extraction (i.e., DHT) and matching process (i.e., SME) is addressed. The main advantage offered by the described approach is the possibility of extracting and matching straight lines directly in the feature space, without requiring complex inverse transformations.

The extraction and integration of multiple 2D image shapes (e.g., circular lines, ellipses, etc.) can notably increase robustness and reliability of the proposed method.

7. REFERENCES

[1]   G.Foresti, V.Murino, C.S.Regazzoni, and G.Vernazza, "Distributed Spatial Reasoning for Multisensory Image Interpretation", *Signal Processing*, Vol 32, Nos.1-2, pp. 217-255, May 1993.

[2]   W.E.L. Grimson, "Object Recognition by Computer: The role of Geometric Constraints", MIT press, 1990.

[3]   J. K. Aggarwal and N. Nandhakumar, "On the computation of motion from sequences of images", *Proc. of IEEE*, Vol. 76, No. 8, pp. 917-935, 1988.

[4]   J. Illingworth and J. Kittler, "A Survey of the Hough Transform", *Computer Vision, Graphics, and Image Processing*, Vol. 44, pp. 87-116, 1988.

[5] J. H. McIntosh and K. M. Mutch, "Matching straight lines", *Computer Vision, Graphics, and Image Processing*, vol. 43, pp. 386-408, 1988

[6]   Y. Liu and T. S. Huang, "Determining straight line correspondences from intensity images", *Pattern Recognition*, vol. 24, no. 6, pp. 489-504, 1991

[7]   G.L. Foresti, C.S. Regazzoni and G. Vernazza, "Grouping of Straight Segment by the Labelled Hough Transform", *Computer Vision, Graphics and Image Processing: Image Understanding* (in press).

Time-Varying Image Processing and
Moving Object Recognition, 3 – V. Cappellini (Ed.)

MOVING OBJECT DETECTION AND TRACKING IN AIRPORT SURFACE RADAR IMAGES

P.F.Pellegrini, P.Palombo, F.Leoncino, S.Cuomo, M.Frassinetti, E.Piazza

Electronic Engineering Department - University of Florence

Work supported by Progetto Finalizzato Trasporti 2, C.N.R. (Italian National Research Council)

1. INTRODUCTION

According to the different problems and techniques related to the detection and recognition of kinematics parameters of moving objects, the present work mainly deals with the processing of images gathered by a high-resolution radar sensor.

The problem with monitoring aircraft movement on airport surfaces is of great importance, especially in the presence of high traffic rates or bad conditions of visibility.

Research trends for future years are oriented towards the development of integrated systems, which collect information coming from various sensors, and distribute it both to ground air-traffic controllers and to moving vehicles. The realization of a system for the automatic control of airport surface traffic can play an important role in this context.

Moreover, in the near future, in those airports in which take-offs and landings are permitted in conditions of very poor visibility and even in conditions of no visibility at all, (categories III B e C), the radar sensor must supply, with a high degree of precision, the identification of the aircraft and it's position on the aerodrome.

These kinds of problems are investigated with respect to effective signals obtained by an operating radar. Here, in particular, we outline some algorithms for tracking moving vehicles on airport surfaces, using the ASMI radar sensor at Fiumicino Airport, Rome.

The basic characteristics of the radar images used for this purpose are:

Antenna elevation (from the ground)	32 m
Carrier frequency	35 GHz
Range resolution	4.5÷7.5 m
Angular resolution	0.4 deg
Maximum range	0.9 Km (0.5 NM)

Images taken into consideration for this work were obtained from a scan converter, in a Standard PAL format with grey levels related to the intensity of the radar echo.

Compared to standard camera images, radar image characteristics present some peculiarities because of sensor properties (see Fig. 1):

- pixel brightness is affected by a radial distortion, due to the nature of radar echo detected;

- the water vapour content in the air (fog, rain etc.) causes the worsening of image brightness and contrast ;

- radar "shadows" in radial direction appear in the proximity of protruding objects in the scene.

Fig. 1 - An example of ASMI radar image

This last characteristic can be properly associated with some target geometrical features: that is, the length and the shape of the shadow can provide useful information about the elevation and shape of aircrafts and vehicles that appear in the scene.

The geometry of the images is corrected for both distortion and difference in sampling x and y direction.

The aim of this work can be outlined in the following points:

- Location of moving aircrafts and vehicles on runways, taxiways and aprons of the aerodrome.
- Determination of the motion of each vehicle in order to automatically determine those manoeuvres which are dangerous and not allowed.

- Identification of the different aircraft classes and recognition of the prow angle.
- Integration of this data with information coming from other sensors (local sensors, GPS etc.) to obtain an integrated system for automatic surface movement control.
- Re transmission to the aircrafts of the positions of moving and standing objects with a scaled, zoomable map of the airdrome in order to supply a perspective synthetic vision even in conditions of no visibility.
- Development of enhanced algorithms and procedures that can be used on new-generation 95 GHz radar systems.

The configuration of the experimental set-up is shown in Fig.2

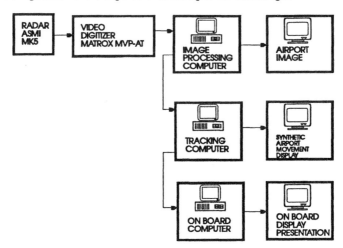

Fig. 2 - Experimental Set-up Configuration

2. ANALYSIS STRATEGY

In order to extract moving objects, we chose to employ feature-based techniques [1] that are especially customized for the sensor in use. In particular, two main issues can be pointed out:

- Image background does not change in different frames (except for possible slight distortions). This allows one to obtain a reference image that can be employed for the detection of moving objects [2].
- Airport planimetry (runways and buildings) is known so that it can be used as a template for image registration by means of correlation or "relaxation" techniques [3][4].

Based upon these points, a method for moving vehicle detection and tracking on airport surfaces has been developed. The analysis strategy includes four main steps:

1. Location of runways and manoeuvring areas by matching a reference template to the recorded images. Runway edges are extracted using Hough Transform based techniques [5]; the matching stage is performed so that it can correct slight distortions that are introduced during image acquisition;

2. Moving object detection, bounded to the zone of interest (runways and manoeuvring areas). In particular, a first estimate of the aircraft position is done by fast and simple algorithms (as profile analysis of the runway) followed by "region growing" techniques [6] applied for object shape estimation;

3. Identification of the position (some meter precision) and aircraft axis direction (some degree precision). Type of the detected vehicles, with template correlation techniques and Hough Transform algorithms, and the evaluation of their displacement with respect to the frame sequence;

4. Display of the obtained results.

These analyses are related to the ASMI-MK5 Radar images, and because of the scan converter processing, they have less information than that of the raw data coming from the video signal of the receiver.

However, the obtained results have allowed for the choice of an analysis strategy which would produce better results on new-generation radar systems.

The process can be summarized in the following phases:

1) Image processing every Δt sec (typical $\Delta t = 0,5 - 1$ sec)
  - quick estimation of the aircraft position
  - recognition of the aircraft and of its shape
  - aircraft classification
  - a more detailed estimation of the aircraft position
  - prow angle estimation

2) Multitemporal analysis of the parameter obtained in 1)
  - determination of non-moving vehicles
  - determination of temporal parameter evolution by a sampling with time step of Δt

3) Periodic analysis of multitemporal images
  - identification of parts belonging to the moving object vehicle every T sec (Typical T = 10 sec)
  - use of techniques even if they are time consuming (e.g. Optical Flow)
  - correspondence check of moving points with parameters obtained in 2)

The output of these results is a parameter set that is exploited to control the automation of the airport surface movement.

These parameters have been used in the development of this work in order to display the kinetic scene of the airport traffic in a synthetic way.

3. HARDWARE CONFIGURATION

The overall system takes into account a structure based upon different computers, as shown in Fig. 2. It corresponds to the actual system configuration used for the test.

The images from the scan converter were recorded on a magnetic tape (VHS standard) in different operation modes at the Fiumicino Airport for a total time of 40 minutes.

The system operates in real time with signal acquisition from the play back video recorder. The image is sampled by using a MATROX MVP-AT image processing board with dimension of 512 x 512 pixel (8 bit/pixel).

The host computer is a PC with a 386DX CPU (like the other computers shown in Fig. 2.).

The other computer also allows for the driving of the monitor of the synthetic display of the airport surface movement presentation.

The on-board computer operates in a similar way to the system placed on the cockpit of the given airplane to which the position data of the aircraft in the airport are sent (e.g. via Mode S radio link).

4. FAST ESTIMATION OF AIRCRAFT POSITION

For a fast determination of points belonging to the airplane (or vehicle) images, different algorithms have been used that were based mainly upon the scanning of the airport surface along the runways, taxiways and terminal areas.

Valid results have been obtained through the projection computation along the runway axis. Taking into account the expected aircraft width, the results obtained with the previous method are quite similar to the information extracted by the section of the runways or taxiways along their axes.

In Fig. 3 a typical obtained profile of a taxiway is shown. Due to the different effects on the radar signal (radar distance, gain control with the distance, atmospheric and weather conditions. etc.), the energy related to the scene can be considered as a function of the position.

Local properties of the signal must be taken into account. Good results have been obtained by evaluating the pulse area of the profile (high pass filtered) along the runways, as shown in Fig. 3b.

Fig. 4 gives an example of the system performance in detecting the airplanes. Here we can see how the radar echo sometimes disappears and the tracking algorithm, therefore, must process this possibility.

For the first analysis of the aircraft position , while considering a portion of about 1 Km², a processing time of about 0.1 sec is employed on a 386 Personal Computer with a MATROX MVP-AT image processing board.

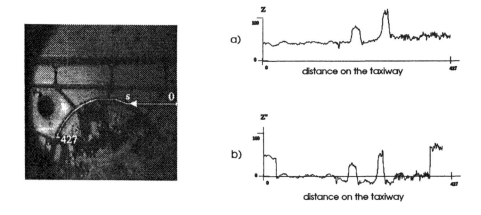

Fig. 3 - Typical Obtained Taxiway Profile: a) actual, b) high pass

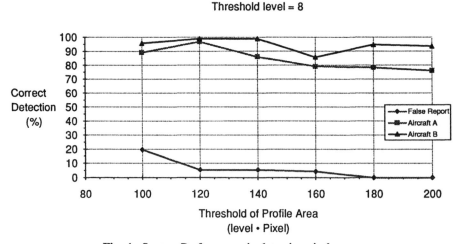

Fig. 4 - System Performance in detecting airplanes

5. FORM EXTRACTION

The extraction of the aircraft echo radar behaviour is obtained by growing the neighbourhood of the points previously found.

Thresholds related to average local intensity of the signals and maximum of the signal are considered (the maximum of the signal is only slowly effected by noise due to integration connected with the image acquisition by the PPI presentation). This threshold can be refined by "learning" the results of different aircraft types in a final operating structure of the system.

Form parameters, such as the centroid, area, inertial moment, direction of principal inertial moment etc., are extracted. These parameters can be related to different classes of aircrafts.

Due to the electromagnetic shadow of the illuminated part of the target, the shape of the airplane is usually not symmetrical. This problem is strongly connected to the determination of the axis of the aircraft.

6. ASPECT ANGLE OF THE AIRCRAFT

For a correct procedure of traffic control, the orientation of an aircraft or a vehicle is very important. The determination of this parameter can be done in different ways:

-by matching the aircraft template;
-by using the Hough transform for the most significant part of the extracted form;
-by using the velocity vector in multitemporal analysis.

The first methods can be used on each single image.

It must be observed that for the image obtained by the radar, the echo shape matches better with the geometric shape of the aircraft for the zone first illuminated by the electromagnetic wave that arrives from the radar sensor, than for the other zones. Taking this into account for a given aircraft, the useful reference portion of the image profile changes with respect to the angle between the axis of the aircraft and the direction from the radar signals that light the aircraft itself.

The Hough transform can be used for the matching of the contour template with the shape.

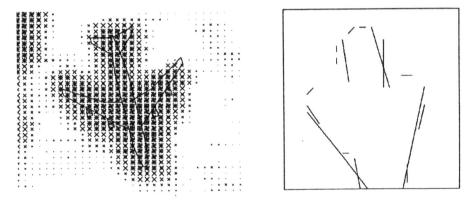

Fig. 5 - Example of detected B 747 aircraft shape and the obtained Hough Transform.

7. THE MOTION TRAJECTORY

The trajectory estimate is determined by a location sequence in the X, Y airport space, and when the aircraft echo is lost in some frames, an interpolation is made. In addition, a filtering process based on the laws of motion is considered.

From these elaborations we detect events that conflict with given constraints.

In particular, the minimum distance between aircrafts can be controlled, as well as other constraints.

8. PERIODIC CONTROL OF MOVING OBJECT

This function is considered as redundant information and as in control of the previously used algorithms.

If some mismatching appears on the pixel involved with the motion, a message gets sent to the image processing computer (Fig. 2) and an image analysis is performed in the neighbourhood of these zones.

These techniques taken into consideration can be simple techniques, such as the evaluation of the difference of sequential images, or more sophisticated, such as the optical flow procedures.

Because of the following topics, optical flow can be an interesting tool for our analysis:

a) the space temporal noise in the images is small;

b) the motion of the objects in the scene is smooth;

c) the motion of the objects under consideration varies continuously over the sequence.

9. SYNTHETIC PRESENTATION OF THE MOTION

The synthetic presentation of the motion in an experimental set-up is obtained through the tracking computer that is driven by the position data and other parameters supplied by the image processing computer.

The system has a library of different aircraft shapes which are presented with their correct position and orientation on the display screen.

The airport image is geometrically drawn with a given scale, and it uses the actual dimension of the runways, taxiways and aircraft shapes.

For a correct presentation, the data coming from the processing computer is geometrically corrected to compensate for the radar presentation distortion (bilinear correction formula).

REFERENCES

[1] J.K.Aggarwal, N.Nandhakumar, "On the computation of motion from sequences of images"- A review", Proceedings of the IEEE, Aug. 1988

[2] K.S.Fu, R.C.Gonzalez, C.S.G.Lee, "Robotica", McGraw-Hill, 1989

[3] A.Goshtasby, "Template matching in rotated images", IEEE Transactions on PAMI, 1985

[4] S.A.Kuschel, C.V.Page, "Augmented relaxation labelling and dynamic relaxation labelling", IEEE Transactions on PAMI, 1982

[5] J.Illingworth, J.Kittler, "A Survey of the Hough Transform", Computer Vision Graphics and Image Processing, Vol.44, 1988

[6] W.K.Pratt, "Digital image processing", Wiley Ed., 1978

K
TRAFFIC MONITORING

Time-Varying Image Processing and
Moving Object Recognition, 3 – V. Cappellini (Ed.)
401

A MONITORING SYSTEM FOR LOCAL TRAFFIC CONTROL BASED ON VEHICLE TRACKING

M.Annunziato, I.Bertini, S.Giammartini, F.Pieroni

ENEA, Diagnostic Technologies Division,
Via Anguillarese 301, 00060 Rome, Italy

1. INTRODUCTION

This paper describes a methodology for the traffic images analysis, which allows functionality of recognition and correlations among vehicle flows. A prototypal system, named SMART (System for Monitoring, Analysis and Regulation of vehicle Traffic) has been realized and widely experimented in several real traffic contexts. It is based on a central elaboration unit that receives and processes images acquired by fixed or mobile sensor units, that are located throughout the area to be controlled. The study provides an algorithm for the vehicles recognition with the aim of monitoring the correlated flows between two or more measuring points. In view of the generality of the proposed system, its possible applications are manifold: assessment and correlation analysis the vehicle flows in pre-determined sections or locations along streets, roads and motor ways; monitoring and possibly regulating traffic at cross-roads; reconstruction of local origin-destination matrices by vehicle classes, which are the data base of the computer codes for the simulation of the traffic and its subsequent planning.

2. IMAGE ACQUISITION

The equipment for the experimental study consists of CCD sensors which, suitably located in the area to be monitored, take continuous pictures of the vehicles in motion. The storage of the data relative to these pictures is not made on the whole image but by acquiring only pixels along pre-determined reference lines (targets) the location of which on the video is selected by the operator oriented in perpendicular sense with respect to the vehicle flow . The number and the size of the targets are selected according to monitored object: one-way street, two-way street, dual carriage way road, cross-roads etc. During the acquisition, the intensity of the pixels which varies with the modification of the scene is read at video frequency and stored maintaining the same chronological succession as in the real scene. The end product is a data file which can be considered as spatial-temporal diagram or a new image where on one axis the pixel lines sampled from each frame and on the other axis the time sampling are plotted in sequence. This image is characterized by a small dimension corresponding to the number of sampled pixels and by a larger time dimension (similar methodology in [1],[2],[7]).

Fig. 1 shows an example of acquisition on two targets, 50 pixels each (first two columns), 25 frame/sec, where for ease of visualization the successive frames are indicated in adjacent columns. Thus an image of the vehicle is obtained by taking advantage of its motion; the apparent width of the vehicle will be proportional to its actual width, while the apparent length is related to the residence time on the target during its passage. For this reason, the vehicles on the image can appear deformed as a consequence of any speed variation during the passage on the target. By allocating two targets close to each other (called *In* and *Out*) on a lane to be monitored, it is possible to count the vehicles, to determine the speed, and to classify of the vehicle on the basis of its actual dimensions.

3. VEHICLE SEGMENTATION

3.1 The distance parameter

The principal step of the analysis for each traffic monitoring system consists in the segmentation of the vehicles with respect to the road background ([3],[4],[8]). We propose a method based on the definition of a *reference line* L_R defined as the target time sample, which statistically represents the road background in absence of vehicles. Subsequently for each instant a *distance parameter* D comparing each time sampled line with the reference line is computed. In this way a distance time diagram is obtained. Several attempts have been carried out for the distance parameter definition. The best results have been obtained using the following two definitions:

$$D_t = \sum_{i=0}^{n} \left[(L_R(i) - L_{Rm}) - (L_t(i) - L_{tm}) \right]^2 \tag{1}$$

$$D_t = \sum_{i=0}^{n} \frac{\left[L_R(i) * L_t(i) \right]}{(L_{R\sigma} * L_{t\sigma})} \tag{2}$$

where $L_t(i)$ represents the value of the i-th pixel of the line sampled on the target at the time t. The R subscript is related to the reference line, the m and σ subscript are related respectively to the mean value and standard deviation computed on the whole line. Finally n represents the number of sampled frames. The normalization, which has been introduced in the definitions (1) and (2), is necessary in order to make the distance function independent from variations of the illumination of the global road during the measuring time.

The module also provides a second method to carry out the distance vector calculation: it minimizes the spurious shapes bearing due to local noise, and resolves also more complex situations, like when the vehicles take up only a fraction of the target line (multi-lines roads). In this case the sum of the distance differences isn't calculated on the whole target line, but on the inside of a *traveling window* (whose dimension is M) of pixels from the i-th pixel by the following equation:

$$D_t = \max \left[D_t(i, i + M) \right] \tag{3}$$

The time diagram of D function is characterized by an almost constant value, representing the background, and by a set of marked peaks, which shows the presence of remarkable variation with respect to the background. This event is associated by the passage objects on the targets *In-Out* (Fig.3). In order to build the reference line $L_R(i)$, a PDF (Probability Distribution Function) analysis of the distance time history with respect an arbitrary line is performed. As reference line, it is taken that line, which corresponds to the distance value nearest to the value corresponding to the PDF maximum. By avoiding drift problems of the reference, this is calculated again every 1000 frames of the spatial-temporal diagram. Finally, after the identification of the real reference lines, the PDF of the distance function is again computed.

3.2 Vehicle extraction

The vehicle locating in the spatial-temporal diagram is carried out by fixing a binarization threshold on the distance vector. It has been clear, that a fixed value of the threshold hasn't been effective in conditions different from these where the threshold has been optimized.

In order to solve this problem an adaptive threshold has been defined on the base of distance PDF. This function is characterized by "half Gaussian" zone near the origin corresponding to the background noise and vehicles noise at higher distance values. From this function, the standard deviation σ_d of the background noise is computed. The threshold T is defined as the product of the standard deviation per a *sensibility coefficient* S which has been pointed out through a wide experimentation on different traffic contexts:

$$T = S \; \sigma_d \qquad\qquad 4)$$

By this definition the threshold becomes adaptive with respect to the variations connected to the sensor noise or the environment noise. Before the analysis, in order to reduce potential instantaneous noises or to solve eventually part of the vehicles which have the same pattern in respect to the road pattern, a median filter on the distance vector is performed.

Another kind of disturbances are related to real variations of the reference line because of the presence of vehicles parked on the road sides or of slowly moving sun shadows. A filter in order to check these disturbances which generally introduce a platform on the distance vector is provided. When a suspected reference modification is identified (i.e. very long vehicles), a new reference line from that instant is computed.

At the end of this phase, all the frames, where almost a *moving object* is present on the target, are selected. In order to extract only the area actually occupied by the vehicles, a new binarization procedure operates on the single points directly into the selected areas. The pixel by pixel binarization is based of an adaptive threshold on the differences between the reference line and the target samples. Finally a linking procedure provides to extract one or more vehicles present in the selected areas producing a series of *AOI* (Area Of Interest), which are related each one to a *moving object*. Fig. 2 shows an example of identified vehicles marked with red blocks.

4. CORRESPONDENCE ANALYSIS

At the end of the segmentation phase the AOIs for each target are identified. In order to obtain speed and dimensional measurements it is necessary to establish the correspondences among the AOIs on the two different targets.

4.1. Correspondence criteria

The proposed algorithm correlates each vehicle from the *inlet* target with the vehicles from the *outlet* target. The selection is based on a coupling square matrix which describes all possible candidates of *outlet* target for each vehicle in the *inlet* target. A candidate group correlated with an *inlet* vehicle is composed by the vehicles which are present after the

correlated with an *inlet* vehicle is composed by the vehicles which are present after the corresponding time frame but included in a limited time interval defined on the base of the minimal expected velocity. In order to distinguish vehicles running on different lanes on a multi-lane target a constraint is fixed on the barycentral and dimensional correlation: a canditate is rejected if the maximum spatial-temporal overlapping of the two zones is lower than 50%. After the computation of the coupling matrix, this is analyzed both from the beginning and from the end of the acquisition and the conflicts are solved by a hierarchical procedure.

The coupling system is able to cut off the vehicles running in prohibited direction and remove the crossing people on the target lines or parking vehicles. Fig.2 shows an example, where the valid coupling is marked by the green line. On the base of the identified correspondences, the vehicle speed and, consequently their dimensions are computed. By using the dimension information, each vehicle in several reference classes is subdivided. At the end the mean flow (vehic/h) and the density (vehic/km) and the speed distribution for each class are available.

4.2. Experimental tests and comparison with other conventional systems

These procedures have been implemented in the SMART system. The system was applied to carry out measurements in different urban and extra urban contexts (single lane, multi-lane, highway, cross-roads). During the measurement, two other conventional measurement systems (pressure tubes and magnetic turns) have been utilized. Furthermore the scene has been recorded by a video-recorder and the vehicle passages have been counted by an operator at video low speed . These data have been used in order to check the errors committed by magnetic turns, pressure tubes, SMART system and by a simplified image analysis system based on only two pixels instead of two target lines (called m2p in the Fig.4 which reports an example of comparison). The global results showed a better accuracy of SMART with respect to the other systems especially at high density of the flows (for example very low speed vehicle lanes). In every cases treated the average error rates were been lower than 5%, with maximum error rates lower than 10%. In the same conditions the conventional systems (magnetic turns, pressure tubes), showed errors up to 20% at high density of the flows.

5. VEHICLE RECOGNITION

In order to apply this system to several traffic problems (like the cross-roads control, the public or special vehicle identification, the correlated flows analysis) it is necessary gives to the system a vehicle recognition capability. In this paragraph the developed methodology is presented. Two or more different measurement points are supposed: for each point, by a double target before described a list of vehicles is identified. The reference recognition problem is *to identify the corresponding image of a specific vehicle monitored (reference vehicle) in a certain measure point inside a list composed by the images of the vehicles monitored in all the other measure points.* This problem requires an algorithm characterized by high robustness to the disturbances connected to the different illumination among the several measure points, and to the lightly different perspective view of the points.

5.1. Basic principles of the method

Because of the required robustness, the algorithms based on contour description or exact image templates have been excluded. The proposed method is based on a two-dimensional transform in order to give to the system some invariance properties (as shown by [11] et al.), which will be discussed. Among the various types of transforms we have chosen the DCT in view of its better performance as compared with others, such as Fourier, Hadamard-Haar or Slant transforms, because although the latter algorithms are effective in terms of calculation speed, the DCT offers a higher capacity for concentrating energy ([5],[6],[10] et al.). The DCT calculation is performed on the transformed image in the spatial-spatial domain by a interpolating procedure based on the speed measurement.

5.2. Definition of the discriminating parameters

It was observed that each coefficient concentrates a different quantity of data, showing more or less significant and recognizable characteristics of the image; therefore, to consider all coefficients in their totality may produce undesirable effects, as the high discriminating capacity of some of them is reduced by casual and spurious values of some other, less significant ones. Instead of use a single matching between the reference vehicle and a possible candidate, matching was operated on particular areas of the DCT matrices, on the only basis of their discriminating capacity. For each block of coefficients a matching parameter has been considered generating n parameters. The classification is performed directly in the *matching space* (n-dimensional) where the origin corresponds to the identity; in fact this approach shows a good selectivity between the different groups of vehicles ([9]).

The DCT areas have been realized by considering groups of coefficients at similar polar distance from the DCT origin (similar spatial frequency). Examples of DCT sub-grouping are reported in Fig.5. This criteria gives the algorithm the vehicle orientation invariance.

The matching parameters are defined as the sum of the squared differences among the correspondent DCT coefficients, which are located in the above mentioned zones of the matrices, by the following equation:

$$P_j = \sum_{i=0}^{N_j} (X_{Rji} - X_{Cji})^2 \tag{5}$$

where X_{Rji} are the DCT coefficients concerning the reference image, X_{Cji} are the DCT coefficients concerning the candidate image, N_j is the number of elements of the j-th area under consideration and P_j is the j-th parameter.

As further discriminating parameter we chose the matching between the histograms of the vehicle images, both reference and candidate.

5.3. Classification, parameter selection and recognition effectiveness

The classification criteria adopted is the minimal Euclidean distance from the origin in the parameter space between all the candidate vehicles and the reference vehicle. Two modalities of parameters normalization in the classification space have been experimented: the standard deviation and the mean value ([6],[12]). The mean value normalization has given the best results in the real data experimentation. This consists to divide each parameter for the mean value of the parameter concerning all the candidates. efficiency decreases with the size of the candidate list, the tests have been carried out on databases built with different list size.

In order to optimize the recognition we studied the possibility to operate a more accurate parameter selection in order to obtain a better effectiveness and in any case a reduction of the necessary quantity of discriminants. A sensitivity study has been conduced in order considering different groups of parameters. The graph in Fig.6 shows an example of effectiveness tests comparing the recognition efficiency for four different groups of parameters. Finally the best parameter set has tested in different conditions (by extracting from the acquisition file a series of couples of corresponding vehicle images on *In* and *Out* targets) and it has shown a mean recognition efficiency of about 80%.

6. CROSS-ROADS MONITORING BY THE VEHICLE RECOGNITION

The previous methodology can be applied to the cross-roads monitoring. In this paragraph a methodology for the cross-roads monitoring based on the recognition proposed technique is presented. The structure of four-way cross-roads may be represented as in Fig.7, where $\Phi_{0..7}$ indicate respectively the incoming (0,2,4,6) and the outgoing (1,3,5,7) vehicle flows. We are interested in the definition of a equations model to fully represent the characteristics of a cross-roads, that is to express in a mathematical way the relationship between the individual incoming, outgoing and correlated flows. The describing equation set consists of the following system:

incoming flows	outgoing flows	closing relation on the node
$\Phi_0 = \Phi_{07} + \Phi_{03} + \Phi_{07}$	$\Phi_1 = \Phi_{01} + \Phi_{21} + \Phi_{31}$	$\Sigma i\, \Phi_{In} = \Sigma i\, \Phi_{Out}$
$\Phi_2 = \Phi_{21} + \Phi_{23} + \Phi_{25}$	$\Phi_3 = \Phi_{03} + \Phi_{23} + \Phi_{43}$	
$\Phi_4 = \Phi_{43} + \Phi_{45} + \Phi_{47}$	$\Phi_5 = \Phi_{25} + \Phi_{45} + \Phi_{65}$	
$\Phi_6 = \Phi_{61} + \Phi_{65} + \Phi_{67}$	$\Phi_7 = \Phi_{07} + \Phi_{47} + \Phi_{67}$	(6)

This system, containing nine equations and twenty unknowns. The knowledge of the eight input-output flows (Φ_x) are not enough to define completely the problem, but the measuring of at least three *correlated flows* (Φ_{xx}) is necessary.

6.1. Realization of a cross-roads controller

In order to define the method for the solution of the specific case, the following procedure was used: direct shots were made by means of TV-camera located at a corner of a cross-roads, at a height of 15 meters above ground and provided with a wide-angle lens. For the data acquisition eight double targets were set, one for each incoming and outgoing lane, so that the overall dynamics of the vehicle movement was recorded. The acquisition file consists of a certain number of adjacent columns, in which the images of the vehicles are chronologically distributed.

Suppose that a vehicle comes from lane 0 and goes to lane 3; two images of this vehicle there will be present: one in the vehicles list corresponding to the 0 lane and the other in the vehicles list corresponding to the lane 3. Therefore, if we wish to know in which direction the vehicle has gone through the cross-roads, it will be necessary to find the corresponding image among all the *Out* vehicle lists. In doing this, the following design limitations have been introduced, which allowed the number of vehicles to be compared to be kept within reasonable limits: 1) the time coordinates of the images which form the list of the candidates must be successive to those of the reference image and included in a fixed maximum time interval (necessary to overpass the cross-roads); 2) elimination of an image, if this image has been recognized and coupled with already analyzed vehicle images.

A first verification of the method gave the following results: as for the counting of the vehicle correlated flows in the various lanes a mean rate of error of 10% was observed; the verification of the image coupling showed a rate of error of 26%. To the aim of investigating on the reasons for coupling errors, a simulation was carried out. We analyzed a video section of a double-way street, where four targets, two for each lane were allocated assuming that street lanes, instead of running parallel to each other, crossed each other perpendicularly to form a cross-roads. The application of the proposed algorithm to this particular case gave the following results: for the vehicle correlated flows in the various lanes a rate of error of 4.5% was observed; the verification of the image coupling showed a rate of error of about 9.5%.

The reasons for this considerable reduction in error rates are to be found in the overcoming of some problems due to perspective errors introduced by lateral vision instead of central vision on the cross-roads.

The 10% error on correlated flows (and 5% on the main flows) is considered satisfactory for most of applications (cross-road control, traffic simulator codes...) considering also that this information cannot obtained with much less powerful conventional sensors like magnetic turns. In every case probably a new vehicle *re-scaling* before the DCT analysis and a DCT block redefinition could be studied in order to decrease the influence of perspective effects.

REFERENCES

1. H. Taniguchi, H. Furusawa, A. Seki, S. Ikebata, "Recognition of moving vehicles using directional-temporal plane transform", Proceedings of SPIE, vol. 1607, p.52-61.
2. P.G. Michalopoulos, R.D. Jacobson, C.A. Anderson, J.C Barbaresso, "Field deployment of Autoscope in the FAST-TRAC ATMS/ATIS programme", Traffic Engineering and Control, vol. 25, p.386-391.
3. J.g. Verly, R.L. Delanoy, D.E. Dudgeon, "A model-based system for automatic target recognition", Proceedings of SPIE, vol. 1471, p.266-282.
4. M.A. Vogel, T.C. Chan, "Model-derived multisensor target discrimination", Proceedings of SPIE", vol. 1003, p.247-254.
5. A.K. Jain, "Image Data Compression: A Review", Proceedings of the IEEE, vol. 69, n. 3, March 1981.
6. A. Rosenfield, A. Kack, " Digital Picture Processing ", Academic Press, ed. 1981.
7. A.D. Houghton, G.S. Hobson, N.L. Seed, R.C. Tozer, "Automatic vehicle recognition", Second International Conference on Road Traffic Monitoring, (Conf.Publ.n.299), p.71-78.
8. M. Burkard, N. Rehefeld, "Real-time road surveillance by an optical sensor", Proceedings of SPIE, vol. 860, p. 133-139.
9. A. Houghton, N.L. Seed, R.W.M. Smith, "Real time vehicle recognition", Proceedings of SPIE, vol. 901, p.65-72.
10. W.H. Chen, C.H. Smith, "Adaptive coding of monochrome and color image", IEEE Transaction on Communications, vol. COM-25, n.11, Novembre 1977.
11. L. Capodiferro , R. Cusani, G. Jacovitti, P. Melli, M. Vascotto, "Scale, rotation and shift invariant object recognition by means of correlation techniques", Eight IASTED International Symposium Measurement, Signal Processing and Control, Settembre 1986.
12. R. Gonzalez, P. Wints, "Digital Image Processing", Admission-Wesley Publishing Company, 1987.

Fig. 1 Representation of monitored scene by videocamera and of locating of two targets.

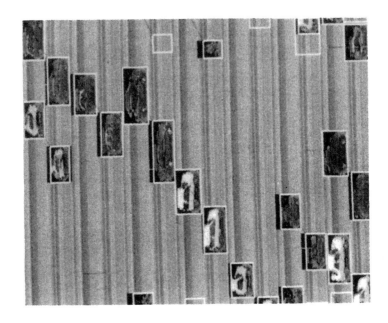

Fig. 2 Visualization of the space-time diagram.

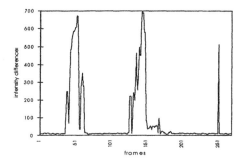

Fig. 3 Graph of the distance parameter.

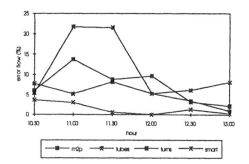

Fig. 4 Comparison of some methodologies : m2p, (monitoring by only 2 pixels), tubes, turnes, SMART (proposed methodology.).

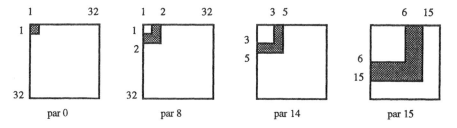

par 0   par 8   par 14   par 15

Fig. 5 Areas of DCT, which are used in the matching.

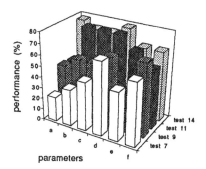

Fig. 6 Comparison of the obtained effectiveness in the four tests, by the use of particular groups of the discriminant parameters.

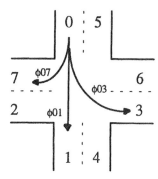

Fig. 7 A four-way crossroad scheme.

Time-Varying Image Processing and
Moving Object Recognition, 3 – V. Cappellini (Ed.)

Automatic vehicle counting from image sequences

G.Nicchiotti and E.Ottaviani

Elsag Bailey S.p.A, Research & Development Department,
Via Puccini 2, 16134 GENOVA

A method for automatic vehicle counting from image sequences is presented. The structure of the algorithm consists of two steps detection and tracking. The detection is based on a background subtraction approach, while the fundamental framework of the tracking is the Directional Temporal Transform. A discussion of the computational efforts and some results in urban and motorway contest are finally presented.

1. INTRODUCTION

To have knowledge about traffic on the road network is getting more and more essential, due to the increasing number of cars travelling along and the importance of the related economical and social impacts. This is presently obtained by means of permanent measurement stations and periodic surveys. The aim is to collect data about traffic conditions in order to get statistical knowledge for traffic management activities (maintenance schedule, traffic flow plans, etc.) and to provide drivers with information about accidents or traffic jams and achieve safe and confortable driving.

Counting vehicles is the main category of measurements for applications concerned with traffic planning and mantenaince. It is frequently carried out using mobile counting stations and it is used to analyse changes in traffic from one year to another and to study peak phenomena (rush hours, major holidays, etc).

Most measurements are currently based on point-like transducers such as photodiodes, ultrasonic sensors, microwave sensors [1], inductive loops and pneumatic tubes [2]. These sensors however are limited in the traffic information they can provide because they could not exploit the 2-D spatial information contained into a traffic scene. This means it is necessary to solve a lot of ambiguity problems in order to correctly measure the traffic parameters. Furthermore construction costs, maintenance access problems and other infrastructure impacts need not be incurred if the video camera can be used as sensors.

Recently computer vision has reached a level of maturity which has led to a variety of applications in industry and security systems. Traffic monitoring seems to be a new interesting field for the application of computer vision techniques [3-5]. This paper describes a simple method for analyzing image sequences and counting the passing vehicles for real time traffic measurement purposes.

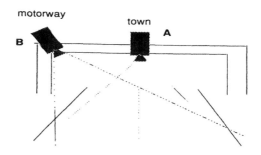

Figure 1: Camera set-up

2. EXPERIMENTAL SETUP

A standard b/w CCD camera was used to record the image sequences. The position of the camera and its focal length were mainly determined by the need of framing about 25 meters of roadway.

Indeed, as we shall see, the structure of the algorithm includes a tracking step and we need the car to stay in the observed sequence for at least three frames. Considering a maximum speed of 120 km/h and a sampling rate of the sequence of 4 Hz, a depth of field of 25 meters is required. So we used a 25 mm focal lenght camera installed at about 8 meters of height from the ground.

We have recorded data in two different situations: at first we analyzed sequences taken on a motorway with four lanes of traffic and the vehicles going off the camera; afterwards we framed an urban traffic situation again with four lanes but with cars both approaching and moving away from the camera.

The camera set-up in the two situations and the typical images we got are shown in figures 1 and 2.

In both circumstances the camera was mounted on an overpass, but we have to point out that in the former case, due to installation problems , we were forced to get a partially side view. As we shall see we got the best results with a top view as reported by many authors [6], mainly because we could overcome occlusion problems due to perspective.

The image sequences were then digitalized with a 4 Hz or 8 Hz rate in 256×512 pixels grey level images with 8 bit/pixel.

3. ALGORITHM STRUCTURE

After an interactive phase at the beginning of the process, the algorithm mainly consists of two steps: the detection and the tracking procedures.

Indeed, as the camera is stationary with respect to the environment, a part of the image corresponds to the static background, which is only subject to illumination changes, while

412

Figure 2: Typical images

the events we are interested in, take place only in a fixed subimage (foreground).

Therefore at the beginning of the process it is possible to select the monitored area, also reducing the computational requirements. and the detection step extracts moving objects only in this user-defined subimage. The image area outside this window, as we will see later, is used to perform the background updating.

In this phase we as well select the lanes we want to monitor and define for each lane the segment the traffic flux will be projected on in the tracking step. The flow chart in figure 3 illustrates the detailed framework of the whole algorithm.

3.1 Detection

Moving object detection from an image sequence is a key task in computer vision. Essentially we can distinguish between two main methodologies: intra-frame difference [7] and background subtraction [8].

As we are interested into analyzing the usually caothic scenario of the city traffic and, in addition, with sub-sampled image sequences (sampling rate lower than standard video rate), we decided to adopt the later approach and successive frames in the image sequence are subtracted from a reference image (background) depicting the scene in its undisturbed state. The background reference, in the monitored subimage, needs to be continuosly updated in order to compensate the gradual changes due to the variable light conditions of the outdoor scenes.

Indeed the reference updating cannot be performed using the pixel grey values of the selected area because of the temporary disturbance due to the passing vehicles we want to count, thus we use the time statistical properties of the image luminance in a small (about a thousand pixel) window outside the monitored region.

These data allow us to estimate frame by frame two parameters (substantially gain

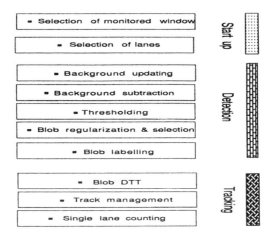

Figure 3: Algorithm flow chart

and offset) of a linear transformation that connects the luminance of the previous frame with the current luminance. This transformation is then applied to the measured grey levels in the monitored area to compensate variable light conditions

This method gives good perfomances and less computational burden with respect to more sophisticated background updating techniques such as multi-model Kalman filtering, operating pixel by pixel in closed loop with the moving objects detection [9], and it represents a trade-off between computational costs and results.

Then the difference image is generated and thresholded with a constant threshold empirically stated, giving a binary map of detected raw blobs. We tried as well a dynamic threshold consisting of a constant bias and a variable offset added or subtracted in a way dependent on the previous neighobour pixel classification [10], but the improvements in performance did not enough counterbalance the computational costs growth.

In this step the main sources of errors are due to shadows and acquisition noise. We partially overcome these effects processing the raw blobs extracted in this phase in order to select only the ones large enough to cover a significant area.

3.2 Blob processing

In this phase we elaborate the raw blobs extracted in the detection step aiming: to regularize the blob shapes, to reduce the number of blobs both suppressing the little ones and merging the close ones and finally to label the connected blob regions in order to assign in the tracking procedure each blob to a one and only one lane.

At first a morphological filter is applied to eliminate little blobs due to noise, to merge the close blobs and to regularize the shape of the selected ones. Good results are obtained with a 3×3 binary modal filter. It simply counts detected pixels in a fixed neighbourhood of a point and assigns this point the blob only if detected pixels are more than the non-detected ones.

Figure 4: An example of the detected blobs

An area thresholder is then applied to definitively suppress small blobs, as vehicles size must be large enough. The remaining blobs are then labelled and the intersection between each blob and the *lane lines* interactively predefined during the start-up phase is considered: let us recall that the *lane line* is a 1-D linear path which crosses each lane along its center inside the monitored window.

The projection of the blobs of vehicles on a directional axis roughly parallel to the lane in order to get a 1-D data stream, allows us to transform a 2-D image sequence into a more suitable representation to perform tracking of moving objects. This transformation, introduced in [11], is named Directional Temporal plane Transform (DTT). We carry out one DTT transform for each lane on which we are interested. DTT then generates for each lane a 2-D image by placing the 1-D data stream side by side in temporal order. Blob DTT represents a terse and effective representation for the following tracking step.

3.3 Tracking and counting

The tracking procedure works on the data generated by DTT transform. A vehicle track is created, updated or estinguished depending on the results on the present frame related with the previous ones via DTT transform (on the same lane).

We introduce in the tracking procedure some a priori knowledge about the traffic direction by means of a *coherence criterium*. According to this rule a blob DTT can award only the tracks in a coherent position with respect to the blob itself taking into account the traffic direction (see fig.4 for example).

A new track is created when an enough large DTT blob can not update, satisfying the coherence criteria, an existing track.

Each track can be updated by one and only one blob DTT.

But due to the structure of the processing it can be possible that a single blob projection updates more than one track. In this case we will get two (or more) coincident tracks.

This fact allows us to overcome problems due to the viewing geometry which can influence the gaps between adjacent vehicles and produce fusion of successive blobs.

As we told each track can be updated by one and only one blob projection but a new track can be created at any position of the lane if the DTT of the blob overcomes a dimension threshold. This eliminates problems of blob fission (one blob splitting in

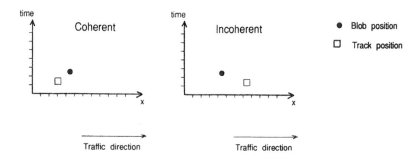

Figure 5: Coherence criterium:an example

Routine	Time (ms)
BCK	100
THR	70
BLOB	450
DTT	40
TRK	10
TOTAL	670

Table 1:Execution times for the single routines

more), resolving a fission with one track updating and one new track creation.

A track is estinguished when no one of the new data can update it. In this latter case, if track life is quite long a vehicle (or more in case of coincident tracks) is counted onto the lane.

The counting logic keeps into account the possibility of vehicle paths crossing two or more lanes only if the lane lenght is short enough (or the vehicle speed high enough) to not allow a vehicle to stay for many frames on more then one lane.

4. TESTS AND RESULTS

The algorithm is implemented on a PC-486 DX 33 Mhz in C language. Because only a window of the image is fully processed and more because the algorithm does not require floating point operations, the average computational effort is about 0.7 sec/frame to examinate a typical urban scene with four lanes of traffic. Most of the time,as shown in table 1, is spent in the low-level processing (blob detection), and mainly in the morphological filtering. However, real-time performances can be achieved with image-oriented multiprocessor architectures like the EMMA2^{TM} computer developed by Elsag Bailey (for example via a data parallelization on the single lanes). EMMA2 can be provided with a front-end image processing unit capable to acquire and process image sequences with built-in standard image processing functions and programmable DSP devices to execute more complex elaborations.

We tested our algorithm on several traffic sequences (256×512 images with 8 bit/pixel at a sampling rate of 4 and 8 Hz) both on motorways and city roads, with satisfactory results.

Results show a good robustness of resulting traffic statistics, with a detection probability of 97% and a false alarm of 4%. This happens even if the detected blobs sometimes do not look exactly like the true vehicle shapes. In addition, the 1-D tracking scheme seems to be more robust and cost-effective with respect to a fully 2-D tracking algorithm, leading also to a simple way to count vehicles.

5. CONCLUSIONS AND FUTURE TRENDS

We have presented an algorithm for automatic vehicle counting on a four lane traffic way at a rate slightly less than 2 hertz on a PC-486 with a few percent of error. We test the procedure on a day-light set of images both in urban and motorway context. The algorithm structure fits a data parallelization on a multiprocessor machine. Miscounting are essentially due to shadows, and, most of all, drivers poor lane discipline. Speed estimation and vehicle classification, on the basis of their size, can be easily obtained with a light further postprocessing and will be future objects of investigation.

REFERENCES

1. K. Uetakaya, H. Fukui Microwave vehicle detector, 24th ISATA, Florence 1991 pg 757.
2. S.Espiè and F.Lenoir The future of road traffic measurements, Recherche Transports Securitè 6, pg.51 (1991).
3. A.T.Ali and E.L.Dagless Computer vision-aided traffic monitoring, IEE ISATA 91, pg 55, (1991).
4. P.Bouthemy and P.Lalande Detection and tracking of moving objects based on statistical regularization method in space and time, Proc. Conf. on Computer Vision, Antibes (1990), pg 307.
5. R.M.Inigo Application of machine vision to traffic monitoring and control, IEEE Trans. on Vehicular Technology, Vol.38, N.3, pp.112-122, (1989).
6. A.Yoshizawa and M.Aoki Road traffic measurement at an intersection by image sequence analysis, Proceedings ISATA 91, pp.427-434, (1991).
7. T.S.Huang (ed.) Image sequence Analysis, Springer NY (1981).
8. S. brofferio et al. A backgruond updating algorithm for moving objects scenes Time varying Image Processing and Recognition 2, pg.277, ed. V.Cappellini, Elsevier Pub. (1990).
9. K.P.Karmann and A.von Brandt Moving object recognition using an adaptive background memory, Time varying Image Processing and Recognition 2, pg.289, ed. V.Cappellini, Elsevier Pub. (1990).
10. T.J.Ellis, P.Rosin and P.Golton Model Based Vision for Automatic Alarm Interpretation, IEEE Aerospace and Electronic Systems Magazine, 6 (1991).

11. H.Taniguchi, H.Furusawa, A.Seki and S.Ikebata Recognition of moving Vehicles using directional temporal plane transform Personal correspondance.

Time-Varying Image Processing and
Moving Object Recognition, 3 – V. Cappellini (Ed.)

418

Matching Image Processing Technologies with GPS system for Vehicle Tracking

Drago Torkar, Rudolf Murn, Dušan Peček

Department of Computer Science and Informatics, Jožef Stefan Institute, Jamova 39, 61111 Ljubljana, Slovenia✢

GPS has become lately a very popular satellite system for positioning and navigation. It can also be used in real time vehicle tracking. In the paper we present a prototype of a system that uses information of the GPS system, passes it to a computer that contains an electronic map of large dimensions and a program that can use these data to tell the user the position of its vehicle and visualise this position on map. To prepare maps for vehicle tracking some image processing has to be done. Combining this with GPS data pre-processing produces a satisfying preciseness for vehicle tracking anywhere, even in concentrated cities.

1. INTRODUCTION

In spite of all technologies and the modern equipment, that has been developed in recent years, orientation and navigation in unknown territory, specially in concentrated cities, seems still to be the problem. Maps and city plans are usually satisfactory if the time is not limited. Otherwise different approach should be chosen. One of the worlds most advanced and, in the last few years, most popular positioning system, that allows real *global* positioning and navigation, is Global Positioning System (GPS), also known as NAVSTAR GPS. Quick, precise and effective positioning in any territory can be built of two parts: GPS system and computerised maps obtained by image processing technologies merged with application dependant overhead. From here to vehicle tracking is only a step. Every vehicle determines its own position and passes it to a *command* centre, that gathers this information, visualises it on a computer and passes certain command or information through a bi-directional communication channel back to the vehicle.

Creating and updating digital maps is a very demanding and sophisticated task. A variety of image processing technologies can be involved to provide computer based maps of sufficient quality. If electronic maps should have been of any practical use in positioning, navigation or vehicle tracking, many attributes must be pined to them. These attributes compose a huge data base that demands correct and fast enough handling. Special care must be taken to insure fast and clear enough readability of electronic maps overheaded with GPS position and other attribute data.

Many difficulties should be overcome in developing such a system. GPS receiver, that is available to civil users, uses unprecise C/A code, that can obtain position to a precision of 30-50 m. This is certainly not enough for navigation in cities or even on the roads outside towns.

✢ This work was supported by the Slovenian Ministry of Research and Technology

Some pre-processing of data must be done before applying it onto a map. To determine behaviour of a GPS receiver and GPS data, some data collection in different conditions should be done and comparison to a known, precise position must be made. This can lead to a selection of a correcting technique.

2. GPS SYSTEM OVERVIEW

GPS was introduced in 1973 by US Department of Defence. Two programs of satellite navigation, TIMATION that was being developed by US Navy and PROGRAM 621B that was being developed by Air Forces, were joint into a common project NAVSTAR GPS. The system was very good planed. Main characteristics were settled in the beginning and were not changed until now. It was planed to put 21 primary space vehicles and 3 spares into 6 twelve hours, 63^0 inclination orbits. Later the number was reduced to 21 space vehicles because of financial troubles. The inclination was also changed to 55^0 that causes some troubles with reception in Mediterranean. Today there are 19 satellites launched in 6 orbit planes at height 20 231 km. 2-D navigation was first possible in October 1978 and 3-D was made possible in December 1978, both only a few hours a day [4]. Today 2-D coverage is provided 24 hours a day and 3-D navigation is possible for about 21 hours a day. The system is expected to be declared fully operational in 1995 with guaranty of US government not to charge direct users' fees for ten years from that point.

GPS provides precise military service, using P-Code signals with accuracy under 1m, and civil service, using C/A code with accuracy of 100 m. Besides GPS there are also other satellite positioning services, like GLONASS (Russia), Navsat and Granas (Europe).

2.1. Principles

Space vehicles transmit spread spectrum signals on two bi-phase right-circular modulated carriers, denoted as L1 = 1575,42 MHz and L2 = 1227,60 MHz. Both frequencies are an integer multiple of atomic standard, present on each satellite. L1 carrier is modulated by both a 10.23 MHz clock rate *P-signal* and a 1.023 MHz *C/A-signal*. The L2 carrier is normally modulated by a P-signal identical to that on L1, but it can also be modulated by the C/A-signal to support special testing applications [3]. On L1, the C/A signal has twice the power of P code signal. The period of the P-code is 267 days. Each space vehicle contributes 7 day period code. This way 36 satellites can transmit its position simultaneously and receiver can detect the code from each satellite unambiguously. The period of C/A code is 1 ms and contains so called Handover word that is used by P code receiver to synchronise itself to the very long P-code. To civil users only C/A code is available.

Satellite sends its position, status, information of other space vehicles in the system, time base correction and some other data, 16 all together to the receiver. Receiver calculates so called pseudo range to each space vehicle by estimating time that signal used to travel from satellite to receiver. This range is composed of real range and error ranges due to difference between satellites and receiver's time base and ionosphere delay.

$$\rho_k = t_k c = \sqrt{(x - x_k)^2 + (y - y_k)^2 + (z - z_k)^2} + cdt$$

Receiver knows (from received data) the position of a certain space vehicle (x_k, y_k, z_k). It must calculate its own position (x, y, z) and time difference (*dt*) due to ionosphere delay and time base difference. These are four unknowns and therefore it needs to know at least pseudo ranges to four space vehicles to find out 3-D information. If it doesn't "see" four satellites, only

2-D information is available or if the number is less than 3 it can not estimate accurate position at all. How to avoid such an outage is one of the problems in using GPS data for positioning and navigation.

2.2. Differential GPS

To increase accuracy of GPS position determination an approach called differential GPS should be used. A reference receiver on known location calculates corrections for satellite data and passes them to user receivers through radio or other data link. User receivers make corrections of position calculated by satellite data. As stated in [1] accuracy 3-5 m can be reached this way. The problem is that user receivers should receive the same constellation of satellites as reference one. The area that is "covered" by one reference receiver is limited. To cover large distances a network of reference receivers must be used.

2.3. Error sources

The main error sources that affect GPS accuracy are [6]:

- *Selective availability (SA)*: The Department of Defence of the USA corrupts GPS when using C/A mode. The characteristics of this degradation are classified, but it is known that SA involves a combination of signal "dithering" and ephemeris data manipulation. In the first case SA introduces a slowly varying delay into the time of satellite signal transmission. The gross features of this dithering can be approximated by a second order Markov process [2]. In the second case, SA introduces a bias error into satellite ephemeris data. All this results in 100 m error for single user and about 0.8-2 m error if differential corrections are used.

- *Troposphere:* Troposphere can introduce a delay at very low satellite elevation angles (below 10°). This can result in 2-3 m error for a single user and less than 1 m for differential GPS.

- *Ionosphere*: Ionosphere error depends on sun activity. P code receiver can simply correct this problem using data from satellites. C/A receiver can use a model to reduce the error, which is about 30 m in worst case for a single user and 2-3 m for differential GPS.

- *GPS receiver:* Error consists from background noise, thermal noise, multiple access interference, measurement quantisation error and code tracking loop noise. All together contributes to 1 m error, which can not be reduced using differential corrections.

- *Multipath*: Multipath error is usually less than a meter. It increases in concentrated cities, mountains...

- *Satellite clock:* Unpreciseness of satellite clock can result in about 1 m error.

- *Bad geometry* of space vehicles that are "seen" by the receiver can result in so called dilution of precision (DOP). The position range accuracy can be calculated through DOP by multiplying pseudo range accuracy by DOP.

3. MAPS AND DATABASE PREPARATION

To prepare suitable surrounding for vehicle tracking first a CAD computer program was made. The program is able to handle raster and vector images of large dimensions and to do semi-automatic vectorisation of maps. Through vectorisation some useful features of map data can be collected. The program is also able to connect to relational database and to interchange data with geographic information systems (GIS). It can also read GPS receiver's data in real time and do some basic pre-processing. Important is that there is no limit in largeness of maps and so tracking can be made on any area and on map of any measuring scale.

3.1. Map preparation

Maps were imported in computer through scanner in resolutions 200 - 400 dpi. We found these resolutions suitable for vehicle tracking because some details on maps are not lost and the amount of data is not too big. Then binarisation was done, using variable local threshold [8].

$$T_n = [C_n \sigma_n + \mu_n], \qquad n = 1,2,3,...,N$$

T_n is threshold value, σ_n is standard deviation of nine (the discussed pixel and eight neighbours) pixels from its' mean value μ_n. C_n is *threshold coefficient*, defined as $C_n = \dfrac{kH_n}{N}$, where k is a value between 0.65 an 1, H_n is the number of pixels on whole image that have the same grey level as pixel n.

Due to scanner and paper errors the image of a rectangular map that is imported in computer is distorted to a general quadrangular.

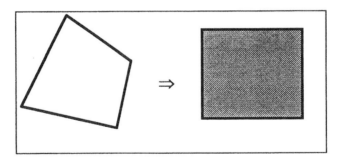

Figure 1. Transforming general quadrangular to regular rectangle

So a hyperbolic mathematical transformation was done to every map image to get as regular data as possible.

3.2. Attribute data base

The data base, holding data, that are "pined" to a certain location on raster maps, was created. Same attributes (street, building, bridge outlines), were collected through vectorisation [5], [7] , the other were added manually (street, city, building, names, numbers, messages, etc.).

Some digital images of the places were also added. When data base query is triggered all or selected attributes of a certain zone are displayed. Query can be triggered manually or by approaching of the tracked vehicle to a certain location.

4. GPS DATA PROCESSING

To determine behaviour of GPS system in the area of Slovenia, some static and dynamic measurements were done. The whole system, we intended to build, is planed on using single GPS receiver (no differential GPS) for vehicle tracking. Two reasons caused such decision: there is no network of reference receivers available yet and the complexity (and price) of a system would be increased by using communication channels for differential corrections. To do measurements low-cost GPS receiver (LOWRANCE) was used.

4.1. Static measurements

Many static measurements on different terrain were made to state the visibility of satellites and sensibility to some error sources. All position data were transformed to a national grid system (Gauss-Krüger system). This is a metric system that uses projection of the earth shape to plane. Due to this projection a static X and Y offset were stated. This offsets must be compensated when absolute position is calculated. Figure 2 shows a typical static measurement. All positions were moved to minimum x and minimum y component of all measured positions to get the impression of distances.

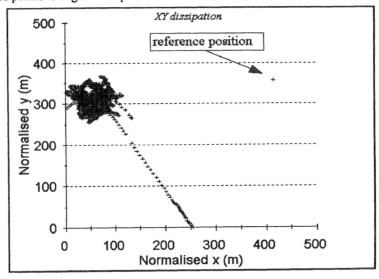

Figure 2. XY static dissipation measurement

Figures 3 and 4 shows time presentation of this data. All data are valid (3 or more satellites). SA activity is clearly visible. All other error remains within the limit of 100 m. Reference point is placed in coordinate origin.

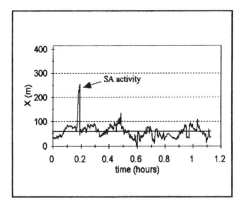

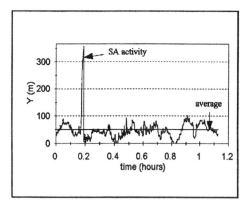

Figure 3. Time dissipation of X coordinate of position measurements from fig. 2

Figure 4. Time dissipation of Y coordinate of position measurements from fig. 2

4.2. Dynamic measurements

To establish behaviour of GPS receiver in dynamic conditions, accuracy to speed dependency was observed. Worst case of these measurements is shown on figure 5. This problem is successfully solved by dedicated receivers. Test drives were made in different places: in a city, in an open road, through the woods, etc. Three main problems were observed through these measures: satellite failure, SA and "shadowing".

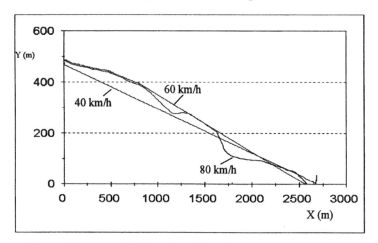

Figure 5. Low cost GPS receiver is sensible to it's speed

4.3. Satellite failure

When a certain space vehicle disappears from receiver's horizon and another comes in (or not), must receiver recalculate its select strategy and adapt to new conditions. This causes momentary troubles in calculating position. Recovery is typically fast (10 sec). We applied no special processing to solve this condition. When receiver doesn't "see" enough satellites (3 or more), positioning is rather impossible. GPS data have no meaning and position determination is completely wrong. When the pre-processor notes that there are not enough space vehicles present (this is reported by the receiver), all incoming data are ignored and tracking is suspended.

4.4. "Shadowing" problem

In concentrated cities, in woods, mountains, etc., some or all satellites can be shadowed by buildings, trees, mountains. Shadowing can happen in a moment, so the receiver still reports satisfying number of satellites, but calculated position is totally wrong. Data pre-processor (it inspects data before being applied to the map) keeps track of vehicle's average speed and direction. Every next position is defined relatively to the latest one. When comes to a sudden change of speed or direction - greater than dynamically defined threshold - pre-processor holds all incoming data, calculates new threshold and observes some next data samples. If new positions follow the predicted changes and no new change happens in a few next seconds, then a vehicle has speeded up or has changed the direction. Otherwise data is observed until the old conditions are fulfilled. Then the last "good" point is directly connected with a new "good" point.

Figure 6. Shadowing during city drive. Straight lines connect the last "good" point
before and the first "good" point after shadowing (measuring scale 1 : 25 000).

4.5. Selective Availability problem

The changes of relative position and speed are small, when SA takes place. The pre-processor can not successfully recognise it. If signal "dithering" is applied alone (without ephemeris distortion), than it can be modelled by second order Markov process [2]. SA could be removed by using differential corrections. Fig. 7 shows how SA appears in vehicle tracking.

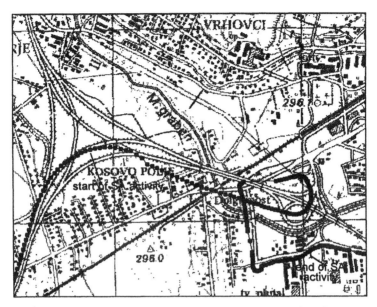

Figure 7. Changes due to SA are small and continuos. Duration is limited to a few minutes (measuring scale 1 : 25 000).

5. VEHICLE TRACKING

To make vehicle tracking possible, data link between a vehicle and so called command centre was established through short wave receiver/transmitter with data connector. Each vehicle can observe its position on a local computer and all vehicles in a system are connected by bi-directional data link with command centre. Time multiplexed and multi-channel data processing in command centre were tested. Second approach is less exposed to errors but needs much more data processing. To establish stable data connections a good infrastructure should cover the area to be tracked.

6. CONCLUSIONS

GPS system has become very popular and wide spread. Vehicle tracking is one of its possible applications. In the near future further improvement should be expected. P code service will become available to civil users. Navstar GPS will be combined with other satellite positioning services to achieve greater accuracy. Single receiver will provide satisfactory preciseness for vehicle tracking anywhere in the world. Cars will be equipped with computer, electronic maps and GPS receivers to help driver in navigation through thicker and thicker traffic.

REFERENCES

[1] Per K. Enge, R. M. Kalafus, M. F. Ruane, "Differential Operation of the Global Positioning System", *IEEE Communication Magazine*, vol. 26, no. 7, (1988)

[2] R. M. Kalafus, N. Knable,, J. Kraemer, J. Vilcans, "NAVSTAR GPS Simulation and Analysis Program", *available through the National Technical Information Service as DOT-TSC-RSPA-83-2,*, (1983)

[3]. Stansell, "Civil GPS from a Future Perspective", *Proceedings of IEEE*, vol. 71, no. 10, (1983)

[4] Philip Mattos, "GPS", *Electronic World+Wireless World*, December 1992-April 1993

[5] Abdulah Dedić,Rudolf Murn, Dušan Peček, "Map Data Processing in a Geographic Information System Environment", *Computer Techniques in Environmental Studies IV - Proceedings of Fourth International Conference on Envirosoft '92,Southampton UK,* (1992)

[6] L.Schuchman, B. D. Elrod, A. J. Van Dierendonck, " Applicability of an Augmented GPS for Navigation in The National Airspace System", *Proceedings of the IEEE*, vol. 77, no. 11, (1989)

[7] A. Dedić, Rudolf Murn, Dušan Peček, " Digitalization of Large Area Drawings and Maps", *Proceedings of Mediterranean Electrotechnical Conference - Melecon '91*, (1991)

[8] A.Dedić, "Računalniško podprta digitalizacija in vektorizacija črtnih risb velikih dimenzij", *Master Degree Work*, Faculty of Electrical Engineering and Computer Science, Ljubljana, (1991)

AUTHOR INDEX

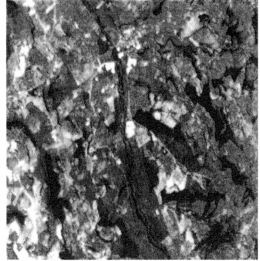

Montespertoli area - June 1991
SAR data: L band HH polarization
Pixel: 6m (range) x 12m (azimuth)

Montespertoli area - June 1991
TMS data: True color image
Pixel: 25m x 25m

Data fusion of SAR and TMS data:
HIS transformation

Data fusion of SAR and TMS data:
HIS+PC transformation

Printed and bound by CPI Group (UK) Ltd, Croydon, CR0 4YY

03/10/2024

01040328-0010